D0204703

# PROJECTIVE AND EUCLIDEAN GEOMETRY

# PROJECTIVE AND EUCLIDEAN GEOMETRY

Second Edition

W. T. FISHBACK

Professor of Mathematics
Earlham College

John Wiley & Sons, Inc.

New York | London | Sydney | Toronto

10 9 8 7 6 5 4 3 2

Library of Congress Catalog Card Number: 76-81329

SBN 471 26053 3

Printed in the United States of America

# PREFACE

The major change in this revised edition is an expansion of the former last chapter into three chapters to consider in detail the descent from real projective geometry to Euclidean, hyperbolic, and elliptic geometry; thus the title of the book may no longer be completely appropriate. There should now be enough material in the book for a two-semester course. By the addition of a new section in Chapter 8, the final three chapters can now be treated independently of Chapter 10 by those who wish to omit this material; Chapter 8 and Sections 9.4 and 9.5, give all of the results needed in the last chapters.

Many of the corrections and simplifications that have been made in the first ten chapters were suggested by users of the book. I thank all of those who have made suggestions.

W. T. Fishback

*Richmond, Indiana*
*April 1969*

# PREFACE TO THE FIRST EDITION

This book contains materials suitable for a one-semester course in projective geometry or a one-semester course in elementary foundations of geometry. It is addressed to upperclassmen, beginning graduate students, and high school mathematics teachers with little or no experience in modern abstract mathematics and assumes only knowledge of elementary synthetic and analytic Euclidean geometry. The materials are a fusion of materials used in a one-semester projective geometry course taught at Ohio University and of a one-semester course in foundations of geometry taught at National Science Foundation Institutes for secondary school teachers at Clark University, Harvard University, and Ohio University.

The projective geometry course is taught with the conviction that synthetic and analytic treatments of the subject are both important and should be given appropriate emphasis even if this involves limitation of the amount of material covered. The course begins with a synthetic treatment since this synthetic approach is the one most familiar and natural for the student. Introduction of the projective plane is accomplished by the addition of ideal elements to the Euclidean plane rather than through an axiom system because it is felt this is more natural to the student and helps him to build on his previous geometry experience. This course contains the materials of Chapters 4 to 10 of this book. The course involves both plane and solid projective geometry, but the solid geometry parts have been isolated in the starred sections of the text and may be omitted without loss of continuity. Linear algebra is introduced in Chapter 7 at the point where it is needed. Classes with appropriate training in this area might well omit this chapter.

The foundations of the geometry course contain the materials of Chapters 1 to 6, 10, and 11, with starred sections omitted. Again the initial approach is synthetic to allow the student to use his previous geometry training. The goal of such a course is an indication of the type of synthetic treatment needed to "repair" Euclidean geometry as embodied in Hilbert's axioms, followed by a sufficient development of projective geometry to enable the student to appreciate an axiomatic development of geometry at the projective level and to see how such a development creates order out of projective, Euclidean, non-Euclidean, and other related geometries. Analytic projective geometry is pursued just far enough to enable the student to appreciate the power and beauty of the material of Sections 10.1 and 10.2.

Other teachers may find other choices of topics appropriate for their classes. The text starts with Euclidean geometry and develops projective geometry sufficiently far to show the central position of projective geometry with respect to affine, Euclidean, and non-Euclidean geometries.

Exercises are included throughout the book to enable the student to test his mastery of the material and to indicate directions of further development of the topics being discussed. Exercises involving solid geometry are starred while those involving Euclidean use of projective methods are indicated by #. Answers to selected numerical problems and hints for some of the proofs given as exercises will be found at the end of the book. Every chapter contains a list of selected references considered appropriate to the level of this book.

I wish to record my appreciation for the help given me by my students at Ohio University, Clark University, and Harvard University in the development of the materials of this book. I also wish to acknowledge the valuable criticisms of early versions of the work by Professors Carl H. Denbow and Rodney T. Hood.

W. T. Fishback

*Athens, Ohio*
*September 1962*

# CONTENTS

# PROJECTIVE AND EUCLIDEAN GEOMETRY

## CHAPTER ONE

# EUCLID REEXAMINED

It is widely believed that geometry, as contained in Euclid's *Elements*, is perfect and complete; there are no flaws in the text, and all of geometry is to be found here. This is not the case. The geometry of Euclid is only one of many possible geometries and is customarily spoken of as Euclidean geometry. The *Elements* is not a complete exposition of Euclidean geometry, for many theorems have been added during the centuries since it was published. Finally, Euclid's work is not logically perfect; this is not meant to belittle the work, for it was a gigantic step forward and marks the real beginning of axiomatic, deductive mathematics. The fact remains that many of Euclid's proofs can no longer be accepted as proofs and the sets of axioms and definitions he gave are incomplete. In this chapter we indicate some of the logical difficulties in the very beginnings of Euclidean geometry.

## 1.1 UNDEFINED TERMS

Perhaps the greatest contribution of the Greeks to mathematics was the recognition of the necessity for axioms or postulates, statements to be assumed and not proved. When we prove a theorem, the steps in the proof must be justified in terms of previously established theorems; these in turn appeal to still other theorems for justification of the steps of their proofs. Since this tracing backward cannot continue forever, we must begin somewhere and assume certain basic statements. This the

Greeks realized, and they introduced the idea of axiom into mathematics.

A similar situation holds for definitions, and, unfortunately, the Greeks were not so perceptive in this matter. When we define "triangle," we are agreeing to substitute this single word in place of a long phrase involving the words "point" and "line segment." This agreement is a matter of convenience, the single word "triangle" being more easily used than the phrase. The word "triangle" has precise mathematical meaning in terms of the expressions "point" and "line segment." These terms in turn require the assignment of a precise mathematical meaning in terms of definitions which in turn will use mathematical terms. Euclid makes the definition, "A point is that which has not part." The mathematical term involved here is "part." If we attempt to define it, we may perhaps define it in terms of "dimension." We must then define "dimension." Such a process cannot go on forever; either we introduce an infinite number of terms, or we start a circular sequence, defining "dimension," say, in terms of "point." Neither outcome is acceptable.

Just as some of our statements must be axioms and assumed without proof, so must some of our terms be left undefined. Euclid's definitions, "A point is that which has no part" and "A line is that which lies evenly among all its points," contribute nothing to the clarity or rigor of geometry. We agree to leave "point" and "line" undefined in what follows. We may agree intuitively among ourselves what the standard pictorial representations for these undefined terms will be, but this understanding is no more a part of geometry than a picture is a part of the proof of a theorem. Indeed, we shall see in Chapter 4 that a man from Mars could confuse the two terms without any loss of understanding or clarity in the geometry we discuss in that chapter.

Mathematics is concerned with relations between elements of two sets. "Equals," "is greater than," "is congruent to," and "is similar to" are all examples of a relation between numbers or between geometric objects. Some relations must also be undefined in mathematics. One which we shall leave undefined is the relation implicit in the statement, "A point lies on a line." The reader will find it instructive to attempt to define this relation. Some may argue that the trouble could be avoided by having a line not be an undefined term, as we shall do here, but by having a line be a set of points; then a point would be defined to be on a line if it were one of the set of points which is the line. This is a perfectly acceptable way of regarding a line, but it does not resolve the problem at hand. It does not tell which sets of points are to be lines, and it does not tell what it means to say a point is an element of a certain set of points. Undefined relations are still present in this approach to the concept of line. The Greeks unfortunately also failed to perceive the necessity of having certain relations undefined.

## 1.2 PROBLEMS OF EXISTENCE

There is a current professional joke which states that the way to distinguish a mathematician from an engineer or physicist is to give the suspect an equation to solve. The engineer or physicist will proceed to seek a solution; the mathematician will first try to prove that a solution exists. The joke is usually told at the expense of mathematicians. The history of mathematics is replete, however, with tales of hours of wasted efforts attempting to solve problems ultimately proved to be unsolvable. The mathematician knows by now that problems of existence are not to be taken lightly!

One instance in which the Greeks overlooked such a problem is in the need to postulate the existence of points and lines. Failure to do so may mean that we are working in a vacuous system. Most modern axiom systems contain assumptions of existence, and we should like such to be the case in Euclidean geometry. We need not postulate the existence of all the desired elements; the existence of some elements together with other axioms may imply the existence of other elements. Thus if we assume "There exists a line" and "On every line there is at least one point," we would be sure that at least one point exists. In the next chapter we introduce a system of axioms for Euclidean geometry avoiding the pitfalls the Greeks encountered. In that system of axioms we have some axioms of existence to avoid the possibility of a vacuous system.

A second type of existence problem betrayed Euclid in his very first proposition of Book I of the *Elements*. This proposition considers the problem of constructing an equilateral triangle given its base $AB$. The construction employs the circles with radius $AB$ and centers at $A$ and $B$. If these circles meet at a point $C$, then $\triangle ABC$ is the desired triangle. How do we know the circles meet? Intuitively we are certain they do, but Euclid's system of axioms and postulates does not guarantee the existence of such a point $C$. Failure to prove the existence of intersection

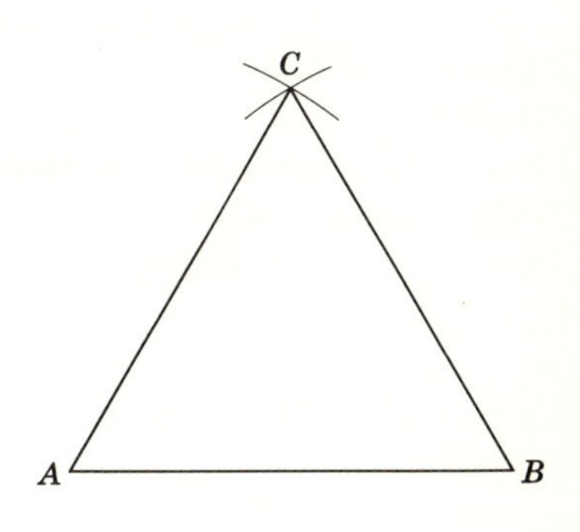

**Figure 1.1**

of loci is a common error in the *Elements*; indeed, the first three propositions all have false proofs, with the fallacy in each being failure to prove existence of the point of intersection of two circles or of a line and a circle.

The problem of proving the existence of such intersections is not to be undertaken lightly. It requires very subtle reasoning and use of the last of the axioms of the next chapter. We do not give such proofs here. They are simple if we have the weapons of analytic geometry present (see the exercises at the end of the section), but the use of analytic methods requires the rigorous development of the real number system, either from its own set of axioms or as a consequence of the geometric axioms used. We assert without proof that the necessary justifications do follow from the axiom system given in Chapter 2.

To illustrate the analytic approach, let us show that a line through the center of a circle and in the plane of the circle meets the circle in two distinct points. For a suitable choice of coordinates we may take the equation of the circle as

$$x^2+y^2=a^2 \qquad a>0,$$

and the equation of the line as

$$y=0.$$

We find at once by simultaneous solution of these equations that the distinct points $(a, 0)$ and $(-a, 0)$ lie on both the circle and the line.

## Exercises 1.2

1. Prove analytically that two circles, each passing through the center of the other, meet in two distinct points.

2. Prove analytically that a line meets a circle in two distinct points if and only if its distance from the center of the circle is less than the radius of the circle.

3. Prove analytically that two circles with radii $a$ and $b$, $a \geqslant b$, and distance between centers $c$ meet in two distinct points if $a-b<c<a+b$, in one point if $c=a\pm b$, and in no point if $c>a+b$ or if $0 \leqslant c<a-b$.

## 1.3 PROBLEMS OF CONGRUENCE

In the preceding sections we have considered difficulties in the axioms, definitions, and first three theorems of the *Elements*. Let us now consider the fourth theorem. This is the "side-angle-side" triangle congruence theorem and is proved by the method of superposition; one triangle is placed on top of the other and words follow which supposedly show that the triangles coincide.[1]

Just what is superposition? Euclid does not define it, so it must be an undefined relation. How does this relation "operate"? We still need axioms that tell us how. It is conceivable that the act of "moving" a triangle makes it shrink in size—at least we must deny this by suitable axioms if we want size and shape (whatever they mean) to be preserved. There is a failure in the *Elements* to recognize the presence of an undefined relation and a need for axioms concerning this relation. The undefined relation could be accepted as superposition, in terms of which congruence could be defined, or we might accept congruence as undefined and use it to define superposition. In any event the classical approach to the problem of congruence and superposition is deficient.

## 1.4 PROBLEMS OF ORDER

One of the best known fallacious geometry proofs is the one that all triangles are isosceles. Let $M$ be the midpoint of side $BC$ of $\triangle ABC$. Let the angle bisector from vertex $A$ meet the perpendicular bisector of side $BC$ at point $P$. From this point drop perpendiculars to $AB$ and $AC$, meeting these sides at $R$ and $S$, respectively. Since $\angle RAP = \angle SAP$

[1]Such "proofs" are sometimes said to be "proofs by hand waving." Euclid himself must have been suspicious of the principle of superposition, for he used it sparingly despite the fact that it could be used in many of the propositions of Book I with as much validity (or better lack of validity!) as in its use in Proposition 4.

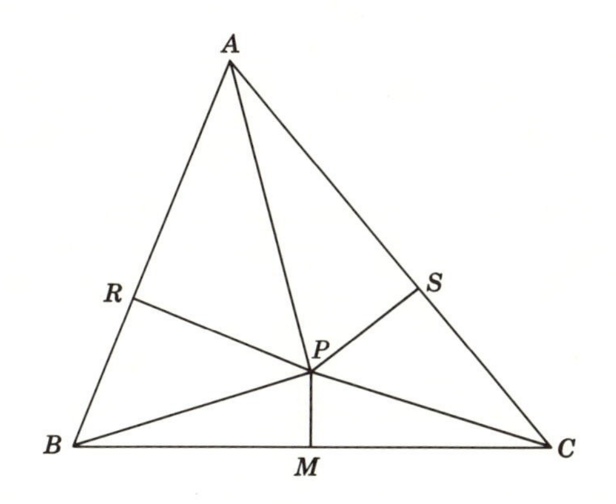


**Figure 1.2**

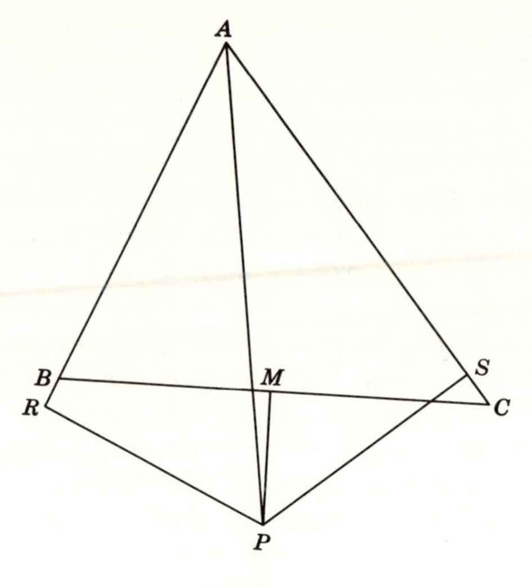


**Figure 1.3**

and $AP = AP$, the right triangles $RAP$ and $SAP$ are congruent. It follows from this that $RA = SA$ and that $RP = SP$. The latter equality, together with the fact that $BP = CP$, implies that the right triangles $BPR$ and $CPS$ are congruent. Hence $BR = CS$. We conclude that $AR + RB = AS + SC$ or $AB = AC$.

The fallacy lies in the assumption that $P$ lies in the interior of the triangle. In fact, if the triangle is not isosceles, $P$ will be outside the triangle, and one of the two points $R$ and $S$ will be exterior on a side while the other will be interior. Thus we actually have $AR - RB = AS + SC$, as shown in Figure 1.3, or $AR + RB = AS - SC$.

The difficulty here from the point of view of the *Elements* is that no account has been taken of the meaning of "exterior" and "interior." These are examples of the concept of *order*, a concept missing in Greek geometry. The axioms and postulates of the *Elements* do not tell us anything about the relation, "A point is between two other points." We cannot adequately define line segment in view of this difficulty. We cannot distinguish between the interior and exterior of a triangle. We cannot even prove the seemingly obvious fact that a line divides a plane into two regions. What is missing is another undefined relation, order, and the axioms concerning this relation. Any reconstruction of the foundations of Euclidean geometry will have to rectify this error of omission.

## 1.5 OTHER AXIOM SYSTEMS

The foregoing sections make it apparent that some reconstruction of the foundations of geometry as conceived by the ancient Greeks is needed. Such reconstructions occupied the attention of many mathematicians during the latter part of the nineteenth and the early part of the twentieth century, Pasch (1843 to 1931), Peano (1858 to 1932), and Hilbert (1862 to 1943) among others. Perhaps the best known of the reconstructions is Hilbert's, a variation of whose axioms for plane geometry is discussed in the next chapter. While more modern and sophisticated approaches have been developed in recent years, Hilbert's axioms do avoid many of the errors cited above and constitute one of the landmarks in the history of Euclidean geometry. They are statements in the spirit of Euclid, using similar language and taking a similar approach to the subject.

It should be noted that the difficulties involving order and congruence could be avoided if use were made of the assumption of all properties of the real number system and of the presence of a "number scale" or coordinate system on a line. There is a growing tendency in modern secondary school geometry texts to employ this approach rather than the classical one. The interested reader is referred to the pioneering text of Birkhoff and Beatley and to the School Mathematics Study Group text cited at the end of the chapter. One text of recent date by Brumfiel, Eicholz, and Shanks has used a modified form of Hilbert's axioms as a foundation for a first course in geometry. It seems clear that the various logical difficulties in the classical approach are in the process of being reduced or eliminated in better, more rigorous introductions to Euclidean geometry.

## REFERENCES

Eves, Howard, *A Survey of Geometry, Vol. I*, Allyn and Bacon, Boston, 1963, Chapter 8.

Klein, Felix, translated by E. R. Hedrick and C. A. Noble, *Elementary Mathematics from an Advanced Standpoint*, vol. 2, Dover Publications, New York, 1939, Part III, II, 3.

Meschkowski, Herbert, translated by A. Shenitzer, *Noneuclidean Geometry*, Academic Press, New York, 1964, Chapter 1.

Wolfe, H. E., *Introduction to Non-Euclidean Geometry*, The Dryden Press, New York, 1945, Chapter 1.

# TEXTS

Birkhoff, G. D., and R. Beatley, *Basic Geometry*, 3rd ed., Chelsea Publishing Company, New York, 1959.

Brumfiel, C., R. E. Eicholz, and M. E. Shanks, *Geometry*, Addison-Wesley Publishing Company, Reading, Mass., 1960.

Curtis, C. W., P. H. Daus, and R. J. Walker, *Euclidean Geometry Based on Ruler and Protractor Axioms*, Studies in Mathematics, vol. II, School Mathematics Study Group, New Haven, 1959.

School Mathematics Study Group, *Mathematics for High School, Geometry*, Yale University Press, New Haven, 1960.

CHAPTER TWO

# HILBERT'S AXIOMS

In this chapter we examine the foundations of Euclidean geometry using Hilbert's axioms. This system of axioms was first presented by the great German mathematician David Hilbert in a series of lectures at the University of Göttingen in 1898 and 1899. The system has undergone refinement in the ensuing years and many variations of Hilbert's original axioms are in existence, all of which are loosely called Hilbert's axioms. We use one such variation of the original system here. Although other approaches to an axiomatic foundation of Euclidean geometry are possible and perhaps even advisable at the secondary school level, this system stands as one of the first and one of the most significant of the modern reformulations of the foundations of Euclidean geometry. It has the advantage of being phrased in terms analogous to those of the classical Greek approach to the subject and of carefully distinguishing the concepts of incidence, order, and congruence, which is of value in subsequent chapters.

## 2.1 AXIOMS OF EXISTENCE AND INCIDENCE

We restrict ourselves here to plane geometry, saving for Section 2.6 consideration of an axiomatic approach to three-dimensional geometry. We take as undefined terms *point* and *line*. As an undefined relation we take *incidence*: the statement that a point lies on a line or that a line passes through a point. The first set of Hilbert's axioms is concerned with this incidence relation and with the postulation of existence of sufficient points and lines to prevent us from talking about a vacuous system. We assume the following:

1. *There exists at least one line.*
2. *On each line there exist at least two points.*
3. *Not all points lie on the same line.*
4. *There is one and only one line passing through two distinct points.*

The first of these axioms assures us that we are not talking about a vacuous system. The second asserts that the incidence relation is actually fulfilled. The third asserts that we are in at least a "two-dimensional" geometry; that is, we are concerned with a system consisting of more than one line and the points it contains.

Meager as this set of axioms seems to be, it is already sufficient for the proof of some basic theorems. We state several of these below, leaving the proofs to the reader.

**Theorem 2.11.** *Through each point there pass at least two distinct lines.*

**Theorem 2.12.** *Not all lines pass through the same point.*

**Theorem 2.13.** *Two distinct lines meet in at most one point.*

## Exercises 2.1

1. Prove that there exists at least one point.
2. Prove Theorem 2.11.
3. Prove Theorem 2.12.
4. Prove Theorem 2.13.
5. What is the minimal number of points consistent with Axioms 1 to 4?

## 2.2 AXIOMS OF ORDER

The second group of axioms is concerned with the concept of order or "betweenness," a concept that we have found missing in Euclid's approach to geometry. The first four of these are one-dimensional and establish the basic relations of order of points on a line.

We take as undefined the relation *between* as used in statements of the form, "Point $B$ is between points $A$ and $C$." For convenience we write $ABC$ when $B$ is between $A$ and $C$. We assume the following:

5. *If $ABC$ then $A$, $B$, and $C$ are distinct and collinear.*

6. *If $ABC$, then $CBA$.*

7. *If $A$ and $C$ are distinct, there is a point $B$ such that $ABC$ and there is a point $D$ such that $ACD$.*

8. *Given three distinct collinear points, one and only one of the points is between the other two.*

9. *Given four distinct collinear points, it is possible to name them $A$, $B$, $C$, and $D$ in such a way that $ABC$, $ABD$, $ACD$, and $BCD$.*

We write $ABCD$ when the points $A$, $B$, $C$, and $D$ fulfill the order relations of Axiom 9. There is a temptation to prove this axiom as a theorem. The reader will find it instructive to try such a proof! The axiom states that if a person is given four distinct collinear points,[1] two are "end points" and two are "interior points"; that is, in the four betweenness statements we can make using three of the four points, only two of the points can be between pairs of points, and each of these two points must be between exactly two different pairs chosen from the remaining three points. Thus if $P, Q$, $R$, and $S$ are distinct and collinear and if $QPS$, $QPR$, $QSR$ are given, the remaining statement must involve $R$, $S$, and $P$ and must have $S$ between $R$ and $P$; that is, we must have $QPSR$. If we have distinct collinear points $T$, $U$, $V$, and $W$ with $VTW$ and $VUW$, Axiom 9 asserts we must have either $VTUW$ or $VUTW$.

With the concept of order or betweenness now introduced, we are able to define carefully the concept of segment.

**Definition 2.21.** *The* segment *determined by the points $A$ and $B$, denoted by $A$-$B$ or $B$-$A$, is the set of all points between $A$ and $B$.*[2]

The final order axiom makes use of the concept of segment. Although part of Hilbert's system of axioms, it predates the system, being ascribed to the German geometer Pasch in 1882. It is usually known as Pasch's axiom.

10. *Given three noncollinear points $A$, $B$, and $C$ and a line not passing through any of these points, if a point of the segment $AB$ lies on the given line, a point of $AC$ or a point of $BC$ also lies on the given line.*

[1]It is possible to prove Axiom 9 once Axiom 10 has been assumed. Thus the first ten axioms are not *independent*, since one can be proved as a theorem. The proof of Axiom 9 is intricate, and for the sake of simplicity we shall retain it as an axiom rather than taking the time to prove it.

[2]As we shall use the term here, the segment does not contain its end points. Analysts often call such a segment *open*. We could just as well agree to include the end points in a segment, obtaining what is called a *closed* segment. We shall consistently exclude end points here.

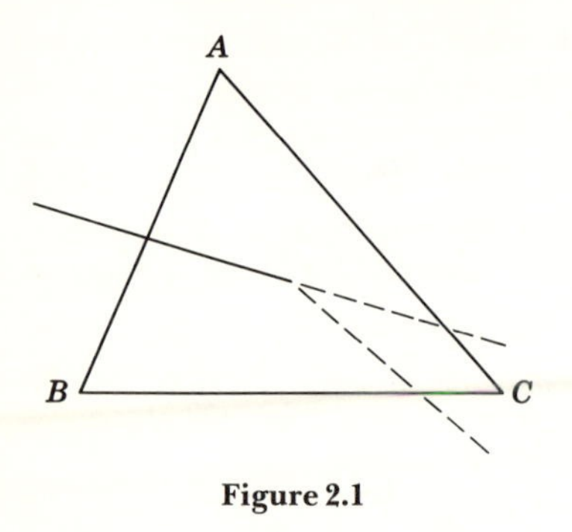


Figure 2.1

If we define triangle in the natural manner (Exercise 2.21), this axiom states that if a line meets one side of a triangle and passes through no vertex, it meets a second side of the triangle.

Many of the important properties of plane geometry that Euclid overlooked and that we tacitly assume can now be proved using these order axioms. Some of these are stated as exercises at the end of the section. We consider here two theorems that lead to a concept needed for the next section.

**Theorem 2.21.** *If O is a point on a line l, then O divides all the other points on l into two classes, two points being in the same class if and only if O is not between them. Two points in the same class are said to be on the same side of O on l.*

PROOF.[3] We choose a point $A$, other than $O$, on $l$. One class $S_1$ consists of all points $B$ for which $ABO$, $BAO$, or $A = B$. A second class $S_2$ consists of all points $C$ for which $AOC$. The existence of such points $B$ and $C$ is assured by Axiom 7. We claim that we must have the order relation $BOC$: this is immediate if $B = A$; if $ABO$, this together with $AOC$ implies $BOC$ in view of Axiom 9; a similar argument holds if $BAO$. If $C'$ is another point in $S_2$, then $AOC$ and $AOC'$ imply $OCC'$ or $OC'C$ in view of Axiom 9, showing that $O$ is not between $C$ and $C'$. If $B'$ is a second point in $S_1$, it is again impossible to have $BOB'$, since the given order possibilities for $A$, $B$, and $O$ and for $A$, $B'$, and $O$ all indicate that $O$ is an "end point" in terms of Axiom 9.

[3]It is not customary in more advanced mathematical writing to use the two-column "statement" and "reason" procedure of higher school geometry. Proofs are customarily written in expository style. The reader must then beware of false steps, and frequently he will be expected to mentally supply missing steps or the reasons for statements made. A proof can thus rarely be read as we read a novel. It may require several readings and considerable mental effort by the reader before it is thoroughly understood. The reader could profit by reexamining some of the two-column high school geometry proofs. There are many missing steps and concealed reasons in almost all such proofs.

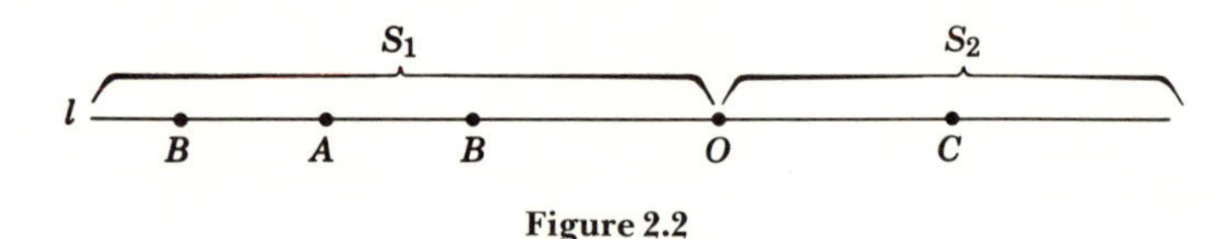


Figure 2.2

The argument shows that $S_1$ and $S_2$ have the properties claimed in the theorem. We assert, finally, that these classes are independent of the choice of point $A$. Suppose that we choose a different $A'$ distinct from $O$. If $A'AO$, any point $B$ in $S_1$ is either $A'$ or is such that $A'BO$ or $BA'O$ (again $O$ must be an "end point" in the application of Axiom 9 to the points $A$, $A'$, $B$, and $O$). Moreover, if $A'AO$ and $AOC$, so that $C$ is in $S_2$, Axiom 9 forces the relation $A'OC$. A similar argument can be carried out if $AA'O$. If $AOA'$, Axiom 9 leads to the conclusion that for any point $B$ in $S_1$, $BOA'$, while for any point $C$ in $S_2$, $A'CO$, $CA'O$, or $C = A'$.

We can also prove a two-dimensional version of this theorem which we shall use frequently in the remainder of this chapter.

**Theorem 2.22.** *Any line divides the set of all points not on the line into two classes, two points being in the same class if and only if the segment they determine does not contain a point lying on the line.*

PROOF: Let $l$ be the given line and let $A$ be a point not on $l$. Let $S_1$ be the class of points $B$ for which $B = A$ or $A$-$B$ does not contain a point of $l$. Let $S_2$ be the class of points $C$ for which $A$-$C$ contains a point of $l$. We first claim that $B$-$C$ must contain a point of $l$. This follows from Axiom 10 if $A$, $B$, and $C$ are noncollinear. If these points are collinear, let $O$ be the point of $A$-$C$ on $l$; we then have $AOC$ and $ABO$ or $BAO$, and in either case we must have $BOC$ by Axiom 9. If $B'$ is a second point

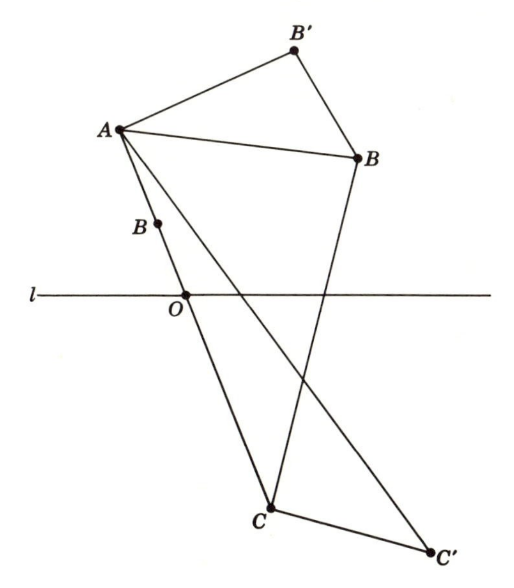


Figure 2.3

of $S_1$, then $B$-$B'$ cannot contain a point of $l$, by Axiom 10 if $A$, $B$, and $B'$ are noncollinear and by Axiom 9 if they are collinear. If $C'$ is a second point in $S_2$, then $C$-$C'$ cannot contain a point of $l$, by Exercise 2.25 if $A$, $C$, and $C'$ are collinear and by Axiom 9 if they are collinear. Finally, we assert that $S_1$ and $S_2$ are independent of the choice of $A$. The proof, similar to that above and using Axioms 9 and 10, is left to the reader.

This theorem is usually called the *plane separation theorem.* It is equivalent to Axiom 10, and many prefer to use it as an axiom, rather than using Pasch's axiom. The statement that Axiom 10 and Theorem 2.22 are equivalent means that either can be assumed and used to prove the other (in the presence of Axioms 1 to 9). Here we have assumed Axiom 10 and have used it to prove Theorem 2.22; we could just as well have assumed Theorem 2.22 and used it to prove Axiom 10 as a theorem (See Exercise 2.211). It is customary to call each of the classes of Theorem 2.22 a *half plane* determined by the given line.

With Theorem 2.21 proved we may now define the concept of ray.

**Definition 2.22.** *If $O$ is a point on line $l$, a* ray *on line $l$ with* terminal point *$O$ is the set of all points on the same side of $O$ together with $O$.*

This definition now enables us to define angle precisely.

**Definition 2.23.** *If $h$ and $k$ are distinct rays with a common terminal point, the pair of rays constitutes an* angle, *denoted by $(h, k)$ or $(k, h)$. The rays are the* sides *of the angle, and the common terminal point of the rays is the* vertex *of the angle.*

The order concepts of this section are sufficient to prove that an angle divides all points of the plane not on the rays into two classes. If the rays are on the same line (in which case the angle is called a *straight angle*), the classes are those into which the line divides the set of all points not on the line (Theorem 2.22). If the rays are not on the same line, each ray is on a line that divides the points not on itself into two

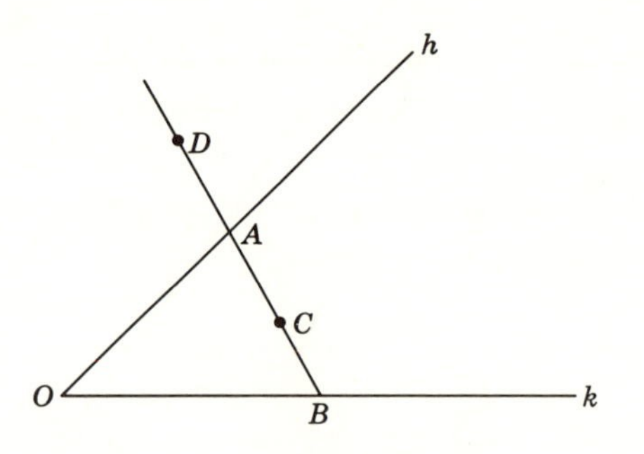


**Figure 2.4**

classes. In each case the ray not on the line of division has all of its points except one in one of these classes. If the ray $h$ is in the class $\mathscr{H}$ and the ray $k$ is in the class $\mathscr{K}$, the set of points common to $\mathscr{H}$ and $\mathscr{K}$ forms a class called the *interior* of the angle $(h, k)$, while all other points not on $(h, k)$ form a second class called the *exterior* of $(h, k)$. These two classes are nonempty. If $(h, k)$ has vertex $O$ and if we choose $A$ and $B$, each distinct from $O$, on $h$ and $k$, respectively, it is clear that any point $C$ for which $ACB$ is in the interior of $(h, k)$ while any point $D$ for which $DAB$ or $ABD$ is in the exterior of $(h, k)$. We can readily prove, using Theorem 2.22, that if a ray with vertex at the vertex of $(h, k)$ has a point in the interior of $(h, k)$, all its points except the vertex are in the interior. Thus it is sensible to speak of a ray in the interior of an angle.

## Exercises 2.2

1. Define a triangle.
2. Define a quadrilateral.
3. Prove that Axioms 1 to 7 imply there are at least three points on each line.
4. Prove that if $A$ and $B$ are distinct, there is a point $E$ such that $ABE$ and there is a point $F$ such that $FAB$.
5. Prove that a line cannot meet all three sides of a triangle. (In keeping with Definition 2.21 we note that a vertex of a triangle is not a point of a side of the triangle.)
6. Prove that a line meeting one side of a quadrilateral must meet a second side.
7. For each of the following, supply the missing order relations:
   *a.* $CAD$ and $CBD$;
   *b.* $RPT$ and $RQT$;
   *c.* $VUW$ and $TWU$.
8. Prove Theorem 2.22.
9. Prove that if a ray with vertex at the vertex of $(h, k)$ has a point in the interior of $(h, k)$, then all its points except the vertex are in the interior.
10. Let $A$ and $B$ be points on the two sides of an angle. Prove that if $ACB$ then $C$ is in the interior of the angle and that if $DAB$ then $D$ is in the exterior of the angle (see Figure 2.4).
11. Assuming Axioms 1 to 9 and Theorem 2.22, prove Axiom 10.
12. A set of points is said to be *convex* if the segment determined by any two points of the set contains only points of the set. Prove that a half plane and the interior of an angle are convex sets.

## 2.3 AXIOMS OF CONGRUENCE

The order concepts of the preceding section have enabled us to define precisely the concepts of segment and angle. We now consider a new undefined relation of *congruence* as applied to segments or angles; we write $A\text{-}B \cong A'\text{-}B'$ for segments and $(h, k) \cong (h', k')$ for angles. Logically speaking, the relation is undefined. Ultimately, if we were to recreate all of Euclidean geometry, we would introduce *measure* concepts for segment and angles, a measure being a real number (length or angle measure) associated with a segment or angle. The measure concept would be tied to the congruence concept by the stipulation that segments or angles will have the same measure if and only if they are congruent. In short the important *metric* ideas of length and measure of angle are being anticipated by the introduction of congruence axioms.

The first of the congruence axioms apply to segments and are the following:

11. *Given a segment* $A\text{-}B$ *and a point* $A'$ *on a line* $a$, *then on each ray on* $a$ *with terminal point* $A'$ *there is one and only one point* $B'$ *such that* $A\text{-}B \cong A'\text{-}B'$.

12. $A\text{-}B \cong A\text{-}B$.

13. *If* $A\text{-}B \cong A'\text{-}B'$, *then* $A'\text{-}B' \cong A\text{-}B$.

14. *If* $A\text{-}B \cong A'\text{-}B'$ *and* $A'\text{-}B' \cong A''\text{-}B''$, *then* $A\text{-}B \cong A''\text{-}B''$.

15. *If* $A\text{-}B$ *and* $B\text{-}C$ *are segments on a line* $a$ *with no common point, if* $A'\text{-}B'$ *and* $B'\text{-}C'$ *are segments on line* $a'$ *with no common points, if* $A\text{-}B \cong A'\text{-}B'$, *and if* $B\text{-}C \cong B'\text{-}C'$, *then* $A\text{-}C \cong A'\text{-}C'$.

Axioms 12, 13, and 14 state that the congruence relation is *reflexive*, *symmetric*, and *transitive*, respectively. A relation that is reflexive, symmetric, and transitive is said to be an *equivalence relation*. Such a relation divides the set on which it is defined into mutually exclusive and exhaustive subsets, each subset consisting of all elements related to each other under the given relation. In the case at hand these subsets, or *equivalence classes*, consist of all segments congruent to each other. Axiom 15 is a precise version, as applied to congruence of line segments, of the assertion often stated as "Equals added to equals give equals."

The next four axioms state properties analogous to those above for angles.

**Figure 2.5**

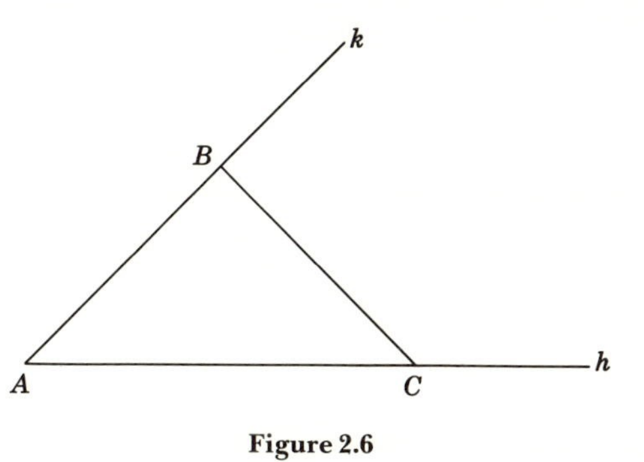


**Figure 2.6**

16. *Given an angle $(h, k)$, not a straight angle, and given a ray $h'$ on a line $a$, then on each side of $a$ there is one and only one ray $k'$ such that $(h, k) \cong (h', k')$. A straight angle is congruent to those and only those angles that are straight angles.*

17. $(h, k) \cong (h, k)$.

18. *If $(h, k) \cong (h', k')$, then $(h', k') \cong (h, k)$.*

19. *If $(h, k) \cong (h', k')$ and $(h', k') \cong (h'', k'')$, then $(h, k) \cong (h'', k'')$.*

The remaining congruence axiom deals essentially with triangles. The triangle can be defined using the order concepts of Section 2.2, and we assume it has been so defined (Exercise 2.21). We note that, given $\triangle ABC$, the segments $A$-$B$ and $A$-$C$ are parts of rays $h$ and $k$ with common terminal point $A$; thus the triangle vertex is associated with the angle $(h, k)$. We follow custom and denote $(h, k)$ by $\angle BAC$, $\angle CAB$, or when there is no ambiguity by $\angle A$.

20. *If in triangles $ABC$ and $A'B'C'$, $A$-$B \cong A'$-$B'$, $A$-$C \cong A'$-$C'$, and $\angle A \cong \angle A'$, then $\angle B \cong \angle B'$.*

This, of course, is the beginning of the "side-angle-side" congruence theorem for triangles, and some authorities prefer to use the "side-angle-side" statement as the final congruence axiom. Actually, the axiom given is sufficient, for we can use it to prove the triangle congruence theorem, after triangle congruence is defined.

**Definition 2.31.** *Triangles $ABC$ and $A'B'C'$ are* congruent $(\triangle ABC \cong \triangle A'B'C')$ *if $A$-$B \cong A'$-$B'$, $A$-$C \cong A'$-$C'$, $B$-$C \cong B'$-$C'$, $\angle A \cong \angle A'$, $\angle B \cong \angle B'$, and $\angle C \cong \angle C'$.*

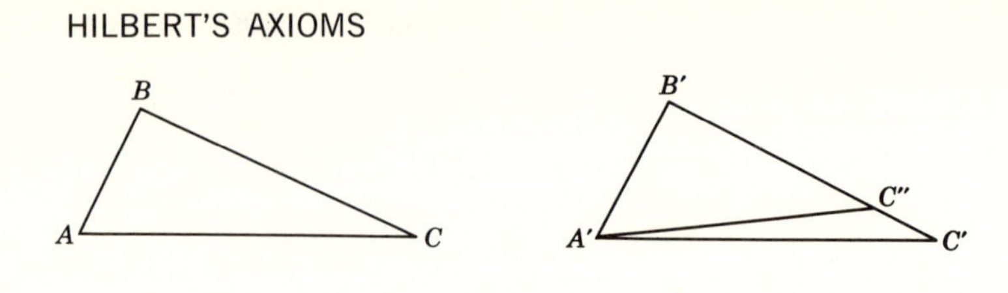


**Figure 2.7**

**Theorem 2.31.** *If in triangles $ABC$ and $A'B'C'$, $A\text{-}B \cong A'\text{-}B'$, $A\text{-}C \cong A'\text{-}C'$, and $\angle A \cong \angle A'$, then $\triangle ABC \cong \triangle A'B'C'$*[4].

PROOF. It follows at once from Axiom 20 that $\angle B \cong \angle B'$ and by symmetry (just change labels on the triangles) that $\angle C \cong \angle C'$. It will thus be sufficient to prove that $B\text{-}C \cong B'\text{-}C'$. On the ray with terminal point $B'$ containing the segment $B'\text{-}C'$ we choose a point $C''$ such that $B\text{-}C \cong B'\text{-}C''$, using Axiom 11. If $B\text{-}C$ and $B'\text{-}C'$ are not congruent, then $C'' \neq C$. If we apply Axiom 20 to triangles $ABC$ and $A'B'C''$, we conclude that $\angle B'A'C'' \cong \angle BAC$. However we have also $\angle B'A'C' \cong \angle BAC$ in contradiction of Axiom 16. We thus conclude that $B\text{-}C \cong B'\text{-}C'$.

In a like manner we can prove

**Theorem 2.32.** *If in triangles $ABC$ and $A'B'C'$, $A\text{-}B \cong A'\text{-}B'$, $\angle A \cong \angle A'$, and $\angle B \cong \angle B'$, then $\triangle ABC \cong \triangle A'B'C'$.*

The proof is left as an exercise for the reader.

Most of the initial theorems of Euclidean geometry make use of the three basic congruence theorems for triangles. Two of these have now been established. We indicate below a sequence of theorems leading to the third of these and now provable in Hilbert's system of axioms.

**Theorem 2.33.** *If $(h, k) \cong (h', k')$ and if $l$ is a ray with vertex of $(h, k)$ as terminal point, and points in the interior of $(h, k)$, then there exists a ray $l'$ with vertex of $(h', k')$ as terminal point and with points in the interior of $(h', k')$ such that $(h, l) \cong (h', l')$ and $(k, l) \cong (k', l')$.*

[4]Note that proper practice in stating congruence of triangles requires that corresponding vertices be written in corresponding positions in designations of the triangles; thus, if we are told that $\triangle PQR \cong \triangle XYZ$, this should mean $\angle P \cong \angle X$, $P\text{-}Q \cong X\text{-}Y$, and so on.

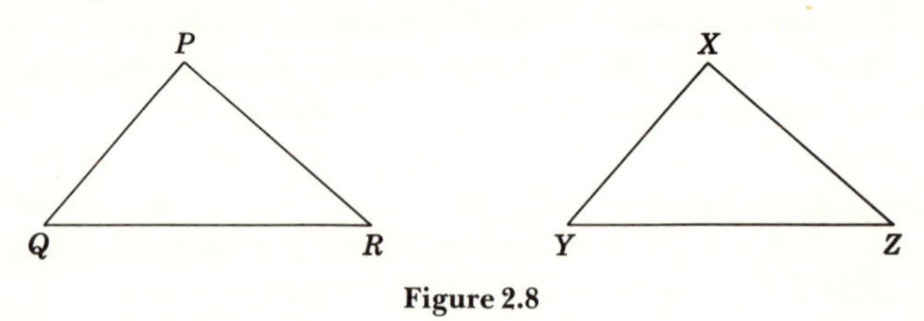


**Figure 2.8**

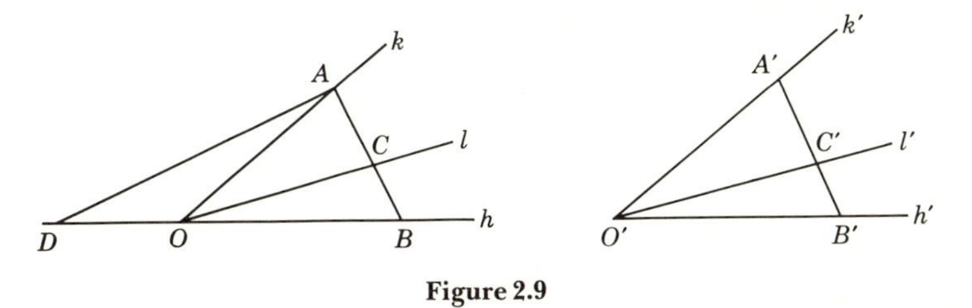


**Figure 2.9**

PROOF. We choose points $A$ and $B$, each distinct from the vertex $O$ of $(h, k)$, on $h$ and $k$, respectively. We assert that $l$ has a point $C$ in common with the segment $A$-$B$. To show this, choose a point $D$ on the line of ray $h$ such that $DOB$. The points of the segment $D$-$A$ will then be on the exterior of $(h, k)$ since they do not lie in the class $\mathscr{K}$ of the previous section. Thus the ray $l$ has no point in common with $D$-$A$. The line containing the ray $l$ has no point in common with $D$-$A$ since $D$-$A$ is in $\mathscr{H}$ while the points on the line of $l$ but not on $l$ are not in $\mathscr{H}$. It follows by Pasch's axiom that the line of $l$ must meet $A$-$B$ with the point clearly being on $l$.

Now if $(h', k')$ has vertex $O'$, choose $A'$ and $B'$ on $h'$ and $k'$, respectively, such that $O'$-$A' \cong O$-$A$ and $O'$-$B' \cong O$-$B$. Choose $C'$ on the line of $A'$ and $B'$ and on the same side of $A'$ as is $B'$ such that $A'$-$C' \cong A$-$C$. Using Theorem 2.31 and Exercise 2.29 we can prove that the ray with vertex $O'$ and containing $C'$ has the desired congruence properties and is in the interior of $(h', k')$. The details are left to the reader.

**Theorem 2.34.** *Let $h$, $k$, and $l$ be coterminal rays, let $h'$, $k'$, and $l'$ be coterminal rays, and let $l$ and $l'$ be in the interiors of $(h, k)$ and $(h', k')$, respectively. If $(h, k) \cong (h', k')$ and $(h, l) \cong (h', l')$, then $(k, l) \cong (k', l')$.*

PROOF. If $(k, l) \not\cong (k', l')$, there is a ray $l'' \neq l'$ such that $(h, l) \cong (h', l'') \cong (h', l')$ and $(k, l) \cong (k', l'')$ by Theorem 2.33. This contradicts Axiom 16.

**Theorem 2.35.** *Let $h$, $k$, and $l$ be coterminal rays, let $h'$, $k'$, and $l'$ be coterminal rays, and let $l$ and $l'$ be in the interiors of $(h, k)$ and $(h', k')$, respectively. If $(h, l) \cong (h', l')$ and $(k, l) \cong (k', l')$, then $(h, k) \cong (h', k')$.*

PROOF. If $(h, k) \not\cong (h', k')$, there is a ray $k''$ on the same side of the line of $h'$ as is $k'$ such that $(h, k) \cong (h', k'')$. By Theorem 2.33 we can find a ray $l''$ in the interior of $(h', k'')$ such that $(h, l) \cong (h', l'')$ and $(h, l) \cong (k', l'')$. If $l'' \neq l'$, the distinct congruent angles $(h', l')$ and $(h', l'')$ contradict Axiom 16. If $l'' = l'$, the angles $(k', l')$ and $(k'', l')$ contradict this axiom. (The rays $k'$ and $k''$ are on the same side of the line of $l'$, namely the side containing the ray opposite to $h$.)

**Theorem 2.36.** *The angles opposite the congruent sides of an isosceles triangle are congruent.*

PROOF. Let $\triangle ABC$ be the isosceles triangle with $A\text{-}B \cong A\text{-}C$. The common proof of this theorem uses the angle bisector at $A$ to produce a pair of congruent triangles. Note that we have not defined angle bisector and that we could not be sure without further proof that this bisector meets the opposite side. It is sufficient to note that by Theorem 2.31 $\triangle ABC \cong \triangle ACB$, hence that $\angle B \cong \angle C$.

**Theorem 2.37.** *If in triangles $ABC$ and $A'B'C'$, $A\text{-}B \cong A'\text{-}B'$, $A\text{-}C \cong A'\text{-}C'$, and $B\text{-}C \cong B'\text{-}C'$, then $\triangle ABC \cong \triangle A'B'C'$.*

PROOF. We consider the ray through $B$ and on the opposite side of the line $BC$ from $B$ to $A$ such that $\angle A''BC \cong \angle A'B'C'$. The point $A''$ on this ray is chosen so that $A''\text{-}B \cong A'\text{-}B'$. It follows that $\triangle A'B'C' \cong \triangle A''BC$. We then use Theorem 2.36 to prove that $\angle BAA'' \cong \angle BA''A$ and $\angle CAA'' \cong \angle CA''A$. Since $A$ and $A''$ are on opposite sides of $BC$, we know that $A\text{-}A''$ contains a point $D$ on $BC$. The congruence of $\angle BAC$ and $\angle B'A'C'$ follows from Axiom 19 and Theorem 2.34 if $DBC$ or $BCD$, and from Axiom 19 and Theorem 2.35 if $BDC$.

## Exercises 2.3

1. Prove that if $ACB$, $A\text{-}B \cong A'\text{-}B'$, $A\text{-}C \cong A'\text{-}C'$, $C'A'B'$ is false, and $A'$, $B'$, and $C'$ are collinear, then $A'C'B'$ and $B\text{-}C \cong B'\text{-}C'$.

2. The angles $(h, l)$ and $(k, l)$ are supplements if they contain no common points in their interiors and if $(h, k)$ is a straight angle. Prove that supplements of congruent angles are congruent.

3. Prove that vertical angles are congruent.

4. An angle is a right angle if it is congruent to its supplement. Prove that an angle congruent to a right angle is a right angle.

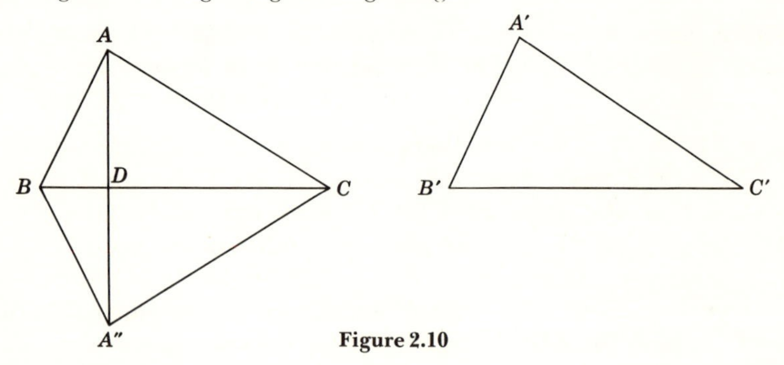


**Figure 2.10**

5. Prove Theorem 2.32.
6. Complete the proof of Theorem 2.33.

## 2.4 AXIOM OF PARALLELS

Our next group of axioms contains a single axiom. The relation involved is *parallel*, which we can define. It might seem more consistent to leave it undefined as we did for incidence, order, and congruence. In fact we can define it, and the axiom of this group could properly be placed among the axioms of incidence, for we are here dealing with that concept. It is given a separate class because of its historical importance and relation to the non-Euclidean geometries.

**Definition 2.41.** *Two lines are* parallel *if they have no common point.*

The axiom is not stated in the form Euclid used[5] and is known as Playfair's axiom, after John Playfair, who published in 1795 a geometry text widely used during the first half of the nineteenth century; the axiom in this form actually dates back to the era of Greek geometry.

21. *Given a line and a point not on the line, there is one and only one line containing the given point and parallel to the given line.*

An immediate consequence of this axiom is

**Theorem 2.41.** *Distinct lines parallel to the same line are parallel to each other.*

If we define alternate interior angles in the usual manner, we can use the triangle congruence theorems to prove

**Theorem 2.42.** *Two lines are parallel if and only if alternate interior angles formed by a transversal on the given lines are congruent.*

From this we obtain at once

**Theorem 2.43.** *Given a triangle ABC, if angles* $(h, k)$, $(k, l)$, *and* $(l, m)$ *are congruent to angles A, B, and C, respectively, and if the former angles have no common points in their interiors, then* $(h, m)$ *is a straight angle.*

## Exercises 2.4

1. Prove Theorem 2.41.
2. Prove Theorem 2.42.
3. Prove Theorem 2.43.

[5]See page 220 of the work by Wolfe cited at the end of Chapter 1.

4. Why can one not state Theorem 2.43 at this point in the form, "The sum of the angles of a triangle is a straight angle"?

## 2.5 AXIOMS OF CONTINUITY AND COMPLETENESS

Our intuition tells us that we can put a number scale on a Euclidean line and that there is room on the line for each real number. More precisely, there is a one-to-one correspondence between the real numbers and the points on a line. This use of a real number scale plays an important role in Euclidean geometry. It can be used to proceed from the congruence relations of Section 2.3 to the ideas of length and angle measure. These in turn are essential in any study of similarity of figures and in the study of area.

The preceding axioms do not imply the presence of the real number scale on a line. A "rational number geometry" would be consistent with these axioms. We could assume that the only points of the Euclidean plane present are those with rational coordinates and that the only lines are those expressible in terms of linear equations with rational coefficients. Such a system of points and lines can be shown to be consistent with the axioms above. We might be tempted to claim that nonparallel lines could fail to meet by crossing at a common "hole" (point with irrational coordinates). Such is not possible, however, for the solution of a pair of linear equations with rational coefficients is rational.

Many modern approaches to secondary school geometry face up squarely to the importance of a real number scale on a line and build their axiom systems around this concept.[6] Hilbert forced the issue by adding two axioms to his system. We state them here, but do not undertake all the analysis required to affirm that they do indeed imply the presence of the real number scale.

22. *Given points $A$ and $B$ and a point $A_1$ between $A$ and $B$, let points $A_2$, $A_3, \cdots$ be chosen so that $A_1$ is between $A$ and $A_2$, $A_2$ is between $A_1$ and $A_3$, etc., and such that $A\text{-}A_1 \cong A_1\text{-}A_2 \cong A_2\text{-}A_3 \cong \cdots$. Then there is a positive integer $n$ such that $B$ is between $A$ and $A_n$.*

This is a geometric statement of the principle of *Archimedean order*. If we assume the idea of length for the moment and let $a$ be the length of $A\text{-}A_1$ and $b$ the length of $A\text{-}B$, we can restate the principle in its usual analytic form

*Given arbitrary $a > 0$ and $b > 0$, there is a positive integer $n$ such that $na > b$.*

[6]See the texts of Birkhoff and Beatley and of the School Mathematics Study Group cited at the end of the chapter.

$A \quad A_1 \quad A_2 \quad A_3 \qquad B \quad A_n$

**Figure 2.11**

A "number system"[7] is said to be *ordered* if there is defined a relation $>$ between pairs of elements of the system which is transitive,

*If* $a > b$ *and* $b > c$, *then* $a > c$,

and obeys the *trichotomy*,

*Given* $a$ *and* $b$, *one and only one of the following holds:* $a > b$, $a = b$, *or* $b > a$.[8]

The system is said to be *Archimedean-ordered* if in addition it fulfills the analytic principle of Archimedean order stated above.

It is possible to construct non-Archimedean-ordered "number systems." It can be shown, however, that any Archimedean-ordered "number system" can be taken to be a part of the real number system. The sense of Axiom 22, therefore, is to require that any "number scale" ultimately used on a line be a part of the real number system. The effect of the final axiom of completeness is to require that the "number scale" so used be all of the real number system.

23. *No additional points or lines can be added to the system without violating one of the preceding axioms.*

These twenty-three axioms are sufficient for a rigorous development of Euclidean geometry. We have indicated some of the early theorems in this chapter but do not pretend to have recreated completely the foundations of Euclidean geometry based on this axiom system. The reader is referred to the references at the end of the chapter for further details in this development. In succeeding chapters we assume that Euclidean geometry has been placed on a firm footing and investigate some of the extensions of this geometry.

## Exercises 2.5

1. Consider the set of all ordered pairs $(a, b)$ of real numbers. Define $>$ by $(a, b) > (c, d)$ if $a > c$ or if $a = c$ and $b > d$. Show that the order is transitive and

[7]We do not here define what we mean by "number system." The proper term is *field*, which is defined and considered in detail in Chapter 10.

[8]This order can be related to the order of Section 2.2 as follows: Given segments $A$-$B$ and $A'$-$C$ with associated "numbers" $b$ and $c$, choose a $B'$ on the line of $A'$ and $C$ on the same side of $A'$ as $C$ such that $A$-$B \cong A'$-$B'$. We say $b > c$ if $C$ is between $A'$ and $B'$. The axioms of order can be shown to imply that the order so defined is transitive and obeys the trichotomy.

obeys the trichotomy but is non-Archimedean. We define $(a, b) + (c, d) = (a + c, b + d)$.

## 2.6* HILBERT'S AXIOMS FOR SOLID GEOMETRY

The preceding sections have contained the axioms pertinent to a development of plane geometry. We outline here briefly the additions and changes required to obtain an axiom system for three-dimensional geometry.

Most of the changes needed are in the axioms of existence and incidence. We take *plane* as an additional undefined term and presume that the undefined incidence relation also applies to statements that a point or a line is on a plane. We must add the following incidence axioms:

24. *On every plane there are at least three points not on a line.*
25. *Not all points lie on the same plane.*
26. *Three points not on a line lie on one and only one plane.*
27. *If a point lies on a line and the line lies on a plane, then the point lies on the plane.*
28. *If two points of a line lie on a plane, the line lies on the plane.*
29. *If two planes have a point in common, they have a second point in common.*

Axiom 3 is now superfluous, having been replaced by Axioms 24 and 25.

No additional order axioms are required, but Axiom 10 must be restated to require that the three given points and the given line all be coplanar (lying in the same plane).

No additional congruence axioms are required. In the definition of angle the points in the interior and exterior of the angle are to be in the plane of the two rays determining the angle. In Theorem 2.43 the rays $h, k, l$, and $m$ must be coplanar.

Finally, Definition 2.41 must have the word "coplanar" inserted to avoid making skew lines parallel by definition.

### Exercises 2.6

1. Prove that there are at least three planes through every point.
2. Prove that not all planes contain a common point.
3. Define parallel planes.

4. Prove that two nonparallel planes meet in a line.

5. Prove that any plane divides the set of all points not on the plane into two classes, two points being in the same class if and only if the segment they determine does not contain a point of the plane.

## REFERENCES

Birkhoff, G. D., and R. Beatley, *Basic Geometry*, 3rd ed., Chelsea Publishing Company, New York, 1959.

Blumenthal, Leonard M., *A Modern View of Geometry*, W. H. Freeman and Company, San Francisco, 1961, Chapter 1.

Golos, Ellery B., *Foundations of Euclidean and Non-Euclidean Geometry*, Holt, Rinehart, and Winston, Inc., New York, 1968, Chapters 5 to 8.

Hilbert, David, translated by E. J. Townsend, *The Foundations of Geometry*, The Open Court Publishing Company, Chicago, 1910, Chapter 1.

Meschkowski, Herbert, translated by A. Shenitzer, *Noneuclidean Geometry*, Academic Press, New York, 1964, Chapter 2.

Robinson, Gilbert de B., *The Foundations of Geometry*, The University of Toronto Press, Toronto, 1940, Chapters 1, 5.

School Mathematics Study Group, *Mathematics for High School, Geometry*, Yale University Press, New Haven, 1960.

Tuller, Annita, *A Modern Introduction to Geometries*, D. Van Nostrand Company, Princeton, 1967, Chapter 2.

## CHAPTER THREE

# THE GROWTH OF GEOMETRY

In the preceding chapters we have considered some of the logical difficulties in the classical development of Euclidean geometry and have seen one possible way of resolving these difficulties. The reexamination of the foundations of Euclidean geometry has taken place gradually over the centuries and reached its climax only in the late nineteenth and early twentieth centuries. This study of the foundations is only one of the currents present in geometry since the publication of Euclid's *Elements*. Throughout the years new theorems have continually been proved, and this process is still going on. In addition to reconstructing the foundations and adding new theorems mathematicians have investigated the relations between axioms and have considered the effect of altering some of these axioms. The result has been the creation of new geometries, related to Euclidean geometry but different from it. The study of these new geometries and the relations among them has considerably broadened our understanding of geometry, and the very discovery of them has had an impact on all of mathematics by showing the value of a strict axiomatic approach and the fruitfulness of considering axioms not necessarily suggested by intuition.

## 3.1 THE GREEK ERA

When we think of ancient Greek geometry, we tend to think of Euclid and the *Elements*. Although this work is perhaps the greatest work in ancient geometry, it certainly is not the only one. Ancient geometry was created by many men and even predates the Greeks. The word "geometry" means "earth measure," and the early growth of the subject was occasioned by the need to measure areas. The Babylonians and Egyptians both developed extensive sets of rules for such measurement, some of them exact and some of them approximate. The entire development by these peoples was completely empirical; the creation of deductive geometry is the gift of the Greeks.

The first Greek credited with the treatment of geometry as a logical system is Thales (640 to 546 B.C.). He is credited with proving such basic theorems as the equality of vertical angles and of the base angles of an isosceles triangle. He spent much time in Egypt, certainly knew the Egyptian empirical geometry, and extended the practice of measurement. It should be noted, however, that many of his results are not of fundamental importance in measurement and mark the beginning of the study of geometry as a branch of knowledge independent of the practical measurement science.

Thales was the originator of a school of geometers which flourished for a hundred years and extended his results. This school was succeeded by that of Pythagoras (580 to 500 B.C.), which produced not only the Pythagorean theorem, but also the theory of parallels, the theory of similar figures, and a fascinating combination of number theory and mysticism. The formalization and intensive study of the logic of geometry is the work of Plato, Aristotle, and their followers.

About 300 B.C. the center of Greek intellectual life shifted to Alexandria, Egypt. In the preceding three hundred years since the appearance of Thales a considerable body of geometric knowledge had been developed. Euclid's great contribution was the collection and synthesis of this knowledge into one logical structure, the *Elements*. This was not accomplished by collecting all known theorems on $3 \times 5$ cards, shuffling them, and writing some appropriate connecting remarks between them. It was necessary to decide which geometric statements to assume as axioms and then decide in what order the known results were to be presented to be a logical sequence. This is not a simple matter. The organization required is overwhelming, and it is almost certainly the case that Euclid created many new proofs in the process of achieving a logical arrangement. He can be criticized for logical incompleteness, as we have done in Chapter 1, and he can be belittled by the claim of nonoriginality of many of the results, but the work is one of the first magnitude for its time and for all time. It could not have been written by a second-rate mathematician!

The Alexandrian school continued into the third century of our era and contributed much to mathematics after the time of Euclid. Certainly the greatest of the men of this school was Archimedes (287 to 212 B.C.). The axiom bearing his name that we encountered in Chapter 2 was an assumption he made about areas; indeed, his work in the problem of finding areas came very close to creating the integral calculus centuries before the time of Newton and Leibniz. Others of note in this school were Apollonius (262 to 190 B.C.), who contributed much to the study of the conics, and Pappus (A.D. 300), who extended the work of the *Elements*.

## 3.2 THE DEVELOPMENT OF NON-EUCLIDEAN GEOMETRIES

Geometry suffered the same fate as most of the knowledge of antiquity and was dormant through the Dark Ages. A revival of interest occurred about 1600, concerned primarily with the development of analytic geometry and its techniques. The *Elements* was translated into the modern European languages beginning in 1492, and many scholars devoted themselves to a critical study of this work.

Most intriguing to them was the axiom of parallels. Men had such a strong intuitive feeling for the inevitability of the parallel axiom that they were convinced it was not independent of the others, but could actually be deduced as a theorem. There is even some suspicion that Euclid was not convinced of its independence, for he avoids it as long as possible and does not use it until Proposition 29 of Book I. (If a transversal cuts two parallel lines, it makes equal alternate interior angles with them.)

We know now that the axiom of parallels is independent of the others. If we replace Playfair's axiom by the axiom:

21′. *Given a line and a point not on the line, there are more than one line containing the given point and parallel to the given line.*

we obtain a consistent, non-Euclidean geometry. This geometry, known as *hyperbolic* geometry, is different in many respects from Euclidean geometry, but the first twenty-eight propositions of the *Elements* do remain valid in it.

If we replace Playfair's axiom by:

21″. *Given a line and a point not on the line, there is no line containing the given point and parallel to the given line,*

we obtain an inconsistent set of axioms, but with suitable modification of the other axioms we obtain a different non-Euclidean geometry, known as *elliptic* geometry (Euclidean geometry is said to be *parabolic*). It contains in common with Euclidean geometry the first fifteen propositions (Proposition 16, an exterior angle of a triangle is greater than either opposite interior angle, contains in its proof as assumption that a line is infinite in extent and does not close up on itself; this assumption cannot be made in elliptic geometry, where lines are similar to the great circles of spherical geometry).

Many mathematicians attempted proofs of the parallel axiom, but failed and produced only false proofs. Notable among these were the attempts of Proclus (410 to 485), Wallis (1616 to 1703), Lambert (1728 to 1777), and Legendre (1752 to 1833). One of the most tragic of these was the attempt of the Italian mathematician Gerolamo Saccheri (1667 to 1733) to prove the axiom by contradiction. In the process of "proving" it, Saccheri actually derived many of the basic theorems of hyperbolic geometry, but he never realized what he had done and concluded by lamely inferring a contradiction which in fact he did not have.

Probably the first man to comprehend the true position of the axiom of parallels was Carl Friederich Gauss (1777 to 1855). Gauss never published his findings, but there is evidence that he made considerable advance in the study of hyperbolic geometry by 1831. Full credit for the creation of hyperbolic geometry must go to Johann Bolyai (1802 to 1860), who published his findings in 1832, and to Nikolai Lobachewsky (1793 to 1856), who published his in 1829. Neither knew of the other's work. The world did not immediately accept the findings of these men, and the modern study of non-Euclidean geometries can be said to date from the great lecture of Bernhard Riemann (1826 to 1866) given in 1854, showing the existence of elliptic and many other non-Euclidean geometries.

## 3.3 THE DEVELOPMENT OF PROJECTIVE GEOMETRY

The famous astronomer Johann Kepler (1571 to 1630) made important contributions to geometry as well as to astronomy. Among other things, he suggested that new points be added to the Euclidean plane, "points at infinity" where parallel lines were to meet. The suggestion did not receive wide and immediate adoption, but it was to bear fruit and ultimately create a new geometry, different from that of Euclid or the non-Euclidean ones described in the previous section.

At the same time additional results in Euclidean geometry were being proved by Gerard Desargues (1593 to 1662), Blaise Pascal (1623 to 1662), and Philippe de la Hire (1640 to 1718). These results, like some of Pappus', were different from much of the material of the *Elements* in that they involved only the incidence relations. As time passed men came to realize that Kepler's suggestion was a fruitful one, leading to a new geometry, and that the results of Pappus, Desargues, Pascal, de la Hire, and others remained valid in this new and more general geometry, which was given the name *projective geometry*.

The first text on the subject is due to Victor Poncelet (1788 to 1867), and the subject developed rapidly in the early part of the nineteenth century with the work of Poncelet, Joseph Gergonne (1771 to 1859), Charles Brianchon (1785 to 1864), and others. The great unification and clarification of the foundations came in the middle of the century with the work of Jacob Steiner (1796 to 1863) and Karl von Staudt (1798 to 1867). The subject flourished throughout the remainder of the century. The climax came at the turn of the century when Hilbert, Oswald Veblen (1880 to 1960), and others made important contributions to an understanding of its foundations and relation to Euclidean and the non-Euclidean geometries.

In succeeding chapters we investigate projective geometry in some detail. It is different from Euclidean geometry, and, as we shall see, it is a geometry of incidence alone in its more elementary aspects. Many consider it to be simpler, more elegant, and more aesthetically satisfying than Euclidean geometry.

## REFERENCES

Blumenthal, Leonard M., *A Modern View of Geometry*, W. H. Freeman and Company, San Francisco, 1961, Chapter 1.

Coxeter, H. S. M., *Non-Euclidean Geometry*, 5th ed., University of Toronto Press, Toronto, 1965, Chapter 1.

Dantzig, Tobias, *The Bequest of the Greeks*, Charles Scribner's Sons, New York, 1955.

Eves, Howard, *An Introduction to the History of Mathematics*, Rinehart and Company, New York, 1953.

Golos, Ellery B., *Foundations of Euclidean and Non-Euclidean Geometry*, Holt, Rinehart, and Winston, New York, 1968, Chapter 9.

Hofman, Joseph E., translated by Frank Gaynor and Henrietta O. Midonick, *The History of Mathematics*, Philosophical Library, New York, 1957.

Meschkowski, Herbert, translated by A. Shenitzer, *Noneuclidean Geometry*, Academic Press, New York, 1964, Chapter 3.

Meserve, Bruce E., *Fundamental Concepts of Geometry*, Addison-Wesley Publishing Company, Reading, Mass., 1955, Chapter 7.

Sanford, Vera, *A Short History of Mathematics*, Houghton Mifflin Company, Boston, 1930.

Sanger, R. G., *Synthetic Projective Geometry*, McGraw-Hill Book Company, New York, 1939, Chapter 9.

Struik, Dirk J., *A Concise History of Mathematics*, 2nd rev. ed., Dover Publications, New York, 1948.

Tuller, Annita, *A Modern Introduction to Geometries*, D. Van Nostrand Company, Princeton, 1967, Chapter 1.

Wolfe, H. E., *Introduction to Non-Euclidean Geometry*, The Dryden Press, New York, 1945, Chapters 2, 3.

CHAPTER FOUR

# PROJECTIVE GEOMETRY

"Parallel lines meet at infinity." Most of us have made this statement or have heard others make it. We did not learn it in our high school geometry classes, and if we ever made the statement there, we were probably told that the statement made no sense. After all, what is infinity to mean in this sentence? Is it not contradictory in view of the fact that parallel lines are defined to be lines that never meet? We must conclude that the statement is nonsense in Euclidean geometry. Although we have no right to talk about "points at infinity," we can add new points to the Euclidean plane in such a way as to remove the exceptions brought about when lines fail to meet. The resulting geometry will not be Euclidean, but will be said to be projective. It is the object of our present study. It is more general than Euclidean geometry and is even incapable of treating some concepts of that geometry, but many believe that it has far more beauty and elegance than does traditional Euclidean geometry.

## 4.1 THE PROJECTIVE PLANE

Let us begin by assuming that we are in the Euclidean plane and that all of its properties have been justified, say by Hilbert's axioms. Our goal is to add additional points to this plane so that the exceptional cases in the incidence relations are removed. If we consider two parallel lines, we presumably must add a point which is to be the point of intersection

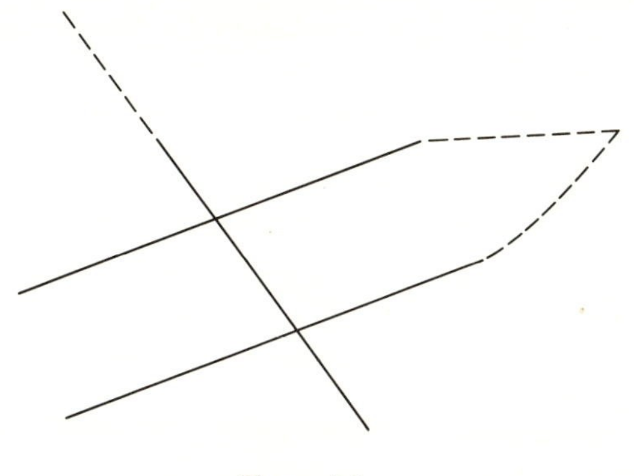

**Figure 4.1**

of these lines. If we now consider a third line parallel to these, the same point might well serve as the point of intersection of this line with either of the original lines. On the other hand, if we consider two intersecting lines, we do not wish to introduce a new common point for these lines, which we wish to continue to have only one point of intersection. It would seem that we would have to add one new point to the plane for each family of parallel lines in the plane. In addition to adding new points we also have to specify the lines on which they lie.

If we agree for the moment that any line is parallel to itself, the relation of parallelism is reflexive, symmetric, and transitive, hence is an equivalence relation (See Section 2.3). The relation thus divides the set of all lines in the plane into mutually exclusive and exhaustive equivalence classes, each class containing the totality of lines parallel to each other or the totality of lines with a given direction. For each class we add a point to the Euclidean plane; the point so added shall lie on all lines of the class with which it is associated and on no other line. We also add one line to the Euclidean plane; all of the new points added shall lie on this line, but no point of the original Euclidean plane shall lie on it.

We call the new points *ideal points;* the original points are called *ordinary points*. The original lines of the Euclidean plane are called *ordinary;* the new line is called the *ideal* line. The resulting system of points and lines is called the *real projective plane*.

## 4.2 THE BASIC INCIDENCE RELATIONS

Using the incidence axiom above, we may now show that the exceptions in incidence relations caused by parallelism have been removed.

In one case we remove the exception in Euclidean geometry brought about by parallelism. In the other case there are no exceptions, but new proof is needed, since we have added new points and lines to the Euclidean plane.

**Theorem 4.21.** *Two distinct points lie on one and only one line.*

PROOF. If the two points are ordinary, the statement follows from Euclidean geometry and the fact that the ideal line contains no ordinary points. Suppose next that one point $A$ is ordinary and one point $P$ is ideal. There is a unique line of the equivalence class determined by $P$ which contains $A$; this line is clearly the unique line containing $A$ and $P$. If both points are ideal, they lie on the ideal line and can lie on no ordinary line, for any such line contains only one ideal point.

Note that Theorem 4.21 holds in Euclidean geometry, yet we have had to consider three cases in the proof in view of the addition of ideal points to the Euclidean plane.

**Theorem 4.22.** *Two distinct lines meet in one and only one point.*

This theorem verifies that we have indeed removed the parallelism-induced exception of Euclidean plane geometry. The proof of the theorem is left as an exercise for the reader.

## Exercises 4.2

1. Prove Theorem 4.22.

## 4.3 THE CONTENT OF PROJECTIVE GEOMETRY

Before we embark on the study of the geometry of the new projective plane we have created let us consider the types of theorems we can anticipate in this new geometry and how they will be related to the theorems of Euclidean geometry.

We created the projective plane by adding ideal points and an ideal line to the Euclidean plane. The goal in this addition was to simplify certain incidence relations. We have verified in the last section, by proving Theorems 4.21 and 4.22, that we have achieved the sought after simplification. This does not mean, however, that ideal points and ideal lines should be labeled as such; indeed, the word "ideal" has no place in projective geometry. If we persist in making a special case of ideal points, we will have achieved no simplification of incidence relations, but rather will have transferred the exceptions caused by parallelism to exceptions caused by the possibility of ideal elements.

Discussion of cases and the distinguishing between ideal and ordinary were necessary in proving Theorems 4.21 and 4.22, but once these theorems have been proved we can use them and not inquire into the nature of the point or line involved; we now know that two distinct lines meet in a unique point—no exceptions—and we do not need to worry about whether one of the lines or the point is ideal. Theorems 4.21 and 4.22 are incidence properties of projective geometry, and since they do not contain any distinction between ideal and ordinary in their statements, neither needs any theorem that is a consequence of these theorems alone.

Hilbert's final axiom states that the system of lines and points cannot be enlarged and remain consistent with the previous axioms. It is pertinent then to inquire which of Hilbert's axioms are no longer applicable in the projective plane.

It is clear that the parallel axiom is no longer valid.

We claim that the order concept of Euclidean geometry creates difficulties if an attempt is made to extend it to projective geometry. The following discussion is not at all rigorous but should indicate the presence of difficulties in any attempt to introduce the order relation in projective geometry.

We can use Hilbert's axioms to introduce the real number scale on the Euclidean line in such a way that order of points of Hilbert's axioms will be consistent with the order of real numbers associated with the points; that is, if points $A$, $B$, and $C$ have as coordinates the real numbers $x$, $y$, and $z$, respectively, and if $ABC$, then $x < y < z$ or $z < y < x$. What is the effect of introducing a new ideal point on the line? Where will it fit in the order of points on the line? We are perhaps tempted to think of this point as a "point at infinity." If we think of the ideal point $I$ of the line as being "infinitely far to the right" on a line and if $A$ and $B$ are distinct ordinary points on the line, we are saying that the order relation must be $ABI$ or $BAI$; that is, $I$ cannot be between two ordinary points. This is inconsistent with Axiom 7, for if we are given an ordinary point $A$ and the ideal point $I$, we can by that axiom find a point $D$ (necessarily ordinary) such that $AID$. Such a "positioning" of the ideal point thus

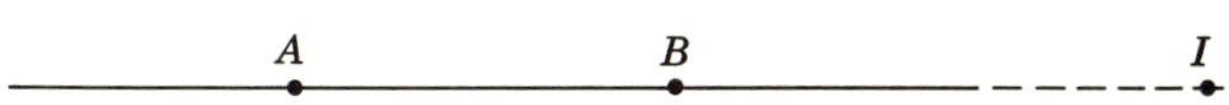

A B I

**Figure 4.2**

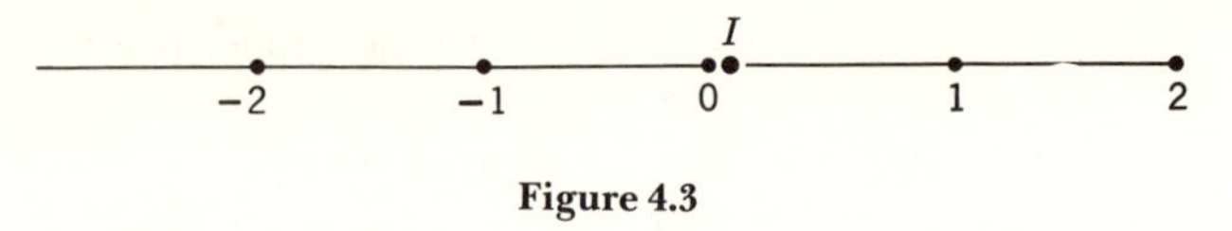


Figure 4.3

leads to difficulties. Putting the ideal point "infinitely far to the left" will of course lead to the same difficulties.[1]

Apparently then we must put the ideal point in the "middle" of the line somewhere. We assert that this also leads to difficulties. Suppose for example, that we try to insert the ideal point between the points with positive coordinates and those with nonpositive coordinates. Then the ideal point will be between two points if and only if one of them has a positive coordinate and the other has a negative or zero coordinate. Let us consider the ideal point and the point with zero coordinate. If Axiom 7 were to apply, there would be a point between these. This point could not have a positive coordinate, for then we would have the ideal point between two points with positive coordinates. It could not have a negative coordinate, for then we would have the point with zero coordinate between two points with negative coordinates. It could not have zero coordinate without violating Axiom 5. A similar difficulty will arise if we attempt to "insert" the ideal point anywhere on the Euclidean line.

If Hilbert's axioms of order were to prevail, they could not be associated with the order of the real numbers used as coordinates in Euclidean geometry. Order is conceivable, but it will not be consistent with the order we used in Euclidean geometry. We accordingly omit order considerations here.

If order is omitted, we cannot define segment, ray, or angle, hence cannot consider congruence. These concepts are also omitted here. This is not to say that it is impossible to introduce metric ideas in projective geometry; we assert only that they are not natural extensions of the same concepts used in Euclidean geometry.

In the projective geometry we study here, we limit ourselves to the use

[1]If "infinitely far to the left" and "infinitely far to the right" were to make sense at all, they would seemingly have to be the same. The parallel lines shown in Figure 4.4 would seemingly have to meet "infinitely far to the left and right" both, but they are to meet in only one point. There is an indication here that the real projective plane is *closed* in some sense. This is indeed the case, but we do not consider the matter here.

Figure 4.4

of Hilbert's axioms of incidence together with Theorems 4.21 and 4.22 and the assumption that the real number scale is present on all but one of the points of a projective line. These in effect are the axioms of real projective geometry, and we list them here for convenience.

1. *There exists at least one line.*

2. *On each line there exist at least three*[2] *points.*

3. *Not all points lie on the same line.*

4. *Two distinct points lie on one and only one line.*

5. *Two distinct lines meet in one and only one point.*

6. *There is a one-to-one correspondence between the real numbers and all but one point of a line.*

## 4.4 THE PRINCIPLE OF DUALITY

There is an interesting relation among the statements above, together with certain elementary consequences of them (Theorems 2.11 and 2.12 and Exercise 2.11). We restate these results in two columns.[3]

[2]The two points of Hilbert's axiom plus the ideal point added to the line.

[3]In the first column we should really include the statement of the one-to-one correspondence between the real numbers and all but one of the points of a line. The second column should then contain the statement that there is a one-to-one correspondence between the real numbers and all but one of the lines through a point. This is readily seen if we consider the points of intersection of these lines with a line not through the given point. We can label each of the lines except one with the real number associated with its point of intersection on the transversal.

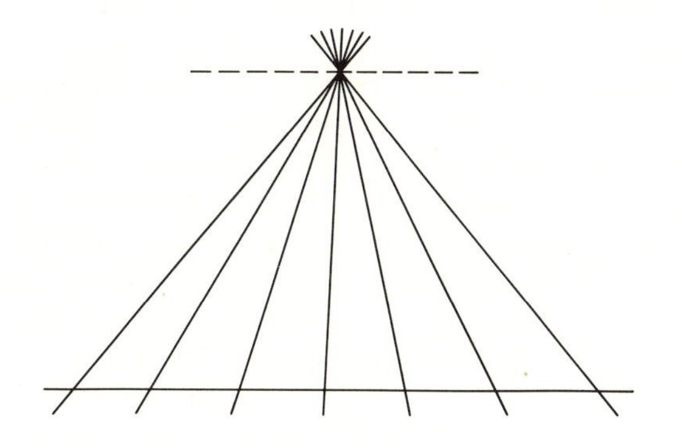

**Figure 4.5**

There exists at least one line.
On each line there exist at least three points.
Not all points lie on the same line.
Two distinct points determine a unique line.

There exists at least one point.
Through each point there pass at least three lines.
Not all lines pass throught the same point.
Two distinct lines meet in a unique point.

If the words "point" and "line," "lie on" and "pass through," and "determine" and "meet in" are interchanged in any of the statements, the statement is changed into its opposite. This property has a very powerful consequence whenever we prove a theorem, for it guarantees that the statement we get from the proved theorem by making the interchanges in words is valid without further proof. In short whenever we prove a theorem we have actually proved two theorems; for example, if we used Theorem 4.21, then Axiom 1, then Theorem 4.22 to justify the steps in the proof of the original theorem, then citation of Theorem 4.22, Exercise 2.11, then Theorem 4.21 will justify the theorem obtained by the interchange of words. Because of the parallelism it is not necessary to write out the details of the proof of the second theorem; we simply note that the principle makes it follow at once from the first theorem. We call the principle the principle of *duality* or *plane duality.* We say the second theorem is the (*plane*) *dual* of the first.

The dual of the statement, "No line can meet a second line in two distinct points," is the statement, "No point, together with a second point, can determine two distinct lines." The dual of the configuration consisting of four noncollinear points is that consisting of four nonconcurrent lines. The dual of a point and a line not through the point is a line and a point not on the line.

Since the ideas implicit in the key words can be expressed in many ways, the dual statement cannot always be obtained by a mechanical change of words. The statement, "Two distinct lines meet in a unique point," can also be stated as "Two distinct lines determine a unique point," "The intersection of two distinct lines is a unique point," or even "Two distinct lines lie on a unique point," if we agree that a line lies on a point if it contains the point. If we do make this latter agreement, it is actually possible to restate the basic statements so that the only change in words required is the interchange of "point" and "line"; thus Theorems 4.21 and 4.22 can be stated as "Two distinct points lie on a unique line" and "Two distinct lines lie on a unique point," respectively. The beginner may wish to alter statements to be dualized to introduce the "on" idea given above, but with practice and experience he should soo reach the point where determining the dual is an easy and automatic process.

There is no logical relation between a theorem and its dual in the sense that one is always a converse, inverse, or contrapositive of the other. In general a pair of dual theorems will not be so related, but on occasion we discover that one is the converse of the other. In such a case, of course, a separate proof of the converse will not be required; it will follow from the duality principle.

Since duality is present in the foundations of projective geometry, we can be sure that it will hold throughout the subject. It should be noted, however, that to preserve this duality it will be necessary to "dualize" each definition made in order to have a name for the dual of the concept defined; thus when we define triangle, quadrilateral, and so on, we have to assign names also to the duals of these configurations. If we define points to be *collinear* when they lie on a common line, we must define the dual concept, the property of lines lying on a common point; this dual concept is defined to be *concurrence*.

## Exercises 4.4

1. Write duals of each of the following:
   *a.* The set of all points on a line.
   *b.* Four points, no three of which are collinear.
   *c.* Two lines, and a point on neither line.
   *d.* The line determined by a given point and the point of intersection of two given lines.
   *e.* All lines in the projective plane.
   *f.* Not all points are on the same line.
   *g.* Distinct concurrent lines have only one common point.
   *h.* Given a line and a point not on the line, distinct lines through the given point meet the given line in distinct points.

## 4.5* PROJECTIVE SPACE

Just as we can remove the incidence exceptions created by parallelism in plane geometry, so we can remove them in the geometry of three dimensions. Although we can extend three-dimensional Euclidean geometry by extending each plane in it in the manner of Section 4.1, it is simpler to make the extension for all of space at once, starting in a manner analogous to that of Section 4.1. It is then readily shown that each plane in the resulting space is a projective plane with all the properties of the previous sections of this chapter.

As in the plane case, the set of all lines in space can be subdivided into equivalence classes of parallel lines where each class contains the totality of lines with a given direction. For each such class we adjoin one point to Euclidean space. Each such point shall lie on each line of the class with which it is associated and on no line of any other class. A point so adjoined shall lie on a plane of Euclidean space if and only if it lies on a line in that plane. If a new point $P$ lies on a plane $\pi$, it then lies on a line (actually many lines) $l$ in $\pi$. If $\pi'$ is now any plane parallel to $\pi$, then $\pi'$ will contain lines parallel to $l$, hence will contain $P$. In a like manner, we see that any point $Q$ on $\pi'$ must lie on $\pi$. We conclude that the set of new points on any two parallel planes, hence on the entire set of planes parallel to each other, is the same. We next adjoin new lines to Euclidean space, one for each family of parallel planes. The set of points on each of these new lines shall be the set of new points on each plane of the family determining the line. Each line so adjoined shall lie on each plane of the family determining it. Finally, we adjoin one plane to Euclidean space. The points and lines on this plane shall be the new points and lines adjoined and no others. The new points, lines, and plane added are called *ideal*; the original points, lines, and planes are called *ordinary*. The resulting system of points, lines, and planes is said to be *real projective space*.

We now prove the incidence relations.

**Theorem 4.51.** *Two distinct planes meet in one and only one line.*

PROOF. If both planes are ordinary and not parallel, their intersection will contain the ordinary points of their ordinary line $l$ of intersection as planes in Euclidean space. It will also contain the ideal point of this ordinary line in view of the manner in which the ideal points were adjoined. The intersection clearly contains no additional ordinary points. We suppose that the intersection contains a second ideal point. Let $m$ and $n$ be lines in the two planes through this second ideal point. These lines are parallel to each other and not parallel to $l$. Because $l$ and $m$, and $l$ and $n$, are coplanar, we know that each pair meets in an ordinary point. Since the ordinary line $l$ can meet the ordinary plane determined by $m$ and $n$ in only one point, this would imply that $m$ and $n$ meet at this point and are not parallel, which is a contradiction. Finally we note that if the planes are ordinary and parallel or if one plane is the ideal plane, the result follows from the manner in which ideal lines were adjoined.

**Theorem 4.52.** *Two distinct points lie on one and only one line.*

PROOF. This is Theorem 4.21 restated. The proof of that theorem is valid except for the case where both points are ideal. If $P$ and $Q$ are ideal, let $l$ and $m$ be the lines determined by them and an ordinary point $O$. Clearly $P$ and $Q$ lie on the ideal line of the ordinary plane determined by $l$ and $m$. If there were a second ideal line through $P$ and $Q$, this would imply the existence of two ordinary planes whose intersection is more than a line, which is a contradiction.

**Theorem 4.53.** *Three noncollinear points lie on one and only one plane.*

PROOF. If all three points are ideal and noncollinear, they can lie only on the ideal plane. We consider the other cases together. Let the points be $A$, $B$, and $C$, and suppose that $A$ is ordinary. Let $l$ and $m$ be the lines determined by $A$ and $B$ and by $A$ and $C$. These lines are ordinary and are distinct if $A$, $B$, and $C$ are noncollinear. The desired plane is the unique ordinary plane containing $l$ and $m$.

**Theorem 4.54.** *If two distinct lines are coplanar, they meet in one and only one point.*

PROOF. If both lines are ordinary, they either meet at an ordinary point and at no ideal point, or they meet at a unique ideal point if they are parallel. If one line is ordinary and one is ideal and if they lie in a common plane, it is clear from the manner in which ideal elements were adjoined that the unique ideal point of the ordinary line must lie on the given ideal line. If both lines are ideal, each is associated with a family of parallel planes. The intersection of a member of the one family with a member of the other family will be an ordinary line, and all intersections thus obtained will produce a family of parallel lines. The unique point of intersection of the ideal lines is the ideal point of the latter family.

**Theorem 4.55.** *Three distinct planes not containing a common line meet in one and only one point.*

The proof is left as an exercise for the reader.

**Theorem 4.56.** *If two distinct lines meet in a point, they lie on one and only one plane.*

PROOF. We consider the case where one line is ideal and one is ordinary and leave the other cases as an exercise for the reader. Since the ideal point of intersection of the lines is on the ordinary line, it must lie on all planes containing the ordinary line. This set of planes will intersect the ideal plane in the set of all ideal lines through the ideal point of intersection. The desired plane is the member of the set of planes meeting the ideal plane in the given ideal line.

**Theorem 4.57.** *A line and a plane not containing the line meet in one and only one point.*

**Theorem 4.58.** *A line and a point not on the line lie on one and only one plane.*

The proofs of Theorems 4.57 and 4.58 are left as exercises for the reader.

## Exercises 4.5

1. Prove that if an ideal point is on an ordinary plane $\pi$ and an ordinary line $l$, then $l$ is either on $\pi$ or parallel to it.
2. Prove Theorem 4.55.
3. Complete the proof of Theorem 4.56.
4. Prove Theorem 4.57.
5. Prove Theorem 4.58.
6. Prove that if two points lie on a plane, the line they determine lies on the plane.

## 4.6* SPACE DUALITY

In Section 4.4 we discussed the important property of duality to be found in plane projective geometry. A similar principle holds in the projective geometry of three dimensions, as evidenced in the following arrangement of the theorems of Section 4.5.

| | |
|---|---|
| Two distinct points determine a unique line. | Two distinct planes meet in a unique line. |
| Three points not determining a line determine a unique plane. | Three planes not meeting in a line meet in a unique point. |
| If two lines determine a plane, they meet in a unique point. | If two lines meet in a point, they determine a unique plane. |
| A line and a plane not containing the line meet in a unique point. | A line and a point not contained on the line determine a unique plane. |

If the words "point" and "plane," and "determine" and "meet in" are interchanged in any of the above, each is changed into the opposite. While the interchange is different, all of the general remarks as to the use of the method made in Section 4.4 still apply. We call the principle *space duality*. Each of the theorems above is the space dual of its opposite member. In the study of three-dimensional projective geometry we must be careful to distinguish between the space and plane duals of a theorem or statement to avoid ambiguity.

## Exercises 4.6

1. Complete the two-column tabulation above by listing Hilbert's space axioms of incidence and their duals in opposite positions. Prove the duals of all axioms not previously proved as exercises.
2. Write the space dual for each of the following:
   *a.* The set of points on a line;
   *b.* The set of points on a plane;
   *c.* The set of planes containing a given line;
   *d.* The set of planes containing a common point;
   *e.* The set of lines passing through a common point;
   *f.* The set of lines lying on a given plane;
   *g.* The set of all planes in space;
   *h.* The set of all lines in space;
   *i.* The set of all points in space.
3. Do the space and plane duals of a configuration have to coincide?

## REFERENCES

Adler, Claire Fisher, *Modern Geometry, An Integrated First Course*, 2nd ed., McGraw-Hill Book Company, New York, 1967, Chapters 4, 5, 7.

Coxeter, H. S. M., *The Real Projective Plane*, McGraw-Hill Book Company, New York, 1949, Chapter 1.

Graustein, William C., *Introduction to Higher Geometry*, The Macmillan Company, New York, 1945, Chapter 2.

Hartshorne, Robin, *Foundations of Projective Geometry*, W. A. Benjamin, New York, 1967, Chapter 1.

Mathews, G. B., *Projective Geometry*, Longmans, Green and Co., London, 1914, Chapters 1 to 3.

O'Hara, C. W., and D. R. Ward, *An Introduction to Projective Geometry*, Oxford University Press, New York, 1937, Chapters 1, 2.

Patterson, Boyd Crumrine, *Projective Geometry*, John Wiley and Sons, New York, 1937, Chapters 1, 2.

Sanger, R. G., *Synthetic Projective Geometry*, McGraw-Hill Book Company, New York, 1939, Chapter 1.

Seidenberg, A., *Lectures in Projective Geometry*, D. Van Nostrand Company, Princeton, 1962, Chapter 1.

Tuller, Annita, *A Modern Introduction to Geometries*, D. Van Nostrand Company, Princeton, 1967, Chapter 3.

Veblen, Oswald, and John Wesley Young, *Projective Geometry*, vol. I, Ginn and Company, Boston, 1910, Chapter 1.

Young, John Wesley, *Projective Geometry*, The Open Court Publishing Company, Chicago, 1930, Chapters 1, 2.

## CHAPTER FIVE

# FUNDAMENTALS OF SYNTHETIC PROJECTIVE GEOMETRY

Just as is so with Euclidean geometry, it is possible to study projective geometry from either the synthetic or analytic point of view. In this chapter we consider some of the basic ideas from the synthetic point of view, and in succeeding chapters we introduce the analytic approach.

## 5.1 THE THEOREM OF DESARGUES

We first consider the famous theorem of Desargues, which will find frequent use in proofs of future theorems. The theorem is named after Gerard Desargues (1593 to 1662), who made extensive contributions to what we now call projective geometry.

In previous exercises we have referred to triangles. We now give a formal definition of a triangle appropriate to projective geometry.

**Definition 5.11.** *A* triangle *is the configuration consisting of three noncollinear points* (vertices) *and the three lines* (sides) *determined by pairs of these points.*

We have already noted that order of points on a line and the concept of line segment are not being considered in projective geometry; hence the sides of a triangle are lines and not line segments (see Figure 5.1).

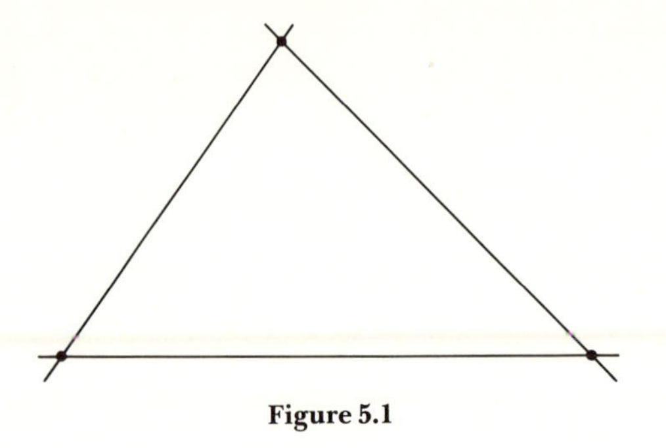

**Figure 5.1**

If we write the dual of Definition 5.11 we obtain

**Definition 5.12.** *A* trilateral *is the configuration consisting of three non-concurrent lines* (sides) *and the three points of intersection* (vertices) *of pairs of these lines.*

It is clear from Figure 5.1 that a triangle and trilateral are the same configuration. We say that a triangle is *self-dual,* and in the future we do not distinguish between a triangle and a trilateral, using only the word triangle.

We are now ready to define two important relations between pairs of triangles.

**Definition 5.13.** *Two triangles are* perspective from a point, *called a* center of perspectivity, *if their vertices can be put into a one-to-one correspondence in such a way that the center of perspectivity is collinear with each pair of corresponding vertices.*

**Definition 5.14.** *Two triangles are* perspective from a line, *called an* axis of perspectivity, *if their sides can be put into a one-to-one correspondence in such a way that the axis of perspectivity is concurrent with each pair of corresponding sides.*

The two relations, which are duals of each other, are illustrated in Figure 5.2. With these relations defined we may finally consider the theorem of Desargues.

**Theorem 5.11 (Desargues).** *If two triangles are perspective from a point, they are perspective from a line, and conversely.*

The theorem is illustrated in Figure 5.3. In this figure it is not necessary to assume the triangles coplanar. We first prove the theorem for

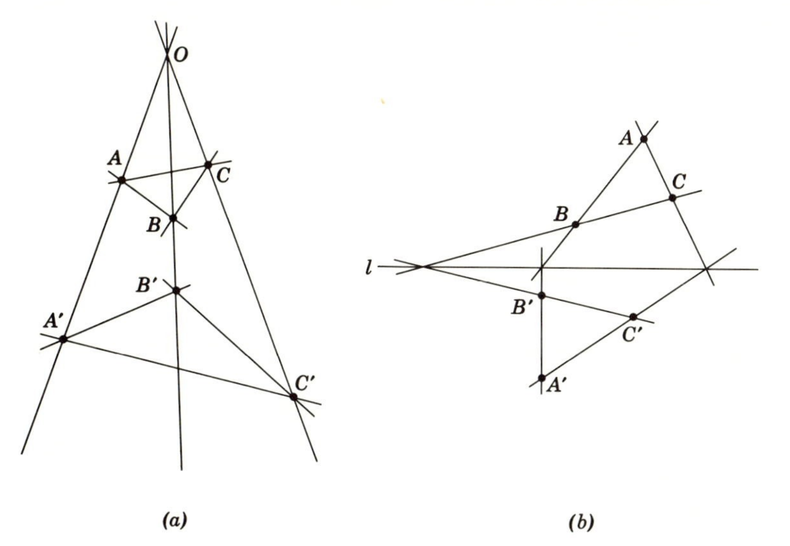


**Figure 5.2** (*a*) Triangles perspective from the point $O$ and (*b*) triangles perspective from the line $l$.

the noncoplanar case, then use this result to prove it in the coplanar case.[1]

[1]We assume the development of projective space, a development occurring in Section 4.5, which the reader may have omitted. The addition of ideal points in space carried out there is similar to that done for the plane in Section 4.1. The resulting space has incidence relations with no exceptions due to parallelism; hence all intersections of planes and of coplanar lines discussed in this proof will certainly exist.

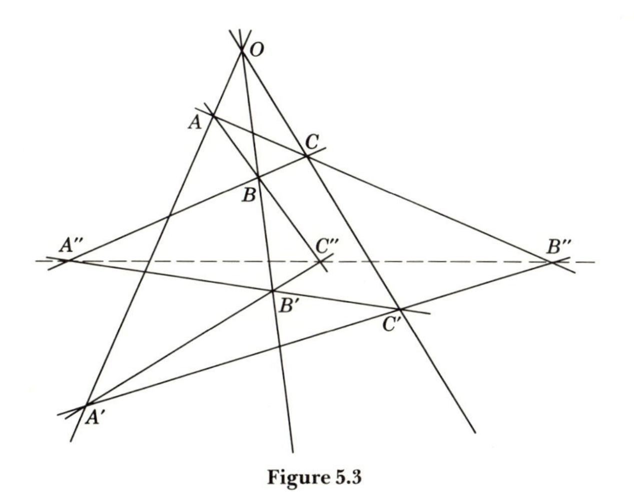


**Figure 5.3**

PROOF. We assume that noncoplanar triangles $ABC$ and $A'B'C'$ are perspective from $O$. We suppose these triangles are in the distinct planes $\pi$ and $\pi'$, respectively. Since the lines $AA'$,[2] and $BB'$ meet at $O$, they determine a plane. The lines $AB$ and $A'B'$ lie in this plane, hence meet at a point, say $C''$. The point $C''$ lies in $\pi$ and in $\pi'$. In a like manner we conclude that $AC$ and $A'C'$ meet in a point $B''$ while $BC$ and $B'C'$ meet in a point $A''$, the points $B''$ and $A''$ both lying in both $\pi$ and $\pi'$. We conclude that $A''$, $B''$, and $C''$ are collinear, lying on the line of intersection of $\pi$ and $\pi'$. This shows that the triangles are perspective from this line. Figure 5.4 illustrates the proof. The converse of the noncoplanar case follows in a like manner, and its proof is left as an exercise for the reader.

We now consider the case where the triangles $ABC$ and $A'B'C'$, lying in a common plane $\pi$, are perspective from a point $O$, which of course is also in $\pi$ (Figure 5.5). We consider a line through $O$ not in $\pi$ and choose distinct points $P$ and $P'$, neither coinciding with $O$, on this line. The lines $PA$ and $P'A'$ lie in the plane determined by the intersecting lines $AA'$ and $PP'$; hence they meet in a point, which we denote by $A''$. By joining other pairs of corresponding vertices to $P$ and $P'$, we obtain

[2]A symbol of the form "$PQ$" here means the line determined by $P$ and $Q$. In Euclidean geometry this symbol on occasion may mean the line determined by the points, the segment determined by the points, the directed segment from $P$ to $Q$, or the length of the segment determined by the points. All of these alternatives are meaningless in projective geometry.

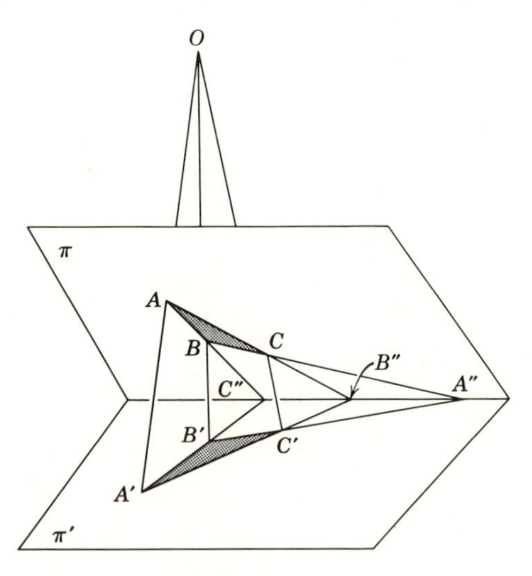


**Figure 5.4**

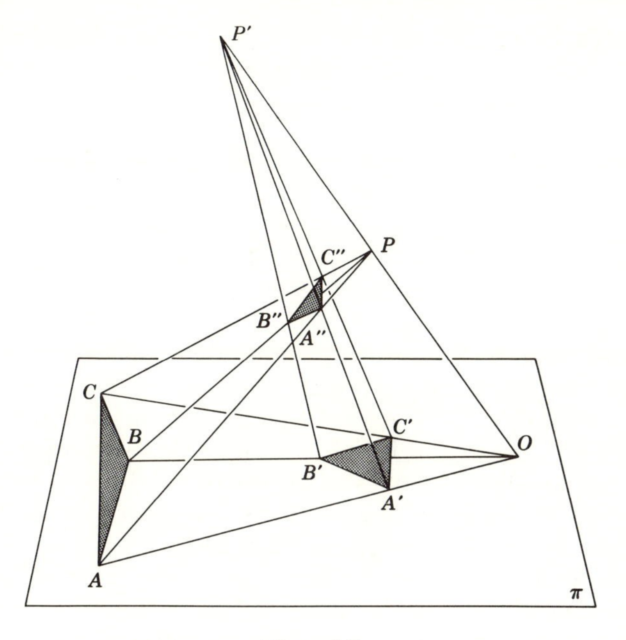


**Figure 5.5**

similar points $B''$ and $C''$. Now the points $A''$, $B''$, and $C''$ cannot be collinear, for if they were, the planes determined by $P$ and the three sides of the triangle $ABC$ would coincide. This single plane would meet $\pi$ in a line containing the points $A$, $B$, and $C$, which is impossible. We see that $A''$, $B''$, and $C''$ determine a triangle and that this triangle is perspective from $P$ with triangle $ABC$ and from $P'$ with triangle $A'B'C'$. The triangle $A''B''C''$ is thus perspective from a line with each of the triangles in $\pi$, and in each case this line is the line of intersection of $\pi$ and the plane of triangle $A''B''C''$. It follows that $A''B''$ meets $AB$ and $A'B'$ at this line of intersection. Since $A''B''$ meets the line of intersection only once, $AB$ and $A'B'$ must meet at a point on this line of intersection. In a like manner we conclude that $AC$ and $A'C'$, and $BC$ and $B'C'$ meet at points on the line of intersection of $\pi$ and the plane of $A''B''C''$. The triangles $ABC$ and $A'B'C'$ are thus perspective from this line.

The converse of the coplanar case above is also its plane dual; however, since we proved the coplanar case in projective space, we cannot argue that the converse portion follows by plane duality. This converse can be proved by choosing a plane distinct from $\pi$ containing the axis of perspectivity and by considering an appropriate triangle $A''B''C''$ in this plane. The proof is left as an exercise.

It should be noted that we used the proof of the noncoplanar case to prove the coplanar case. This was no accident, and it is known that the theorem of Desargues cannot be proved in the plane unless we use lines outside the plane or use the properties of the number scale on all but one point of the projective line. We consider this matter further in Chapter 10.

The theorem of Desargues is valid in plane Euclidean geometry if all the pertinent intersections of lines exist. The theorem is often applicable in cases when this is not so. Thus the theorem assures us that if in the Euclidean plane two triangles have corresponding sides parallel, the lines joining corresponding vertices are either concurrent or parallel; the triangles are *Desarguean* since they are perspective from the ideal line; they are thus perspective from a point, and the lines from it joining the vertices are concurrent if the point is ordinary and parallel if the point is ideal. Further examples of the use of the theorem in Euclidean geometry will be found in the exercises. In general many Euclidean theorems can be proved using projective geometry theorems together with results from Euclidean geometry; we apply a projective result, stating which elements in the statement of the result are to be ideal and which are to be ordinary; this Euclidean interpretation of a projective theorem, together with results of Euclidean geometry, may lead to a simple proof of the Euclidean theorem. We on occasion state exercises which are Euclidean but which are simply done by application of projective results; such exercises are indicated by the symbol #.

## Exercises 5.1

1.* What is the space dual of a triangle? Of a trilateral?

2.* State by number which of Theorems 4.51 to 4.58 is the justification for each of the statements in the preceding proof that noncoplanar triangles perspective from a point are perspective from a line.

3.* Prove that two noncoplanar triangles perspective from a line are perspective from a point.

4.* Prove that two coplanar triangles perspective from a line are perspective from a point.

5.* Prove that two coplanar triangles with a common side are Desarguean.

6.# Verify that each of the following pairs of triangles is Desarguean:

*a.* Two coplanar triangles with corresponding sides parallel;

*b.* Two coplanar triangles with the lines joining corresponding vertices parallel;

*c.* Two congruent triangles in distinct planes with corresponding sides parallel;

*d.* Two coplanar triangles, each with two ideal vertices;

*e.* A triangle and the triangles formed by joining the midpoints of its sides.

7.* State the space dual of the theorem of Desargues.

8. A Desarguean configuration (Figure 5.3) contains ten points. Show that any one of these ten points can be taken as the center of perspectivity for a Desarguean configuration with the same ten points.

9. How can a person plant ten trees in ten rows of three each?

10. Prove that if three triangles are Desarguean in pairs with a common axis of perspectivity, the centers of perspectivity are collinear.

11. Prove that if three triangles are Desarguean in pairs with a common center of perspectivity, the axes of perspectivity are concurrent.

12.# Prove that the medians of a triangle are concurrent.

13. Let $A$, $B$, and $C$ lie on a line $m$ and $A'$, $B'$, and $C'$ lie on a line $n$ such that $AA'$, $BB'$, and $CC'$ are concurrent. Prove that the intersections of $AB'$ and $A'B$, of $AC'$ and $A'C$, and of $BC'$ and $B'C$ lie on a line passing through the intersection of $m$ and $n$.

14.* A tetrahedron is the configuration consisting of four noncoplanar points (vertices) together with the four planes (faces) determined by triples of these points and the six lines (edges) determined by pairs of these points. Two tetrahedra are perspective from a point if their vertices can be put into a one-to-one correspondence in such a way that the point is collinear with each pair of corresponding vertices. Two tetrahedra are perspective from a plane if their faces can be put into a one-to-one correspondence in such a way that the plane is concurrent with (has a common line with) each pair of corresponding faces. Prove that if two tetrahedra are perspective from a point, they are perspective from a plane, and conversely.

## 5.2 HARMONIC SEQUENCES

In the preceding section we defined triangle and then derived an important theorem about triangles. In this section we consider the concepts of quadrangle and quadrilateral and investigate an important relation associated with these configurations.

**Definition 5.21.** *A* complete quadrangle *is the configuration consisting of four (coplanar) points* (vertices)*, no three of which are collinear, together with the six lines* (sides) *determined by pairs of these points. The triangle determined by the points of intersection of opposite sides (sides not containing* *a common vertex) of a complete quadrangle is called its* diagonal triangle, *its vertices being the* diagonal points *of the quadrangle.*

Figure 5.6(*a*) shows a complete quadrangle $P_1P_2P_3P_4$ together with its diagonal triangle $D_1D_2D_3$. Later we shall be able to prove that the diagonal triangle of a complete quadrangle is well defined since $D_1$, $D_2$, and $D_3$ cannot be collinear. If we take the plane dual of this definition, we obtain

**Definition 5.22.** *A* complete quadrilateral *is the configuration consisting of four (coplanar) lines* (sides), *no three of which are concurrent, together with the six points* (vertices) *of intersection of pairs of these lines. The triangle determined by the lines joining opposite vertices (vertices not lying on a common side) of a complete quadrilateral is called its* diagonal triangle, *its sides being the* diagonal lines *of the quadrilateral.*

Figure 5.6(*b*) shows a complete quadrilateral and its diagonal triangle. When we later prove that the vertices of the diagonal triangle of a complete quadrangle are noncollinear, it will follow by duality that the diagonal triangle of a complete quadrilateral is well defined since its sides cannot be concurrent. It is apparent from Figure 5.6 that, unlike the situation with the triangle, a complete quadrangle is not self-dual.

With the complete quadrangle and quadrilateral defined we are now ready to define the concepts of a harmonic sequence of points and a harmonic sequence of lines.

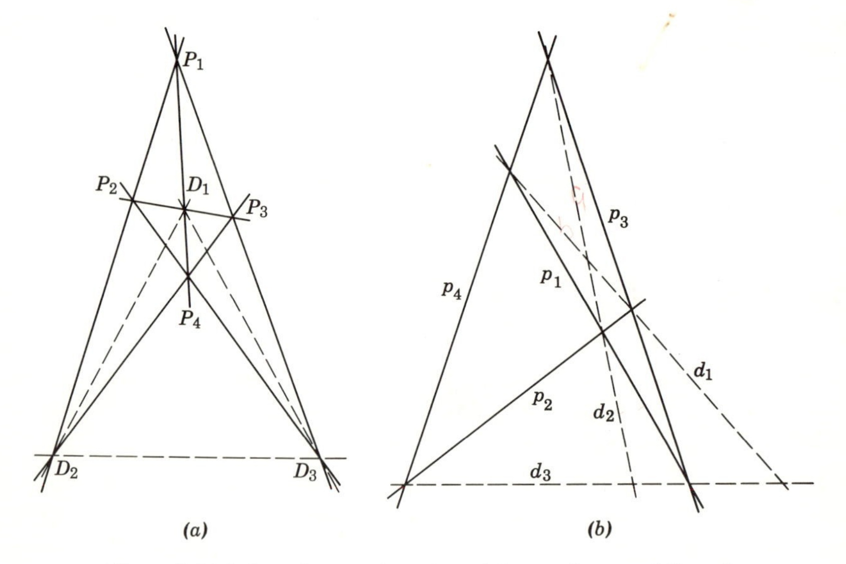


**Figure 5.6** (*a*) Complete quadrangle and (*b*) complete quadrilateral.

**Definition 5.23.** *If four collinear points $A$, $B$, $C$, and $D$ are such that $A$ and $B$ are diagonal points of a complete quadrangle while $C$ and $D$ are on the distinct sides of the quadrangle passing through the third diagonal point, we say that $A$, $B$, $C$, and $D$ form a* harmonic sequence, *$H(A, B; C, D)$, or that $D$ is the* harmonic conjugate *of $C$ with respect to $A$ and $B$.*

**Definition 5.24.** *If four concurrent lines $a$, $b$, $c$, and $d$ are such that $a$ and $b$ are diagonal lines of a complete quadrilateral while $c$ and $d$ pass through the distinct vertices on the third diagonal side of the quadrilateral, we say that $a$, $b$, $c$, and $d$ form a* harmonic sequence, *$H(a, b; c, d)$, or that $d$ is the* harmonic conjugate *of $c$ with respect to $a$ and $b$.*

Note that the order in which the points or lines are stated is pertinent in these definitions; thus if $H(A, B; C, D)$, it does not follow from the definition that $H(A, C; B, D)$, although it is clear that $H(B, A; C, D)$ or $H(A, B; D, C)$.

Given collinear points $A$, $B$, and $C$, it is a simple matter to find a fourth point $D$ such that $H(A, B; C, D)$. We choose a point $O$ not on the line of the given points and choose a point $P$, other than $O$ or $B$, on $OB$. Let $AP$ meet $OC$ at $R$, then let $BR$ meet $OA$ at $Q$. If $D$ is the intersection of $PQ$ and $AB$, it is apparent that $H(A, B; C, D)$, the complete quadrangle being $OPQR$. The dual procedure for finding $d$ such that $H(a, b; c, d)$ is left as an exercise. This construction verifies that a harmonic conjugate of $C$ with respect to $A$ and $B$ exists. We now prove that this harmonic conjugate point is unique.

**Theorem 5.21.** *Given three distinct collinear points $A$, $B$ and $C$, there is a unique point $D$ for which $H(A, B; C, D)$.*

PROOF. We carry out the construction for $D$ twice as indicated in Figure 5.8. We must show that $D$ and $D'$ coincide. To do this, it is

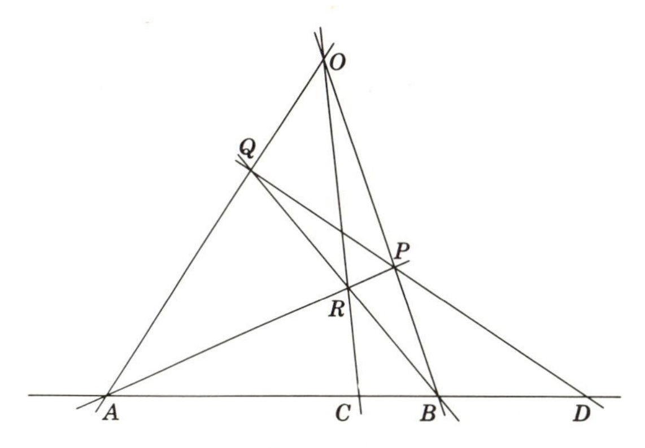


**Figure 5.7**

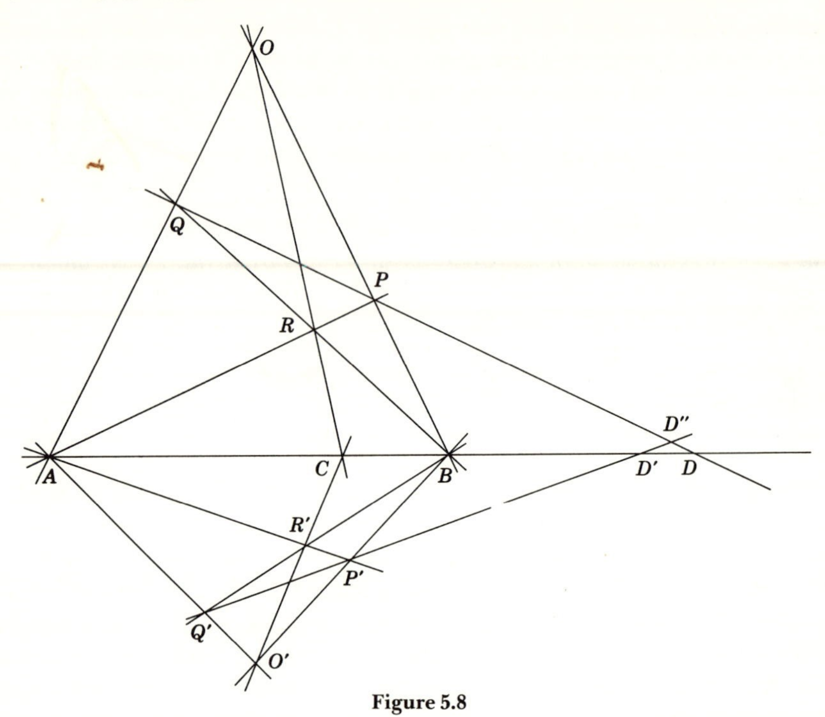


Figure 5.8

sufficient to prove that $D''$, the intersection of $PQ$ and $P'Q'$, lies on $AB$. This is accomplished by repeated application of the theorem of Desargues. We note first that triangles $OQR$ and $O'Q'R'$ are Desarguean, since corresponding sides meet at the collinear points $A$, $B$, and $C$. It follows that $OO'$, $QQ'$, and $RR'$ (not shown in the figure) are concurrent. Similar reasoning applied to triangles $OPR$ and $O'P'R'$ shows that $OO'$, $PP'$, and $RR'$ are concurrent and that the triangles $PQR$ and $P'Q'R'$ are perspective from this common point of concurrence. Corresponding sides of these triangles meet at $A$, $B$, and $D''$, which by the theorem of Desargues must be collinear.

By duality we have

**Theorem 5.22.** *Given three distinct concurrent lines a, b, and c, there is a unique line d for which* $H(a, b; c, d)$.

In our definitions harmonic sequences of points and harmonic sequences of lines are dual relations. A relation between the two concepts is shown in

**Theorem 5.23.** *If a line meets the concurrent lines a, b, c, and d in the points A, B, C, and D, respectively, then* $H(A, B; C, D)$ *if and only if* $H(a, b; c, d)$.

PROOF. We suppose first that $H(A, B; C, D)$ and that the lines meet at a point $O$. Using this point as the point $O$ of the harmonic conjugate construction, carry out the construction. By hypothesis $PQ$ will meet $AB$ at $D$. Consider the complete quadrilateral with sides $AP$, $PQ$, $QB$, and $BA$ and vertices $P$, $Q$, $A$, $B$, $R$, and $D$. Now $OA$ and $OB$ are determined by the respective pairs of vertices $A$ and $Q$, and $B$ and $P$, while $OC$ and $OD$ are determined by $O$ and the remaining vertices, $R$ and $D$. Hence $H(OA, OB; OC, OD)$ or $H(a, b; c, d)$. The converse now follows by duality.

We have previously remarked that the order in which the points or lines are stated is pertinent in a claim that they constitute a harmonic sequence, but that certain changes in order are allowed in view of the definitions given. The following theorem indicates another change which is not trivial.

**Theorem 5.24.** *If $H(A, B; C, D)$, then $H(C, D; A, B)$.*

PROOF. We suppose that $H(A, B; C, D)$ and augment Figure 5.7 by indicating $OD$ and $PC$, meeting at the point $S$, by labeling as $T$ the point of intersection of $OC$ and $QD$, and by indicating $ST$. The resulting configuration is shown in Figure 5.10. We consider the complete quadrangle with vertices $O$, $S$, $T$, and $P$. The sides $OT$ and $PS$ pass through $C$ while the sides $OS$ and $TP$ pass through $D$; moreover, $OP$ passes through $B$. It will thus be sufficient to show that the remaining side $ST$ passes through $A$. Now the triangles $OSP$ and $QTR$ are perspective from the line $CD$. By the theorem of Desargues they are perspective from a point. Since the point must be the intersection $A$ of $RP$ and $OQ$, it follows that $ST$ passes through $A$.

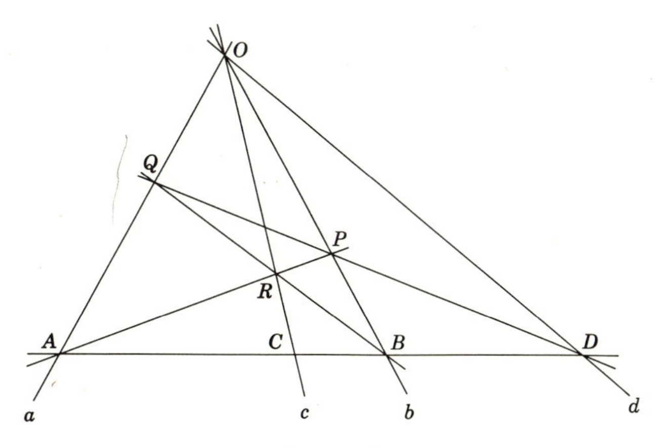


**Figure 5.9**

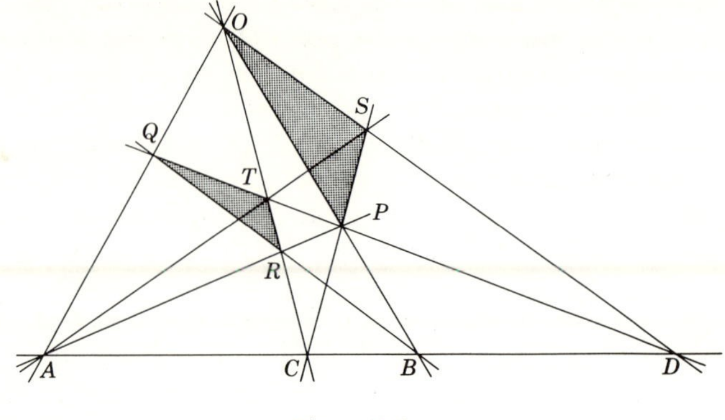


Figure 5.10

## Exercises 5.2

1. Carry out the construction of a line $d$, the harmonic conjugate of a given line $c$ with respect to given lines $a$ and $b$.

2. Assume $H(A, B; C, D)$. Find seven other orders of the points $A$, $B$, $C$, and $D$ for which the harmonic relation holds.

3. Prove that if there is a one-to-one correspondence between the sides of two complete quadrangles such that five pairs of corresponding sides meet in collinear points, then the sixth pair of sides meet in a point on this line.

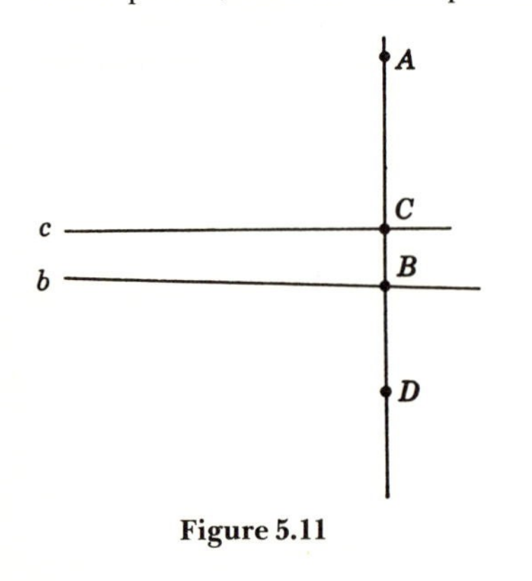


Figure 5.11

4. The point $A$ and the lines $b$ and $c$ are fixed, and $H(A, B; C, D)$. What is the locus of $D$ as the line $AB$ is rotated about $A$?

5. It is desired to extend the line $AB$ beyond the region $R$ with straight edge alone without its crossing $R$. Carry out the construction.

6.# Two lines $a$ and $b$ are coplanar, but their point of intersection is inaccessible. Find the line determined by this point and a given point in the plane of the lines.

7. Construct a complete quadrangle which has a given diagonal triangle. Is it unique?

8. Given the diagonal triangle and one vertex of a complete quadrangle, find the other vertices.

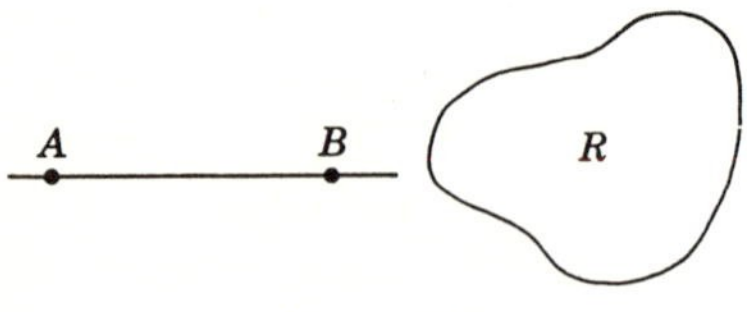


Figure 5.12

9. Let three points, one on each side of a triangle, be collinear. Join the harmonic conjugates of these points with respect to the adjacent vertices of the triangle to the opposite vertices of the triangle. Prove that the three lines so determined are concurrent.

10. Prove that the six sides of a complete quadrangle meet the sides of its diagonal triangle in the six vertices of a complete quadrilateral with the same diagonal triangle.

11.# Prove that if $C$ is the midpoint of the segment $AB$, its harmonic conjugate with respect to $A$ and $B$ is an ideal point.

12.# Prove that the harmonic conjugate of the ideal point on a line with respect to points $A$ and $B$ of the line is the midpoint of the segment $AB$.

13.# Prove that if a line $c$ bisects an angle determined by $a$ and $b$ and if $d$ is the harmonic conjugate of $c$ with respect to $a$ and $b$, then $c$ and $d$ are perpendicular.

14.# Prove that if $c$ and $d$ are perpendicular and $d$ is the harmonic conjugate of $c$ with respect to $a$ and $b$, then $d$ bisects an angle determined by $a$ and $b$.

15.# Let the external and internal bisectors of the vertex angle $A$ of $\triangle ABC$ meet the line of $B$ and $C$ at $D$ and $E$. Prove that $H(B, C; D, E)$.

## 5.3 FUNCTIONS AND RELATIONS

We have already seen the important role played in Euclidean geometry by the concept of congruence. There is a concept that plays an analogous role in projective geometry. In the next section we investigate one specific example of it, and in the next chapter we undertake a more general discussion of the concept. The concept is an example of a function, which we investigate in the present section.

If the triangles $ABC$ and $A'B'C'$ are congruent (Figure 5.13), we understand that *corresponding* sides and angles are congruent. The word "corresponding" is important here, and the congruence is not truly specified unless the correspondence is given. We suppose in Figure 5.13 that the segment $A$-$B$ corresponds to the segment $A'$-$B'$, the segment $A$-$C$ to the segment $A'$-$C'$, and the segment $B$-$C$ to the segment $B'$-$C'$. The correspondence can be made a point-to-point correspondence: we agree that $A$ corresponds to $A'$, $B$ to $B'$, and $C$ to $C'$. The

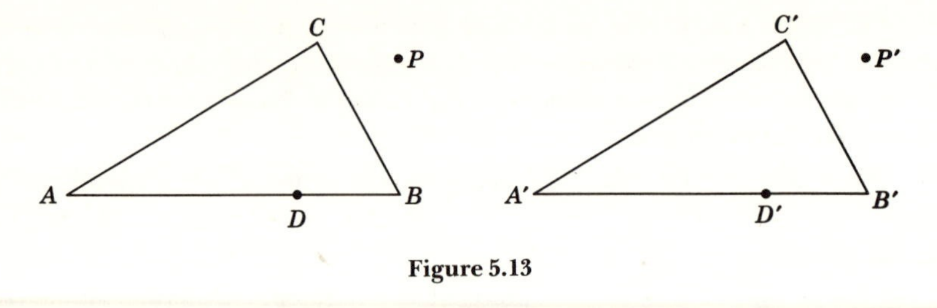


Figure 5.13

point-to-point correspondence can even be extended to points on the sides of the triangle; thus, if $D$ and $D'$ are on $A$-$B$ and $A'$-$B'$, respectively, and if $A\text{-}D \cong A'\text{-}D'$, we may agree that $D$ and $D'$ are corresponding points. The correspondence can even be extended to all points of the plane: any point $P$ will correspond to a point $P'$ where $A$-$P$, $B$-$P$, and $C$-$P$ are congruent to $A'$-$P'$, $B'$-$P'$, and $C'$-$P'$, respectively. We could denote the correspondence in the form $T(A) = A'$, $T(B) = B'$, $T(D) = D'$, $T(P) = P'$, and so forth.

The notation suggests that the function concept is present here just as much as in a statement of the form $y = f(x) = x^2$. In the former case a point $P$ determines a point $P'$ while in the latter case a real number $x$ determines a real number $y$. The function concept is not restricted to numbers, but applies to sets in general. It is a process in which an element of some set ($P$ or $x$ above) determines an element ($P'$ or $y$ in the examples) of some set.

Although the function concept is present, the discussion above does not indicate what the function itself is to be. We note that a pairing process is present in either example. In the geometric case we are concerned with ordered pairs of points $(P, P')$ where $T(P) = P'$; in the real number case we are concerned with ordered pairs of real numbers $(x, y)$ where $y = x^2$. Mathematicians are in general agreement that a satisfactory definition of function is one which makes the function be the set of ordered pairs. We thus make the following definition.

**Definition 5.31.** *A* function, transformation, *or* mapping *from a set $\mathscr{X}$ to a set $\mathscr{Y}$ is a set of ordered pairs $(x, y)$ where $x$ is an element of $\mathscr{X}$, $y$ is an element of $\mathscr{Y}$, and where $(x, y)$ and $(x, y')$ being in the function implies $y = y'$. The subset of $\mathscr{X}$ containing all elements that are first elements of ordered pairs in the function is the* domain *of the function. The subset of $\mathscr{Y}$ containing all elements that are second members of ordered pairs in the function is the* range *of the function. If $(P, P')$ is in the function $T$, we write $T(P) = P'$.*

In the example of the congruence, the domain and range of the transformation are the set of all points in the Euclidean plane. In the example

$y = x^2$ the domain is the set of all real numbers and the range is the set of all nonnegative real numbers.

The requirement that $(x, y)$ and $(x, y')$ being in the function shall imply $y = y'$ makes a function "single-valued"; that is, $f(x)$ is unique. We have not required, however, that a given $y$ come from a unique $x$; that is, in a function we may well have pairs $(x, y)$ and $(x', y)$ where $x \neq x'$. Functions for which such is not the case will be important for us, and we make the following definition.

**Definition 5.32.** *A function is* one-to-one *from a set $\mathscr{X}$ onto a set $\mathscr{Y}$ if its domain is $\mathscr{X}$, its range is $\mathscr{Y}$, and $(x, y)$ and $(x', y)$ being in the function implies $x = x'$.*

Using this definition we can easily prove the following theorem, the proof of which is left to the reader.

**Theorem 5.31.** *If $T$ is a one-to-one function from $\mathscr{X}$ onto $\mathscr{Y}$, then the set of all $(y, x)$ for which $(x, y)$ is in $T$ is a one-to-one function from $\mathscr{Y}$ onto $\mathscr{X}$.*

Suppose now that $T$ and $S$ are functions such that the range of $T$ is a subset of the domain of $S$. If $(x, y)$ is in $T$, then $y$ will be in the range of $T$, hence in the domain of $S$. We see that there will then be a $z$ in the range of $S$ such that $(y, z)$ is in $S$. In such a case an $x$ in the domain of $T$ has determined a $z$ in the range of $S$. We consider the set of all $(x, z)$ for $x$ in the domain of $T$ such that there is a $y$ in the range of $T$ for which $(x, y)$ is in $T$ and $(y, z)$ is in $S$. We assert that we cannot have an $(x, z)$ and an $(x, z')$, $z \neq z'$, in this set, for $z \neq z'$ will imply the impossibility of finding a suitable $y$. We thus have proved:

**Theorem 5.32.** *If $T$ and $S$ are functions for which the range of $T$ is a subset of the domain of $S$, then the set of all $(x, z)$ for $x$ in the domain of $T$, for which there is a $y$ such that $(x, y)$ is in $T$ and $(y, z)$ is in $S$, is a function.*

**Definition 5.33.** *The set of all $(x, z)$ of Theorem 5.32 is called the* product *$ST$ of the functions $T$ and $S$.*

The basic property of this product function is that of the following theorem, the proof being left to the reader.

**Theorem 5.33.** *If $ST$ is defined, $[ST](P) = S[T(P)]$.*

We continue this introduction to functions by stating two definitions and a theorem based on Theorems 5.31 and 5.33.

**Definition 5.34.** *If $T$ is a one-to-one function from $\mathscr{X}$ onto $\mathscr{Y}$, the set of all $(y, x)$ for $(x, y)$ in $T$ is called the* inverse function *of $T$ and is denoted by $T^{-1}$.*

**Definition 5.35.** *The* identity function *on a set $\mathscr{X}$ is the set of all $(x, x)$ for $x$ in $\mathscr{X}$.*

**Theorem 5.34.** *If $T$ is a one-to-one function from $\mathscr{X}$ onto $\mathscr{Y}$, if $I$ is the identity function on $\mathscr{X}$, and if $I'$ is the identity function on $\mathscr{Y}$, then $TT^{-1} = I'$ and $T^{-1}T = I$.*

The proof is left to the reader.

The foregoing is a development of the concept often spoken of as "single-valued function of one variable" and is the concept to be used in this chapter. For completeness and later use let us consider the generalizations needed to remove the restrictions of "one variable" and "single-valued."

An increase in the number of variables is simply handled and does not require any generalization of Definition 5.31. If

$$z = x^2 + 2xy + 3y^3,$$

we often say that $z$ is a (single-valued) function of the variables $x$ and $y$. In the terminology used here the function is a set of ordered pairs $[(x, y), z]$ whose domain is itself a set of ordered pairs $(x, y)$. More generally if $z$ is determined by the $n$ variables $x_1, x_2, \cdots, x_n$, the resulting function is the set of ordered pairs $[(x_1, x_2, \cdots, x_n), z]$. In short, allowance for additional variables can be made by taking for the set $\mathscr{X}$ of Definition 5.31 a set of ordered $n$-tuples. For convenience we write $(x_1, x_2, \cdots, x_n, z)$ rather than $[(x_1, x_2, \cdots, x_n), z]$ for an element of the function; that is, we regard the function as a set of ordered $(n+1)$-tuples rather than a set of ordered pairs, it being understood that the range is the set of last members of the $(n+1)$-tuples. We maintain the single-valued requirement of Definition 5.31; that is, we require that if $(x_1, x_2, \cdots, x_n, z)$ and $(x_1, x_2, \cdots, x_n, z')$ are elements of a function, then $z = z'$.

Many geometric concepts are functions in this more general sense. An example is the harmonic relation of Section 5.2. We know by Theorem 5.21 that for a given set of distinct collinear points $A$, $B$, and $C$ there is a unique $D$ such that $H(A, B; C, D)$. Thus the set of all $(A, B, C, D)$ for which $H(A, B; C, D)$ is a function whose domain is the set of all distinct triples of collinear points.

The set of all $(A, B, l)$ where $l$ is the unique line determined by $A$ and $B$ is a function whose domain is the set of all pairs $(A, B)$ of distinct points.

The set of all $(m, m', P)$ where $P$ is the common point of $m$ and $m'$ is a function. In the projective plane the domain is the set of all $(m, m')$ for which $m \neq m'$; in the Euclidean plane the domain is the set of all $(m, m')$ for which $m \neq m'$ and $m \nparallel m'$.

The metric concepts of Euclidean geometry are easily stated in function form. Any pair of points $A$ and $B$ (not necessarily distinct) determine a unique segment length $m(A,B)$ (where $m(A,B)=0$ if $A=B$). Thus the set of all $(A,B,m)$ where $m$ is the length of the segment determined by $A$ and $B$ is a function. In a like manner three points $A$, $B$, and $C$ determine a unique area $m(A, B, C)$ of the triangle of which they are vertices (where $m=0$ if $A$, $B$, and $C$ are collinear). Thus the set of all $(A,B,C,m)$ is a function.

Playfair's axiom for Euclidean geometry states that there is a unique line $l'$ through a point $P$ not on a line $l$ which is parallel to $l$. The set of all resulting $(P,l,l')$ is a function whose domain is the set of all $(P,l)$ for which $P$ is not on $l$.

To remove the requirement of single-valuedness, we need only remove this restriction in Definition 5.31. In modern usage we reserve the word "function" for the concept frequently called "single-valued function" and use the word "relation" for the more general concept. Thus we have

**Definition 5.36.** *A* relation *from a set $\mathscr{X}$ to a set $\mathscr{Y}$ is a set of ordered pairs $(x,y)$ where $x$ is an element of $\mathscr{X}$ and $y$ is an element of $\mathscr{Y}$.*

It is apparent that a function is a special type of relation but that not all relations are functions. As in the case of functions we can generalize the concept of relation to sets of $(n+1)$-tuples $(x_1, x_2, \cdots, x_n, z)$ where now two $(n+1)$-tuples $(x_1, x_2, \cdots, x_n, z)$ and $(x_1, x_2, \cdots, x_n, z')$ with $z \neq z'$ may be elements of the same relation.

The basic "relations" introduced in Hilbert's axioms are all relations in the sense of Definition 5.36. The incidence relation is just the set of all $(P,l)$ for which $P$ is on $l$. The order relation is the set of all $(A,B,C)$ for which $B$ is between $A$ and $C$. The congruence relations are the sets of all $(A\text{-}B, A'\text{-}B')$ and $[(h,k), (h',k')]$ for which $A\text{-}B \cong A'\text{-}B'$ and $(h,k) \cong (h',k')$, respectively. Finally, the relation of parallelism is just the set of all $(l,l')$ for which $l \parallel l'$. None of these relations is a function.

We find examples of relations also in our previous discussion of projective geometry. As a function the set of all $(A,B,C,D)$ for which $H(A,B;C,D)$ is a relation. The set of all pairs of triangles $(\triangle_1, \triangle_2)$ which are Desarguean is a relation. The set of all $(A,B,C)$ for which $A$, $B$, and $C$ are collinear is a relation; the dual of the relation is the set of all $(a,b,c)$ for which $a$, $b$, and $c$ are concurrent.

## Exercises 5.3

1. Prove Theorem 5.31.
2. Prove Theorem 5.33.

3. Prove Theorem 5.34.
4. Determine which of the following are functions:
   *a.* The set of all $(P, P')$ for which $P \neq P'$;
   *b.* The set of all $(P, P')$ for which $P = P'$;
   *c.*# The set of all $(A, C, B)$ for which $A$ and $B$ are distinct and $B$ is the midpoint of the segment $A$-$C$;
   *d.*# The set of all $(l, P, l')$ for which $P$ is on $l$ and $l'$ is perpendicular to $l$ at $P$;
   *e.** The set of all $(\pi, \pi', l)$ where $\pi \neq \pi'$ and $l$ is in $\pi$ and $\pi'$;
   *f.*# The set of all $(\pi, \pi', l)$ where $\pi \neq \pi'$ and $l$ is in $\pi$ and $\pi'$;
   *g.** The set of all $(\pi, \pi', P)$ where $\pi \neq \pi'$ and $P$ is in $\pi$ and $\pi'$;
   *h.* The set of all $(A, B, C, D)$ where $A$, $B$, $C$ and $D$ form a harmonic sequence in some order;
   *i.* The set of all $(A, B, C, A', B', C')$ where the triangles $ABC$ and $A'B'C'$ are Desarguean.

## 5.4 PROJECTIVE TRANSFORMATIONS

We now define the transformations analogous in projective geometry to the congruence transformations of Euclidean geometry.

**Definition 5.41.** *A* central perspectivity *is a transformation of the points of a line onto the points of a line for which each pair of corresponding points is collinear with a fixed point, called the* center of perspectivity, *in neither the domain nor range of the transformation.*

Figure 5.14 illustrates a central perspectivity. For convenience we shall agree that if $T(A) = A'$, $T(B) = B'$, and so on, we shall write $T(A, B, \cdots) = A', B', \cdots$. Thus in Figure 5.14 $T(A, B, C, D) = A', B', C', D'$. The reader should have little difficulty verifying that the center of perspectivity is unique if the domain and range are distinct, and that if the domain and range are the same, the central perspectivity is the identity function. We tend to associate the idea of projection with central perspectivity as here defined. The projective idea is more general, as indicated in the next definition.

**Definition 5.42.** *A* projective transformation *or* projectivity *is a transformation of the points of a line onto the points of a line which may be expressed as a finite product of central perspectivities.*

Figure 5.15 illustrates a projective transformation $T$ for which $T(A, B, C) = A'', B'', C''$. Not every projectivity is a central perspectivity, but every central perspectivity is a projectivity. In Definition 5.42 we

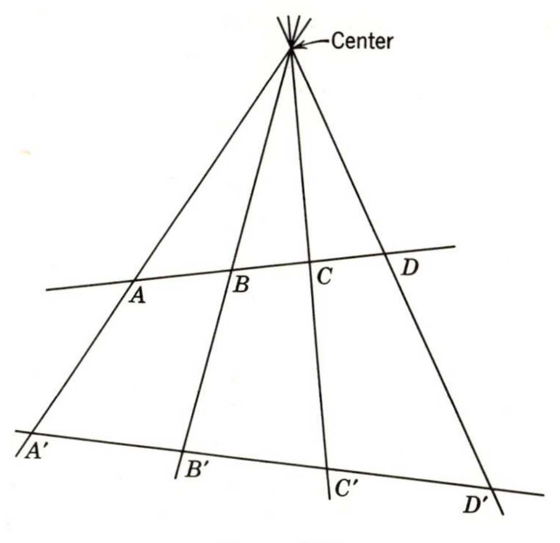


**Figure 5.14**

do not claim that every projectivity can be *uniquely* expressed as a product of perspectivities; this is not the case. Any projective transformation has an inverse which is a projective transformation. We state this as a theorem, leaving the proof to the reader.

**Theorem 5.41.** *If $T$ is a projectivity, $T^{-1}$ exists and is a projectivity.*

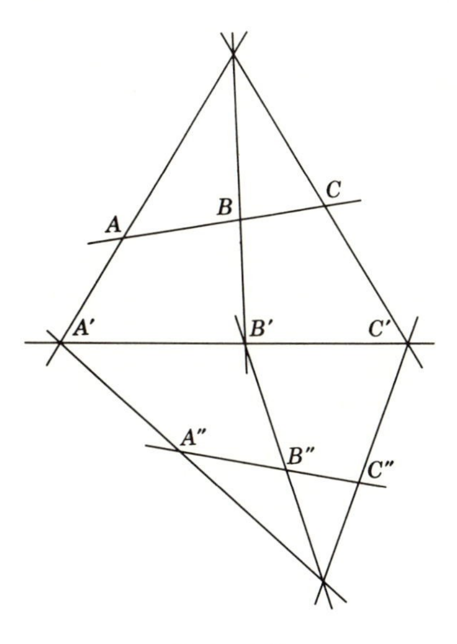


**Figure 5.15**

We can now prove very simply that harmonic sequences of points are preserved under a projective transformation, that is, that a harmonic sequence of points is transformed into a harmonic sequence of points. We have

**Theorem 5.42.** *If $T$ is a projectivity for which $T(A, B, C, D) = A', B', C', D'$ and if $H(A, B; C, D)$, then $H(A', B'; C', D')$.*

PROOF. It will be sufficient to prove the theorem for a central perspectivity, for if a harmonic sequence is carried into a harmonic sequence under a central perspectivity, it will be carried into one by a product of such transformations. In the case of a central perspectivity the result follows at once from Theorem 5.22, for if $H(A, B; C, D)$, then $H(AA', BB'; CC', DD')$, from which it follows that $H(A', B'; C', D')$.

Theorem 5.42 implies a restriction on our ability to construct projectivities between lines, for it says we cannot always take four points $A, B, C$, and $D$ on one line and points $A', B', C'$, and $D'$ on a second line and find a projectivity $T$ for which $T(A, B, C, D) = A', B', C', D'$. This is true since the theorem requires that if $D$ is the harmonic conjugate of $C$ with respect to $A$ and $B$, then $T(D)$ must be the harmonic conjugate of $C'$ with respect to $A'$ and $B'$. If there is a theorem which states that a projectivity is uniquely determined when the correspondence is specified for a given number of points, we see that the given number can be no more that three. On the other hand, it is clear that a central perspectivity can always be found which will transform two given points into two arbitrary points on a second line. Since we have more freedom in a projectivity, it seems reasonable that in this case we can transform three points into three arbitrary points. This is indeed the case, and this result is known as the Fundamental Theorem of Projective Geometry.

**Theorem 5.43 (Fundamental Theorem of Projective Geometry).** *There exists one and only one projective transformation mapping three distinct points on a line onto three distinct points on a line in a given order.*

The domain and range need not be distinct. We shall omit the proof of this theorem[3] and content ourselves with merely outlining the steps of the proof. If $A, B$, and $C$ and $A', B'$, and $C'$ are the given points on the two lines, we seek a projectivity $T$ for which $T(A, B, C) = A', B', C'$. We have already observed that if $H(A, B; C, D)$, then $T(D)$ is uniquely determined. In a like manner if $H(A, C; B, E)$ and $H(B, C; A, F)$, then $T(E)$ and $T(F)$ are uniquely determined. We now have $T$ uniquely determined for six points on the line. We continue by choosing various triples of these points and completing them to harmonic sequences, on

[3]The reader is referred to the works of Mathews, Robinson, and Veblen and Young cited at the end of the chapter for a more detailed discussion of this theorem.

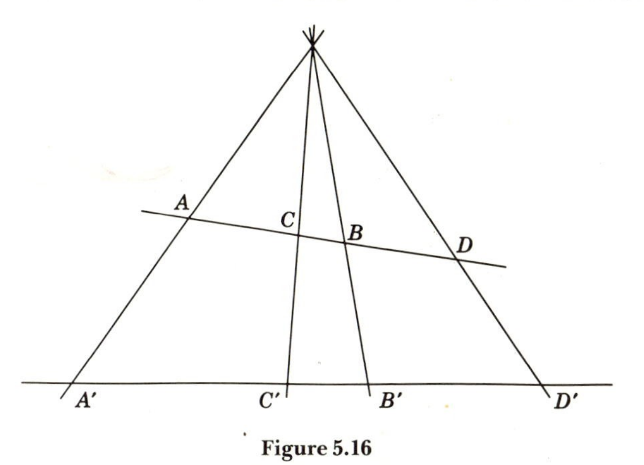


**Figure 5.16**

each point of which $T$ is uniquely determined. This process can be carried on indefinitely. The set of points we obtain on each line if the process is continued indefinitely is said to be a *net of rationality*. If we make use of the real number scale on all but one of the points of a line, we can show that a net of rationality is *dense* on the line; that is, between any two ordinary points of the line there is a point of the net of rationality. The relation between a net of rationality and the set of all points on the line is analogous to the relation between the rational and the real numbers. It is clear that $T$ is uniquely determined on the net of rationality. Finally we can show that the denseness of the net of rationality implies that $T$ is uniquely determined on all of the line.

The Fundamental Theorem is a very powerful tool for proving properties of projective transformations. It promises existence and uniqueness of a projectivity given the relation on three pairs of points. The uniqueness is often of great power in proofs; thus if we have three distinct points $A$, $B$, and $C$, have given a projectivity $T$, and construct a second projectivity $S$ for which $T(A, B, C) = S(A, B, C)$, it will follow from the Fundamental Theorem that $T = S$; that is, $T$ and $S$ are the same projectivity. This principle is illustrated in the final three theorems of this section.

**Theorem 5.44.** *A projectivity between distinct lines can be expressed as the product of two central perspectivities.*

PROOF. Let $T$ be the given projectivity and suppose that $T(A, B, C) = A', B', C'$, where $A$, $B$, and $C$ are distinct. We choose any line through $A'$ different from $A'C'$. Now let $R$ be any central perspectivity from the line $AC$ to this line which maps $A$ onto $A'$, and suppose $R(A, B, C) = A', B'', C''$. We denote by $O$ the intersection of $B'B''$ and $C'C''$. If $S$ is

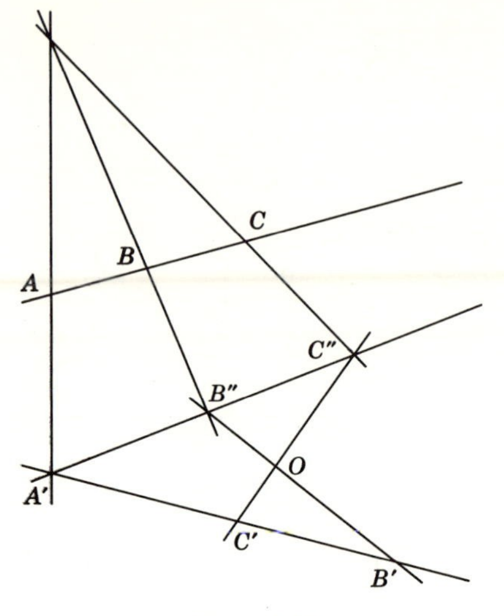


**Figure 5.17**

the central perspectivity for $A'C''$ to $A'C'$ with center at $O$, we have $S(A', B'', C'') = A', B', C'$, and hence $[SR](A, B, C) = A', B', C'$. Finally, it follows from the Fundamental Theorem that $T = SR$.

It is now a simple matter to prove the corresponding theorem for a projectivity of a line onto itself. We state the theorem and leave the proof to the reader.

**Theorem 5.45.** *Any projectivity of a line onto itself can be expressed as the product of three central perspectivities.*

Implicit in the proof of Theorem 5.44 is the following theorem, the proof of which is also left to the reader.

**Theorem 5.46.** *Any projectivity between distinct lines which transforms a point into itself is a central perspectivity.*

## Exercises 5.4

1. Prove that a central perspectivity of a line onto itself is the identity.

2. Show that a projective transformation of a line onto a line is not uniquely expressible as a product of central perspectivities.

3. Show that if $S$ and $T$ are projective transformations for which $ST$ is defined, then $(ST)^{-1} = T^{-1}S^{-1}$.

4. If $S$ and $T$ are projective transformations for which $ST$ is defined, when will $TS$ also be defined?

5. Show that if $H(A, B; C, D)$ and if $T$ is a projectivity for which $T(A, B, C) = A, B, D$, then $T(D) = C$.

6. Prove Theorem 5.41.

7. Prove Theorem 5.45.

8. Prove Theorem 5.46.

9. The dual of the set of points on a line is called a *pencil of lines*. The dual of a central perspectivity is called an *axial perspectivity*, and the dual of the center of perspectivity is called the *axis of perspectivity*. The dual of projectivity is called projectivity. Write the duals of all definitions and theorems of this section.

10. A set of elements on which a multiplication has been defined is said to be a *group* if for any $a$, $b$, and $c$ in the set, (1) $ab$ is an element of the set; (2) $(ab)c = a(bc)$; (3) there is an element $e$ in the set for which $ae = ea = a$ for all $a$; and (4) for each $a$ there is an $a^{-1}$ for which $aa^{-1} = a^{-1}a = e$. Prove that the set of all projective transformations of a line onto itself is a group under the multiplication operation of Definition 5.33.

11. If $A$, $B$, $C$, and $D$ are distinct points on a line, construct projectivities $S$ and $T$ of the line onto itself where $S(A, B, C, D) = B, A, D, C$ and $T(A, B, C, D) = C, D, A, B$.

## 5.5 THE THEOREM OF PAPPUS

Theorem 5.46 can now be used to prove the theorem of Pappus, one of the great classical theorems of projective geometry. The theorem is named after Pappus of Alexandria (A.D. 300 to 370), but it was probably known to Euclid.

**Theorem 5.51 (Pappus).** *If $A$, $B$, and $C$, and $A'$, $B'$, and $C'$ are triples of distinct points on two distinct (coplanar) lines, the points of intersection of $AB'$ and $A'B$, of $AC'$ and $A'C$, and of $BC'$ and $B'C$ are collinear.*

PROOF. We denote these points of intersection by $P$, $Q$, and $R$ as shown in Figure 5.18. Let $PQ$ meet $BC'$ at $D$ and $AC$ at $E$. We must prove that $R$ and $D$ coincide. We denote by $F$ the intersection of $AB'$ and $A'C$ and by $G$ the intersection of $AC'$ and $A'B$. If $T$ is the central perspectivity from $AB'$ to $AC'$ with center $A'$, we have $T(A, P, F, B') = A, G, Q, C'$. If $S$ is the central perspectivity from $AC'$ to $PQ$ with center $B$, we have $S(A, G, Q, C') = E, P, Q, D$. Thus $[ST]$ $(A, P, F, B') = E, P, Q, D$. Since $P$ is self-corresponding, we conclude by Theorem 5.46 that $ST$ is a central perspectivity. Since $AE$ and $FQ$ meet at $C$, $C$ is the center of perspectivity. This implies that $D$ is also the intersection of $PQ$

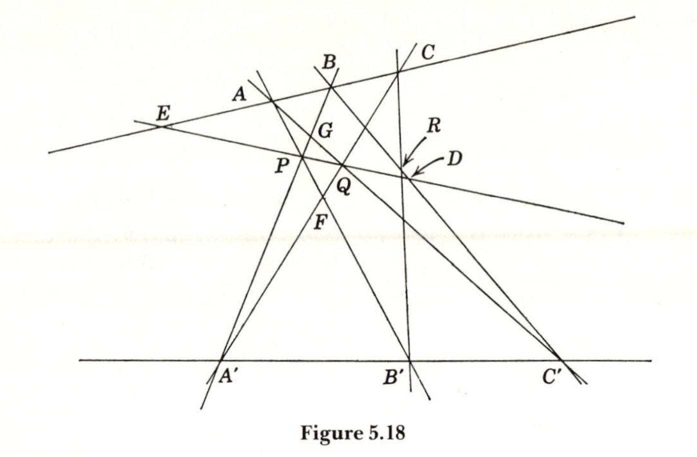


**Figure 5.18**

and $B'C$, whence $PQ$, $BC'$, and $B'C$ are concurrent, and $R$ and $D$ coincide.

If $T$ is a projectivity for which $T(A, B, C) = A', B', C'$, where $A$, $B$, and $C$ are distinct and the domain and range of $T$ are distinct sets, the Fundamental Theorem asserts that $T(D)$ is uniquely determined for any $D$ in the domain of $T$. Pappus' theorem now gives us a method for determining $T(D)$. The line on which the points $P$, $Q$, and $R$ of Figure 5.18 lie is called a *Pappus line*. We assert that $T(D)$ is determined by the condition that $AT(D)$ and $A'D$ (or $BT(D)$ and $B'D$, etc.) must meet on this line. To show this, suppose that $T(A, B, C) = A', B', C'$. Let $U$ be the central perspectivity from $AB$ to $PR$ with center $A'$. We have $U(A, B, C) = A'', P, Q$. Let $V$ be the central perspectivity from $PQ$

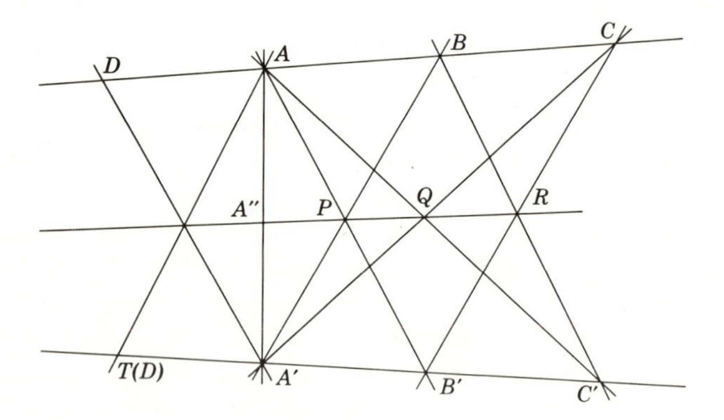


**Figure 5.19**

to $A'B'$ with center $A$. We have $V(A'', P, Q) = A', B', C'$. Hence $[VU](A, B, C) = A', B', C'$ and by the Fundamental Theorem $T = VU$. It is clear then that $AT(D)$ and $A'D$ must meet on $PQ$. We have proved

**Theorem 5.52.** *If $T$ is a projectivity between distinct lines, all points of intersection of the form $PT(Q)$ and $QT(P)$ are collinear.*

We conclude with

**Definition 5.51.** *The* axis of homology *of a projective transformation $T$ between distinct lines is the line containing all points of intersection of the for $PT(Q)$ and $QT(P)$.*

## Exercises 5.5

1. Write the duals of the theorems and definition of this section. (The dual of axis of homology is called the *center of homology*.)

## REFERENCES

Adler, Claire Fisher, *Modern Geometry, An Integrated First Course*, 2nd ed., McGraw-Hill Book Company, New York, 1967, Chapters 4, 5, 7.

Coxeter, H. S. M., *The Real Projective Plane*, McGraw-Hill Book Company, New York, 1949, Chapters 2, 4.

Gans, David, *Transformations and Geometries*, Appleton-Century-Crofts, New York, 1969, Chapters 5, 8.

Graustein, William C., *Introduction to Higher Geometry*, The Macmillan Company, New York, 1945, Chapter 2.

Hartshorne, Robin, *Foundations of Projective Geometry*, W. A. Benjamin, New York, 1967, Chapters 2, 5.

Mathews, G. B., *Projective Geometry*, Longmans, Green and Co., London, 1914, Chapters 4 to 8.

O'Hara, C. W., and D. R. Ward, *An Introduction to Projective Geometry*, Oxford University Press, New York, 1937, Chapters 3, 4.

Patterson, Boyd Crumrine, *Projective Geometry*, John Wiley and Sons, New York, 1937, Chapters 2 to 4, 6.

Rainich, G. Y., and S. M. Dowdy, *Geometry for Teachers*, John Wiley and Sons, New York, 1969, Chapter 3.

Robinson, Gilbert de B., *The Foundations of Geometry*, The University of Toronto Press, Toronto, 1940, Chapters 2, 3, 8.

Sanger, R. G., *Synthetic Projective Geometry*, McGraw-Hill Book Company, New York, 1939, Chapters 2 to 4.

Seidenberg, A., *Lectures in Projective Geometry*, D. Van Nostrand Company, Princeton, 1962, Chapters 1, 2.

Tuller, Annita, *A Modern Introduction to Geometries*, D. Van Nostrand Company, Princeton, 1967, Chapter 3.

Veblen, Oswald, and John Wesley Young, *Projective Geometry*, vol. I, Ginn and Company, Boston, 1910, Chapters 2 to 4.

Young, John Wesley, *Projective Geometry*, The Open Court Publishing Company, Chicago, 1930, Chapters 3, 4.

## CHAPTER SIX

# NATURAL HOMOGENEOUS COORDINATES

The reader is certainly familiar with the use of the real numbers as coordinates of points on a line, of pairs of real numbers as coordinates of points in the plane, and of triples of real numbers as coordinates of points in space. This introduction of coordinates achieves a fusion of geometry and algebra, for associated with a geometric locus is a relation among the appropriate variables, satisfied by the coordinates of those and only those points lying on the locus. In analytic geometry we are able to discover and prove geometric facts by algebraic processes, these processes frequently leading to proofs simpler than the synthetic geometric ones. The object of this chapter is to develop coordinate systems for projective geometry which will permit an analytic study of that geometry.

## 6.1 COORDINATES FOR THE PROJECTIVE PLANE

The coordinate systems of Euclidean geometry would seem to be almost sufficient for projective geometry, since they do assign coordinates to all ordinary points. The assignment of coordinates to the ideal points is the crux of the problem. It would be fortunate if this could be done by adding symbols for the ideal elements. If we consider first the coordinates on a line, this would require the addition of only one symbol for the ideal point on the line. It is difficult to think of the addition of a single element to the real number system that will give us a new system with the familiar properties of the real number system. We know we can add a square root of $-1$ to the system, but we then obtain a number system with many new elements, not just the one new element with which to label the ideal point of the line. Intuitively we may tend to think of the ideal point as a "point at infinity." What is wrong then with the use of the symbol $\infty$ as coordinate of this ideal point? This symbol does not represent a number. If we accept our intuitive ideas of infinity, we might perhaps accept the rules

$$\infty + a = \infty \quad \text{for any real number } a,$$

$$\infty \cdot a = \infty \quad \text{for any real number } a,$$

$$\frac{a}{\infty} = 0 \quad \text{for any real number } a,$$

but our intuition does not tell what number such expressions as $\infty/\infty$ or $\infty - \infty$ are to represent. Even if we could represent coordinates on the line using this symbol, we would be in trouble in the plane when we ask which ideal point is to have coordinates $(\infty, \infty)$.

It seems probable that the addition of a new symbol is not likely to resolve our difficulties and that some other solution might well be sought. We shall see that the coordinates adopted are noticeably different from those of Euclidean analytic geometry and that many of the algebraic techniques adaptable to these coordinates are also different from those of Euclidean analytic geometry.

We consider first the problem of finding the coordinates in the projective plane. To obtain a clue as to the procedure we might well follow, let us consider the Euclidean problem of finding the point of intersection of two parallel lines. If we attempt to find the point of intersection of the parallel lines

$$2x + 3y = 8,$$

$$2x + 3y = 5,$$

by subtracting the second equation from the first, we eliminate both variables, obtain the absurd statement $0 = 3$, and conclude the equations are inconsistent. In general parallel lines have equations which may be written in the form

$$Ax + By + C = 0,$$

$$Ax + By + C' = 0,$$

where $C \neq C'$. If we attempt to solve this system by eliminating the $x$ terms by subtraction, we also eliminate the $y$ terms and get the contradictory statement

$$C - C' = 0.$$

In the projective plane on the other hand we must not come to an absurd statement, for now the two lines meet at an ideal point. One way to avoid the absurdity would be to have the quantity $C - C'$ multiplied by a variable quantity that could assume the value zero. There would be no necessary absurdity if elimination led to

$$(C - C')z = 0.$$

Let us then consider the consequences of introducing a third variable and changing the equation of a line from the form

$$Ax + By + C = 0$$

to the form

$$Ax + By + Cz = 0,$$

where points now have triples $(x, y, z)$ as coordinates. Our discussion would seem to indicate that the third coordinate be zero for an ideal point. If $(x, y)$ are coordinates, old-style, of an ordinary point, we might let $(x, y, 1)$ be new coordinates for this point, for such coordinates would satisfy the new-style equation of any line through this point. What if the third coordinate were neither one nor zero? In this case we may write the coordinates in the form $(ax, ay, a)$ where $a \neq 0$. These coordinates would satisfy the equation of any line through the point with old-style coordinates $(x, y)$, since if

$$Ax + By + C = 0,$$

then

$$A(ax) + B(ay) + Ca = a(Ax + By + C) = 0.$$

This suggests that any triple with coordinates proportional to those of $(x, y, 1)$ should also be taken as coordinates of the point with old-style coordinates $(x, y)$.

Before committing ourselves, let us further consider coordinates of the form $(x, y, 0)$. Any triple proportional to $(1, m, 0)$ would satisfy the new-style equation

$$y = mx + bz,$$

which represents the family of lines with slope $m$. Any triple proportional to $(1, m, 0)$ could thus be coordinates for the ideal point of the family of parallel lines of slope $m$. Vertical lines, furthermore, would have equations of the form

$$x = cz,$$

which would be satisfied by any triple proportional to $(0, 1, 0)$.

The triple $(0, 0, 0)$ is proportional to any triple $(x, y, z)$ and could thus represent coordinates for any point whatsoever. To avoid this ambiguity, we might well agree to discard this particular triple from the coordinate system.

The foregoing indicates that it should be possible to use triples as coordinates in the projective plane, although the resulting correspondence between points and coordinates will not be one-to-one. To avoid confusion with Euclidean space coordinates, we write the triples in the form $(x_1, x_2, x_3)$. We formalize the discussion with the following definition.

**Definition 6.11.** *The set of all real number triples* $(x_1, x_2, x_3)$ *where not all* $x_i$ *are zero constitute* natural homogeneous point coordinates *in the real projective plane, where* $(x_1, x_2, x_3)$ *will be coordinates of*

a. *the ordinary point with Euclidean coordinates* $(x, y)$ *if* $x_3 \neq 0$, $x_1/x_3 = x$, *and* $x_2/x_3 = y$,

b. *the ideal point of the lines of slope* $m$ *if* $x_3 = 0$ *and* $x_2/x_1 = m$,

c. *the ideal point on vertical lines if* $x_1 = x_3 = 0$.

We see that the point with Euclidean coordinates $(3, -2)$ has natural homogeneous coordinates $(3, -2, 1)$, $(-3, 2, -1)$, $(6, -4, 2)$, or any set of numbers proportional to these and not all zero. The ideal point on lines of slope 2 has natural homogeneous coordinates $(1, 2, 0)$, $(3, 6, 0)$, etc. The triple $(7, -4, 3)$ constitutes natural homogeneous coordinates for the point whose Euclidean coordinates are $(\frac{7}{3}, -\frac{4}{3})$ while $(-2, 3, 0)$ are coordinates for the ideal point of lines of slope $-\frac{3}{2}$.

We can develop natural homogeneous coordinates for points on a line in a like manner. Here we need two numbers instead of the one in the Euclidean case. We have

**Definition 6.12.** *The set of all real number pairs* $(x_1, x_2)$ *where* $x_1$ *and* $x_2$ *are not both zero constitute* natural homogeneous point coordinates *of the real projective line, where* $(x_1, x_2)$ *will be coordinates of*

a. *the ordinary point with Euclidean coordinate* $x_1/x_2$ *if* $x_2 \neq 0$,

b. *the ideal point if* $x_2 = 0$.

Thus $(1, 0)$, $(2, 0)$, etc., are coordinates of the ideal point; $(0, 1)$, $(0, 2)$, etc., are coordinates of the origin; and $(-3, 1)$, $(6, -2)$, etc., are coordinates of the point with Euclidean coordinate $-3$.

Our development has implicitly assumed we desired lines in the projective plane to have first degree equations. We can now prove this formally on the basis of Definition 6.11.

**Theorem 6.11.** *An equation in $x_1$, $x_2$, and $x_3$ of the form*

$$u_1x_1+u_2x_2+u_3x_3=0$$

*where not all $u_i$ are zero is the equation of a line, and conversely any line has equation of this form.*[1]

PROOF. We consider the equation

$$u_1x_1+u_2x_2+u_3x_3=0$$

and suppose $u_1$ and $u_2$ are not both zero. If we divide both sides of the equation by $x_3$ and replace $x_1/x_3$ and $x_2/x_3$ by $x$ and $y$, respectively, the equation becomes the "old-style" equation

$$u_1x+u_2y+u_3=0.$$

Using Definition 6.11, we conclude that the ordinary points whose natural homogeneous coordinates satisfy the original equation are those and only those lying on the Euclidean line given by the second equation. We must still show that the original equation is satisfied by the coordinates of the ideal point added to this Euclidean line and by no other ideal points. If $u_2=0$, the corresponding Euclidean line is vertical and the equation is satisfied by the ideal point with coordinates $(0, x_2, 0)$ and by no other ideal point. If the line is not vertical, its slope is $-u_1/u_2$, and the equation is satisfied by the ideal point whose coordinates $(x_1, x_2, 0)$ satisfy

$$u_1x_1+u_2x_2=0$$

or

$$\frac{x_2}{x_1}=-\frac{u_1}{u_2}$$

and by no other ideal point. This is the desired ideal point, and we conclude that

$$u_1x_1+u_2x_2+u_3x_3=0$$

is the equation of a line. If $u_1$ and $u_2$ are both zero, it is clear that the equation is satisfied by the coordinates of all ideal points and only these points; hence the equation is that of the ideal line.

[1]The form of this equation indicates the reason for the use of the word "homogeneous" in our new coordinates. Lines have homogeneous equations; that is, all terms have the same degree. This homogeneity of degree is true for the equations of all algebraic loci studied in analytic projective geometry.

The proof of the converse is similar to that above and is left as an exercise for the reader.

## Exercises 6.1

1. Find natural homogeneous coordinates in the plane for each of the following points:
   *a*. The origin;
   *b*. The intersection of the $x$-axis and the ideal line;
   *c*. The intersection of the $y$-axis and the ideal line;
   *d*. The point $(2, -5)$;
   *e*. The point $(3, -7)$;
   *f*. The point $(1, 0)$;
   *g*. The point $(1, 1)$.

2. Find natural homogeneous coordinates on the line for each of the following points:
   *a*. The origin;
   *b*. The ideal point;
   *c*. The point with coordinate 1;
   *d*. The point with coordinate $-\frac{3}{8}$.

3. Describe each of the following lines:
   *a*. $2x_1 - 3x_2 + 4x_3 = 0$;
   *b*. $7x_1 + 2x_2 + 3x_3 = 0$;
   *c*. $x_1 = 0$;
   *d*. $x_2 = 0$;
   *e*. $x_3 = 0$.

4. Find coordinates of the ideal point or points of each of the lines of the previous exercise.

5. Find the equation of each of the following lines:
   *a*. The extension to the projective plane of the $x$-axis;
   *b*. The ideal line;
   *c*. The line through $(3, 7, 1)$ and $(2, 3, 1)$;
   *d*. The line through $(3, 7, 3)$ and $(2, 3, 5)$;
   *e*. The line through $(3, 7, 1)$ and $(2, 3, 0)$.

6. Prove the converse portion of Theorem 6.11.

7. Show that the equation of the line through the points $(a_1, a_2, a_3)$ and $(b_1, b_2, b_3)$ is

$$\begin{vmatrix} x_1 & x_2 & x_3 \\ a_1 & a_2 & a_3 \\ b_1 & b_2 & b_3 \end{vmatrix} = 0.$$

## 6.2 LINE COORDINATES; DUALITY

Theorem 6.11 indicates the form that the principle of duality takes in plane analytic projective geometry. We consider the equations

$$2x_1 + 3x_2 - 5x_3 = 0$$

and

$$4x_1 + 6x_2 - 10x_3 = 0.$$

It is clear that these are equations of the same line, since they will be satisfied by exactly the same triples of real numbers. Indeed, any triple of numbers proportional to $[2, 3, -5]$ can be taken as the coefficients for the equation of this line. It is also clear that the triple $[0, 0, 0]$ does not lead to coefficients for the equation of any line. The properties of triples used to form coefficients in the equation of a line are strikingly similar to those of natural homogeneous point coordinates, and we make the following definition.

**Definition 6.21.** *The set of all real-number triples* $[u_1, u_2, u_3]$ *where not all* $u_i$ *are zero constitute* natural homogeneous line coordinates *in the real projective plane, where* $[u_1, u_2, u_3]$ *are coordinates of the line with equation*

$$u_1x_1 + u_2x_2 + u_3x_3 = 0.$$

This equation is symmetric in the $u$'s and $x$'s. Theorem 6.11 states that this equation is satisfied by the point $(x_1, x_2, x_3)$ and the line $[u_1, u_2, u_3]$ if and only if the point lies on the line or the line passes through the point. This is a more symmetric statement of the result of the theorem. Now we can interpret this result in two ways. In the original statement of the theorem we were thinking of it in the form, "For fixed $[u_1, u_2, u_3]$

$$u_1x_1 + u_2x_2 + u_3x_3 = 0$$

is the equation of the line with coordinates $[u_1, u_2, u_3]$; that is, it is satisfied by the coordinates of those and only those points lying on this line." Our symmetric statement above asserts that we may also say, "For fixed $(x_1, x_2, x_3)$

$$u_1x_1 + u_2x_2 + u_3x_3 = 0$$

is the equation of the point with coordinates $(x_1, x_2, x_3)$; that is, it is satisfied by the coordinates of those and only those lines passing through this point." Thus

$$2x_1 + 3x_2 - 5x_3 = 0$$

is satisfied by the coordinates of the points on the line with homogeneous line coordinates [2, 3, –5] and is the equation of this line. Similarly

$$2u_1 + 3u_2 - 5u_3 = 0$$

is satisfied by the coordinates of the lines passing through the point with natural homogeneous coordinates (2, 3, –5) and is the equation of this point. We thus have

**Theorem 6.21.** *An equation in $u_1$, $u_2$, and $u_3$ of the form*

$$u_1x_1 + u_2x_2 + u_3x_3 = 0$$

*where not all $x_i$ are zero is the equation of a point, and conversely any point has equation of this form.*

Our introduction of line coordinates has carried over the duality principle to plane analytic projective geometry, for a change from "point" to "line" leads from one set of homogeneous linear equations to another, and these two sets of coordinates have a symmetric appearance in the equation of a line or a point. We note that Theorems 6.11 and 6.21 are plane duals of each other.

## Exercises 6.2

1. Find natural homogeneous coordinates in the plane for the extension to the projective plane of each of the following lines:
   *a.* The coordinate axes;
   *b.* The ideal line;
   *c.* $y = x$;
   *d.* $3x + 2y = 6$;
   *e.* $y = 2x - 7$;
   *f.* $y = 3$;
   *g.* $x = 9$.
2. Find the equation of each of the following points:
   *a.* The origin;
   *b.* The ideal point on the $x$-axis;
   *c.* The point with Euclidean coordinates (1, 1);
   *d.* The point with Euclidean coordinates (3, 7).
3. Prove that the point $(a, b, c)$ can never lie on the line $[a, b, c]$.
4. Why is there no principle of duality for the projective line?

## 6.3 HOMOGENEOUS EQUATIONS

The problems of determining the lines through two points and of finding the point of intersection of two lines are not symmetric in Euclidean analytic geometry, yet they must be so in the projective case since the problems are duals of each other. We observe that this is the case, for the former problem is that of solving

$$u_1x_1 + u_2x_2 + u_3x_3 = 0$$

and

$$u_1x_1' + u_2x_2' + u_3x_3' = 0$$

for $u_1$, $u_2$, and $u_3$, while the latter problem is that of solving

$$u_1x_1 + u_2x_2 + u_3x_3 = 0$$

and

$$u_1'x_1 + u_2'x_2 + u_3'x_3 = 0$$

for $x_1$, $x_2$, and $x_3$.

The problem of solving simultaneous linear equations like the systems above is not the same in terms of the algebra required as in Euclidean geometry, and we investigate this situation in detail.

Suppose we wish to find the coordinates of the point of intersection of the lines $[2, 4, 8]$ and $[1, -1, -1]$; that is, we wish to solve

$$2x_1 + 4x_2 + 8x_3 = 0$$

and

$$x_1 - x_2 - x_3 = 0$$

for $x_1$, $x_2$, and $x_3$. Here we must solve two linear equations in three variables. The constant terms in the equations are all zero, which is not often the case in Euclidean geometry. Simultaneous linear equations in which the constant terms are all zero are said to form a *homogeneous* linear system. It is clear that any homogeneous linear system in $x_1$, $x_2$, and $x_3$ has a solution $x_1 = x_2 = x_3 = 0$. This solution is said to be the *trivial* one, and it can have no geometric meaning since $(0, 0, 0)$ does not represent any point. It is clear that if we can find a nontrivial solution of the system, any set of numbers proportional to the numbers of the solution will also be a solution; for example, $(1, -2, 3)$ satisfies

$$x_1 - x_2 - x_3 = 0$$

as does $(-1, 2, -3)$, $(2, -4, 6)$, etc. This is reassuring since the homogeneous coordinates of a point are unique only to within a nonzero constant of proportionality. The existence of these proportional sets of solutions suggests that we could find a solution by assigning a value arbitrarily to one of the $x_i$ and then solving for the others. In our example we could set $x_3 = 1$. The equations then become

$$2x_1 + 4x_2 + 8 = 0$$

and

$$x_1 - x_2 - 1 = 0.$$

This system has the solution $x_1 = -\frac{2}{3}$, $x_2 = -\frac{5}{3}$. It is clear that any set of numbers proportional to $x_1 = -\frac{2}{3}$, $x_2 = -\frac{5}{3}$, and $x_3 = 1$ is also a solution; a simpler such set would be $x_1 = 2$, $x_2 = 5$, and $x_3 = -3$. In terms of the geometric problem the point of intersection has coordinates $(-\frac{2}{3}, -\frac{5}{3}, 1)$, $(2, 5, -3)$, or any set proportional to these.

This procedure would be embarrassing if the desired point were ideal and had $x_3 = 0$, for our choice $x_3 = 1$ would lead to trouble (the resulting equations would be inconsistent, but the original system would have a nontrivial solution and not be inconsistent). We can avoid this possibility by a second method of solving such a homogeneous system. We eliminate an $x_i$ from the pair of equations. In the example $x_1$ is perhaps most simply eliminated. If we eliminate it, we find

$$6x_2 + 10x_3 = 0.$$

We now find any pair of numbers satisfying this equation. We could take $x_2 = 10$ and $x_3 = -6$. If we substitute these values in either of the original equations, we find $x_1 = 4$. The desired point thus has coordinates $(4, 10, -6)$. The use of the method of elimination until we obtain one equation in two $x_i$, for which a solution can easily be found, avoids the possibility of an unfortunate assignment of nonzero value to an $x_i$ whose value must be zero.

We now consider a third method of solving the system. If we divide each equation by $x_3$, then start to solve for $x_1/x_3$ and $x_2/x_3$ by Cramer's rule, we find

$$\frac{x_1}{x_3} = \frac{\begin{vmatrix} -8 & 4 \\ 1 & -1 \end{vmatrix}}{\begin{vmatrix} 2 & 4 \\ 1 & -1 \end{vmatrix}}$$

and

$$\frac{x_2}{x_3} = \frac{\begin{vmatrix} 2 & -8 \\ 1 & 1 \end{vmatrix}}{\begin{vmatrix} 2 & 4 \\ 1 & -1 \end{vmatrix}}.$$

Since the interchange of two columns in any determinant will change a sign, as will multiplication of a column by $-1$, these solutions could also be written in the form

$$\frac{x_1}{x_3} = \frac{\begin{vmatrix} 4 & 8 \\ -1 & -1 \end{vmatrix}}{\begin{vmatrix} 2 & 4 \\ 1 & -1 \end{vmatrix}}$$

and

$$\frac{x_2}{x_3} = \frac{\begin{vmatrix} 8 & 2 \\ -1 & 1 \end{vmatrix}}{\begin{vmatrix} 2 & 4 \\ 1 & -1 \end{vmatrix}}.$$

It follows readily that a solution of the system is the triple of numbers

$$\left(\begin{vmatrix} 4 & 8 \\ -1 & -1 \end{vmatrix}, \begin{vmatrix} 8 & 2 \\ -1 & 1 \end{vmatrix}, \begin{vmatrix} 2 & 4 \\ 1 & -1 \end{vmatrix}\right) = (4, 10, -6).$$

We now generalize this third procedure. We seek the point of intersection of the lines

$$\begin{aligned} u_1x_1 + u_2x_2 + u_3x_3 &= 0 \\ u_1'x_1 + u_2'x_2 + u_3'x_3 &= 0. \end{aligned}$$

If we solve by Cramer's rule for $x_1/x_3$ and $x_2/x_3$, we find

$$\frac{x_1}{x_3} = \frac{\begin{vmatrix} -u_3 & u_2 \\ -u_3' & u_2' \end{vmatrix}}{\begin{vmatrix} u_1 & u_2 \\ u_1' & u_2' \end{vmatrix}} = \frac{\begin{vmatrix} u_2 & u_3 \\ u_2' & u_3' \end{vmatrix}}{\begin{vmatrix} u_1 & u_2 \\ u_1' & u_2' \end{vmatrix}}$$

and

$$\frac{x_2}{x_3} = \frac{\begin{vmatrix} -u_1 & u_3 \\ -u_1' & u_3' \end{vmatrix}}{\begin{vmatrix} u_1 & u_2 \\ u_1' & u_2' \end{vmatrix}} = \frac{\begin{vmatrix} u_3 & u_1 \\ u_3' & u_1' \end{vmatrix}}{\begin{vmatrix} u_1 & u_2 \\ u_1' & u_2' \end{vmatrix}}.$$

Homogeneous coordinates of the point of intersection are thus

$$\left(\begin{vmatrix} u_2 & u_3 \\ u_2' & u_3' \end{vmatrix}, \begin{vmatrix} u_3 & u_1 \\ u_3' & u_1' \end{vmatrix}, \begin{vmatrix} u_1 & u_2 \\ u_1' & u_2' \end{vmatrix}\right),$$

assuming these determinants are not all zero.

Since
$$\begin{vmatrix} a & b \\ c & d \end{vmatrix} = 0$$

if and only if its rows are in proportion, all the determinants will be zero if and only if their rows are proportional; that is, $[u_1, u_2, u_3] = k[u_1', u_2', u_3']$. This would imply that $[u_1, u_2, u_3]$ and $[u_1', u_2', u_3']$ are line coordinates of the same line. We conclude that for distinct lines not all of the determinants are zero.[2] We have proved

**Theorem 6.31.** *The point of intersection of the distinct lines* $[u_1, u_2, u_3]$ *and* $[u_1', u_2', u_3']$ *has natural homogeneous coordinates*

$$\left(\begin{vmatrix} u_2 & u_3 \\ u_2' & u_3' \end{vmatrix}, \begin{vmatrix} u_3 & u_1 \\ u_3' & u_1' \end{vmatrix}, \begin{vmatrix} u_1 & u_2 \\ u_1' & u_2' \end{vmatrix}\right).$$

By plane duality we also have

**Theorem 6.32.** *The line determined by the distinct points* $(x_1, x_2, x_3)$ *and* $(x_1', x_2', x_3')$ *has natural homogeneous coordinates*

$$\left[\begin{vmatrix} x_2 & x_3 \\ x_2' & x_3' \end{vmatrix}, \begin{vmatrix} x_3 & x_1 \\ x_3' & x_1' \end{vmatrix}, \begin{vmatrix} x_1 & x_2 \\ x_1' & x_2' \end{vmatrix}\right].$$

## Exercises 6.3

1. Find the coordinates of the line determined by each of the following pairs of points:

   *a.* $(3, -5, 7)$ and $(2, 1, -3)$;
   *b.* $(4, -6, 3)$ and $(4, -6, 1)$;
   *c.* $(1, 1, 0)$ and $(0, 1, 0)$;
   *d.* $(2, 5, -3)$ and $(3, -2, 0)$;
   *e.* $(3, -5, 7)$ and $(6, -10, 14)$.

2. Find the coordinates of the point of intersection of each of the following pairs of lines:

   *a.* $[3, -5, 7]$ and $[2, 1, -3]$;

[2]This use of Cramer's rule is valid only if

$$\begin{vmatrix} u_1 & u_2 \\ u_1' & u_2' \end{vmatrix} \neq 0.$$

The determination of the coordinates of the point of intersection as the above-ordered triple of determinants can be justified by an appropriate application of Cramer's rule, since some such determinant can not vanish.

*b*. $[4, -6, 3]$ and $[4, -6, 1]$;
*c*. $[1, 1, 0]$ and $[0, 1, 0]$;
*d*. $[2, 5, -3]$ and $[3, -2, 0]$;
*e*. $[3, -5, 7]$ and $[6, -10, 14]$.

3. Rework parts *c*, *d*, and *e* of Exercise 6.15.

4. Find the coordinates of the line determined by the point $(2, 3, -1)$ and the intersection of $[3, 4, -2]$ and $[1, -1, 2]$.

5. Use the result of Exercise 6.17 to obtain an alternate proof of Theorem 6.32.

## 6.4* SPACE COORDINATES

The point and line coordinates we have been considering are those needed for plane analytic projective geometry. The corresponding coordinates for space geometry are introduced in a similar manner. We give an outline of the development, leaving the details to the reader.

It would be natural to use four numbers as point coordinates in space and keep the assumption that proportional sets are coordinates of the same point. Analogous to the plane case it would be desirable to have the first three coordinates proportional to the Euclidean coordinates in the case of an ordinary point, while an ideal point would be distinguished by a zero fourth coordinate. For an ideal point the first three coordinates would be associated with the direction of the lines through that ideal point. A consistent theory is created if we take the three nonzero coordinates as direction numbers of the lines through the given ideal point. Formally we define space coordinates in

**Definition 6.41.** *The set of all real number four-tuples* $(x_1, x_2, x_3, x_4)$ *where not all* $x_i$ *are zero constitute* natural homogeneous point coordinates *in real projective space, where* $(x_1, x_2, x_3, x_4)$ *are coordinates of*

a. *the ordinary point with Euclidean coordinates* $(x, y, z)$ *if* $x_4 \neq 0$, $x_1/x_4 = x$, $x_2/x_4 = y$, *and* $x_3/x_4 = z$,

b. *the ideal point on the lines with direction numbers* $x_1 : x_2 : x_3$ *if* $x_4 = 0$.

In space we desire planes to have first-degree equations. Definition 6.41 assures that this is the case, and we have

**Theorem 6.41.** *An equation in* $x_1, x_2, x_3$, *and* $x_4$ *of the form*

$$u_1x_1 + u_2x_2 + u_3x_3 + u_4x_4 = 0$$

*where not all* $u_i$ *are zero is the equation of a plane, and conversely any plane has equation of this form.*

The proof is similar to that of Theorem 6.11, and the details are left to the reader.[3]

Next we define the $u$'s in the equation of Theorem 6.41 to be natural homogeneous coordinates of the plane determined by the equation. We have

**Definition 6.42.** *The set of all real number four-tuples* $[u_1, u_2, u_3, u_4]$ *where not all* $u_i$ *are zero constitute* natural homogeneous plane coordinates *in real projective space, where* $[u_1, u_2, u_3, u_4]$ *are coordinates of the plane whose equation is*

$$u_1x_1 + u_2x_2 + u_3x_3 + u_4x_4 = 0.$$

The symmetry of the $x$'s and the $u$'s in the equation of a plane is the manifestation of space duality. The equation either states that a point lies on the plane whose coordinates are the $u$'s, or it states that a plane passes through the point whose coordinates are the $x$'s. We have as space dual of Theorem 6.41.

**Theorem 6.42.** *An equation in* $u_1, u_2, u_3$, *and* $u_4$ *of the form*

$$u_1x_1 + u_2x_2 + u_3x_3 + u_4x_4 = 0$$

*where not all* $x_i$ *are zero is the equation of a point, and conversely any point has equation of this form.*

We note that the problems of finding the plane through three noncollinear points and of finding the intersection of three noncollinear planes are dual. Indeed, they are just the problem of solving simultaneously three homogeneous linear equations in four variables. Our previous discussion of such systems still applies. Again a determinant approach gives a compact and symmetric solution to the problem, and we have the following dual theorems.

**Theorem 6.43.** *The point of intersection of the noncollinear planes* $[u_1, u_2, u_3, u_4]$, $[u_1', u_2', u_3', u_4']$, *and* $[u_1'', u_2'', u_3'', u_4'']$ *has natural homogeneous coordinates*

$$\left( \begin{vmatrix} u_2 & u_3 & u_4 \\ u_2' & u_3' & u_4' \\ u_2'' & u_3'' & u_4'' \end{vmatrix}, -\begin{vmatrix} u_1 & u_3 & u_4 \\ u_1' & u_3' & u_4' \\ u_1'' & u_3'' & u_4'' \end{vmatrix}, \begin{vmatrix} u_1 & u_2 & u_4 \\ u_1' & u_2' & u_4' \\ u_1'' & u_2'' & u_4'' \end{vmatrix}, -\begin{vmatrix} u_1 & u_2 & u_3 \\ u_1' & u_2' & u_3' \\ u_1'' & u_2'' & u_3'' \end{vmatrix} \right).$$

[3]It will be convenient to recall that the Euclidean plane with equation $Ax + By + Cz + D = 0$ has a perpendicular direction with direction numbers $A:B:C$ and that lines with direction numbers $a:b:c$ and $a':b':c'$ are perpendicular if and only if $aa' + bb' + cc' = 0$.

**Theorem 6.44.** *The plane determined by the noncollinear points* $(x_1, x_2, x_3, x_4)$, $(x'_1, x'_2, x'_3, x'_4)$, *and* $(x''_1, x''_2, x''_3, x''_4)$ *has natural homogeneous coordinates*

$$\left[\begin{vmatrix} x_2 & x_3 & x_4 \\ x'_2 & x'_3 & x'_4 \\ x''_2 & x''_3 & x''_4 \end{vmatrix}, -\begin{vmatrix} x_1 & x_3 & x_4 \\ x'_1 & x'_3 & x'_4 \\ x''_1 & x''_3 & x''_4 \end{vmatrix}, \begin{vmatrix} x_1 & x_2 & x_4 \\ x'_1 & x'_2 & x'_4 \\ x''_1 & x''_2 & x''_4 \end{vmatrix}, -\begin{vmatrix} x_1 & x_2 & x_3 \\ x'_1 & x'_2 & x'_3 \\ x''_1 & x''_2 & x''_3 \end{vmatrix}\right].$$

The computation needed to verify these coordinates is left to the reader. The noncollinearity implies that not all the determinants can be zero, as will be proved in Chapter 8.

## Exercises 6.4

1. Find homogeneous coordinates in space for each of the following points:
   *a.* The origin;
   *b.* The intersection of each of the coordinate axes with the ideal plane;
   *c.* The point $(3, 0, -2)$;
   *d.* The point $(4, -6, -1)$;
   *e.* The point $(5, 0, 0)$.

2. Find homogeneous coordinates in space for the extension to projective space of each of the following planes:
   *a.* The coordinates planes;
   *b.* The ideal plane;
   *c.* The plane $x-2y+3z=4$;
   *d.* The plane $2x+7y-8=0$;
   *e.* The plane $y=4$.

3. Find coordinates of the plane determined by each of the following sets of points:
   *a.* $(1, 2, 1, 3)$, $(0, 1, 4, -7)$, and $(2, 1, 3, -3)$;
   *b.* $(3, 4, -1, 1)$, $(2, -3, 3, 5)$, and $(-2, 1, 1, -3)$;
   *c.* $(3, 4, -1, 2)$, $(2, 1, -1, 1)$, and $(1, 0, 3, 0)$;
   *d.* $(1, 1, 1, 0)$, $(1, 1, 0, 1)$, and $(1, 0, 1, 1)$;
   *e.* $(1, 2, 1, 3)$, $(0, 1, 4, -7)$, and $(1, 3, 5, -4)$.

4. Find the coordinates of the point of intersection of each of the following sets of planes:
   *a.* $[1, 2, 1, 3]$, $[0, 1, 4, -7]$, and $[2, 1, 3, -3]$;
   *b.* $[3, 4, -1, 1]$, $[2, -3, 3, 5]$, and $[-2, 1, 1, -3]$;
   *c.* $[3, 4, -1, 2]$, $[2, 1, -1, 1]$, and $[1, 0, 3, 0]$;
   *d.* $[1, 1, 1, 0]$, $[1, 1, 0, 1]$, and $[1, 0, 1, 1]$;
   *e.* $[1, 2, 1, 3]$, $[0, 1, 4, -7]$, and $[1, 3, 5, -4]$.

5. Find the equation of the plane determined by (1, 2, 1, 3), (3, 0, 1, 0), and the intersection of [1, 0, 2, 3], [2, 0, 1, 1], and [1, 1, 1, 1].

6. Prove that the plane through the points $(a_1, a_2, a_3, a_4)$, $(b_1, b_2, b_3, b_4)$, and $(c_1, c_2, c_3, c_4)$ has equation

$$\begin{vmatrix} x_1 & x_2 & x_3 & x_4 \\ a_1 & a_2 & a_3 & a_4 \\ b_1 & b_2 & b_3 & b_4 \\ c_1 & c_2 & c_3 & c_4 \end{vmatrix} = 0.$$

7. Prove Theorem 6.41.

8. Prove Theorem 6.43 (assuming noncollinearity implies not all of the determinants are zero).

## REFERENCES


Adler, Claire Fisher, *Modern Geometry, An Integrated First Course*, 2nd ed., McGraw-Hill Book Company, New York, 1967, Chapter 8.

Artzy, Rafael, *Linear Geometry*, Addison-Wesley Publishing Company, Reading, Mass., 1965, Chapter 3.

Busemann, Herbert, and Paul J. Kelly, *Projective Geometry and Projective Metrics*, Academic Press, New York, 1953, Chapter 1.

Coxeter, H. S. M., *Non-Euclidean Geometry*, 5th ed., University of Toronto Press, Toronto, 1965, Chapter 4.

Dorwart, Harold L., *The Geometry of Incidence*, Prentice-Hall, Englewood Cliffs, N. J., 1966, Chapters 2, 3.

Gruenberg, K. W., and A. J. Weir, *Linear Geometry*, D. Van Nostrand Company, Princeton, 1967, Chapter 2.

Hartshorne, Robin, *Foundations of Projective Geometry*, W. A. Benjamin, New York, 1967, Chapter 1.

Levy, Harry, *Projective and Related Geometries*, The Macmillan Company, New York, 1961, Chapters 2, 3.

Maxwell, E. A., *The Methods of Plane Projective Geometry Based on the Use of General Homogeneous Coordinates*, Cambridge University Press, Cambridge, 1960, Chapter 1.

Rosenbaum, Robert A., *Introduction to Projective Geometry and Modern Algebra*, Addison-Wesley Publishing Company, Reading, Mass., 1963, Chapter 5.

Seidenberg, A., *Lectures in Projective Geometry*, D. Van Nostrand Company, Princeton, 1962, Chapter 1.

Tuller, Annita, *A Modern Introduction to Geometries*, D. Van Nostrand Company, Princeton, 1967, Chapter 5.

## CHAPTER SEVEN

# VECTORS AND MATRICES

We have seen in the preceding chapter that the algebra associated with natural homogeneous coordinates is not quite the same as that of Euclidean analytic geometry. The important techniques of elementary analytic projective geometry come from that part of modern algebra concerned with the theory of vector spaces and matrices, and these techniques lead to very elegant and frequently simple proofs in an analytic study of projective geometry. We consider the applications of these techniques in the next chapter. In the present chapter we consider those aspects of the theory that we use later. The discussion is brief and certainly does not cover all of even the fundamental aspects of these topics. The reader is encouraged to consult the references at the end of the chapter for further background.

## 7.1 REAL VECTOR SPACES

The natural homogeneous coordinates we have introduced are $n$-tuples of real numbers, $(x_1, x_2, \cdots, x_n)$, where $n$ is 2, 3, or 4. We consider the algebra of these $n$-tuples; our results will be valid for all $n$, although we use them in the next chapter only for $n$ of 2, 3, or 4. We define the sum of two $n$-tuples by the simple relation

$$(x_1, x_2, \cdots, x_n) + (y_1, y_2, \cdots, y_n) = (x_1 + y_1, x_2 + y_2, \cdots, x_n + y_n). \tag{7.11}$$

Thus we have

$$(2,-4,3,7)+(1,2,-4,1)=(3,-2,-1,8).$$

If $c$ is a real number, we can also define the product of $c$ and an $n$-tuple by

$$c(x_1, x_2, \cdots, x_n) = (cx_1, cx_2, \cdots, cx_n); \tag{7.12}$$

for example,

$$7(2,-4,3,7)=(14,-28,21,49).$$

If this multiplication is defined for all $c$, we must allow the $n$-tuple $(0, 0, \cdots, 0)$ in our discussion. We denote this $n$-tuple by $\mathbf{0}$.

Let us denote the set of all $n$-tuples (for fixed $n$) by $\mathscr{V}$. It is readily verified that for all $\mathbf{v}_1$, $\mathbf{v}_2$, and $\mathbf{v}_3$ in $\mathscr{V}$ and for all real numbers $c_1$ and $c_2$

1. $\mathbf{v}_1 + \mathbf{v}_2$ is in $\mathscr{V}$;
2. $c_1\mathbf{v}_1$ is in $\mathscr{V}$;
3. $(\mathbf{v}_1+\mathbf{v}_2)+\mathbf{v}_3=\mathbf{v}_1+(\mathbf{v}_2+\mathbf{v}_3)$;
4. $\mathbf{v}_1+\mathbf{v}_2=\mathbf{v}_2+\mathbf{v}_1$;
5. $\mathbf{v}_1+\mathbf{0}=\mathbf{v}_1$;
6. for each $\mathbf{v}_1$ there is a $-\mathbf{v}_1$ such that $\mathbf{v}_1+(-\mathbf{v}_1)=\mathbf{0}$;
7. $c_1(\mathbf{v}_1+\mathbf{v}_2)=c_1\mathbf{v}_1+c_1\mathbf{v}_2$;
8. $(c_1+c_2)\mathbf{v}_1=c_1\mathbf{v}_1+c_2\mathbf{v}_1$;
9. $(c_1c_2)\mathbf{v}_1=c_1(c_2\mathbf{v}_1)$;
10. $1\mathbf{v}_1=\mathbf{v}_1$.

These properties are fulfilled by many sets $\mathscr{V}$ other than the set of $n$-tuples considered here. They would hold, for example, if $\mathscr{V}$ were the set of all functions continuous on the interval $0 < x < 1$; if $f$ and $g$ are such functions and $c$ is a real number, we would define the functions $f+g$ and $cf$ by $[f+g](x)=f(x)+g(x)$ and $[cf](x)=cf(x)$, respectively. The set of all solutions of a given homogeneous linear differential equation with constant coefficients fulfills the above under the same definitions of the operations. The conditions are even fulfilled if $\mathscr{V}$ is the set of real numbers itself. Another example would be the set of all vectors in three-dimensional Euclidean space where addition is carried out by the familiar parallelogram rule. Indeed, if we place all vectors with their initial points at the origin and their terminal points at $(x, y, z)$, then the vectors can be identified with the set of all 3-tuples. It is customary to use this example to give a name for $\mathscr{V}$, and we have

**Definition 7.11.** *A set is a* real vector space *if there is defined an operation of addition* (+) *on the set and an operation of multiplication of elements of the set by real numbers satisfying 1 to 10 above.*

We have at once in view of our discussion.

**Theorem 7.11.** *The set of all n-tuples of real numbers with operations defined by (7.11) and (7.12) is a real vector space.*

We call the elements of a vector space *vectors* and the real numbers *scalars.* The operation of multiplication of a vector by a scalar is called *scalar multiplication.* We note we cannot identify vectors with points in the projective plane in a one-to-one manner since many vectors represent coordinates of the same point while the zero vector **0** represents no point.

## Exercises 7.1

1. Find the sum of each of the following pairs of vectors:
   *a.* $(2,-7), (4,5)$;
   *b.* $(1,-4,9), (-3,5,3)$;
   *c.* $(2,-8,-4), (0,0,0)$;
   *d.* $(3,-1,4,-7), (4,2,-3,-5)$;
   *e.* $(3,-1,4,-7), (-3,1,-4,7)$.
2. Compute each of the following:
   *a.* $3(2,-7)+2(4,5)$;
   *b.* $4(1,-4,9)-5(-3,5,3)$;
   *c.* $3(2,-8,-4)+7(0,0,0)$;
   *d.* $-4(3,-1,4,-7)+2(4,2,-3,-5)$;
   *e.* $-4(3,-1,4,-7)+2(-3,1,-4,7)$.
3. Verify properties 1 to 10 for the set of $n$-tuples.
4. Prove that a vector space is a group under addition (see Exercise 5.410).

## 7.2 LINEAR INDEPENDENCE

Basic to the study of vector spaces are the ideas of linear independence and linear dependence, and we find these ideas extremely useful in the applications of vector space theory to projective geometry. The main concepts and result are those of the next two definitions and the following theorem.

**Definition 7.21.** *The vectors* $\mathbf{v}_1, \mathbf{v}_2, \cdots, \mathbf{v}_m$ *are* linearly independent *if*

$$c_1\mathbf{v}_1 + c_2\mathbf{v}_2 + \cdots + c_m\mathbf{v}_m = \mathbf{0} \tag{7.21}$$

*implies*

$$c_1 = c_2 = \cdots = c_m = 0.$$

*Vectors not linearly independent are* linearly dependent.

We note that if vectors are linearly dependent, equation (7.21) holds with $c_i \neq 0$ for some $i$. Thus (1, 2, 3), (1, 1, 1), and (5, 7, 9) are linearly dependent, since

$$2(1,2,3)+3(1,1,1)-1(5,7,9)=(0,0,0)=\mathbf{0}.$$

**Definition 7.22.** *The vector* $\mathbf{v}$ *is a* linear combination *of* $\mathbf{v}_1, \mathbf{v}_2, \cdots, \mathbf{v}_m$ *if there exist scalars* $c_1, c_2, \cdots, c_m$ *such that*

$$\mathbf{v}=c_1\mathbf{v}_1+c_2\mathbf{v}_2+\cdots+c_m\mathbf{v}_m.$$

**Theorem 7.21.** *A set of vectors is linearly dependent if and only if some one of the set is a linear combination of the others.*

PROOF. Suppose first that $\mathbf{v}_1, \mathbf{v}_2, \cdots, \mathbf{v}_m$ are linearly dependent. Then

$$c_1\mathbf{v}_1+c_2\mathbf{v}_2+\cdots+c_m\mathbf{v}_m=\mathbf{0}$$

where not all the $c_i$ are zero. Without loss of generality we may assume $c_m \neq 0$. Then

$$\mathbf{v}_m=(-c_1/c_m)\mathbf{v}_1+(-c_2/c_m)\mathbf{v}_2+\cdots+(-c_{m-1}/c_m)\mathbf{v}_{m-1},$$

whence $\mathbf{v}_m$ is a linear combination of $\mathbf{v}_1, \mathbf{v}_2, \cdots, \mathbf{v}_{m-1}$.

Conversely, if

$$\mathbf{v}_m=d_1\mathbf{v}_1+d_2\mathbf{v}_2+\cdots+d_{m-1}\mathbf{v}_{m-1},$$

then

$$d_1\mathbf{v}_1+d_2\mathbf{v}_2+\cdots+d_{m-1}\mathbf{v}_{m-1}+d_m\mathbf{v}_m=0$$

where $d_m=-1\neq 0$. We conclude the vectors are linearly dependent.

Suppose we are asked whether the vectors (1, 3, 1), (2, −1, −1), and (1, 2, 3) are linearly independent or dependent. We must determine the nature of $c_1$, $c_2$, and $c_3$ for which

$$\begin{aligned}&c_1(1,3,1)+c_2(2,-1,-1)+c_3(1,2,3)\\&\qquad=(c_1+2c_2+c_3,\,3c_1-c_2+2c_3,\,c_1-c_2+3c_3)=(0,0,0),\end{aligned}$$

or equivalently

$$\begin{aligned}c_1+2c_2+c_3&=0\\3c_1-c_2+2c_3&=0\\c_1-c_2+3c_3&=0.\end{aligned}$$

This system has only the trivial solution $c_1=c_2=c_3=0$, since

$$\begin{vmatrix}1&2&1\\3&-1&2\\1&-1&3\end{vmatrix}=-17\neq 0.$$

## Exercises 7.2

1. Determine whether each of the following sets of vectors is linearly dependent or independent. When the set is dependent, write one of the vectors as a linear combination of the others.

   *a.* $(2, 4)$ and $(3, 7)$;
   *b.* $(2, 3, 5)$, $(4, 2, 1)$, and $(4, 6, 9)$;
   *c.* $(2, 3, 5)$, $(4, 2, 1)$, and $(0, -4, -9)$;
   *d.* $(2, 3, 5)$, $(4, 2, 1)$, and $(4, 6, 10)$;
   *e.* $(2, 3, 5)$, $(4, 2, 1)$, and $(0, 0, 0)$;
   *f.* $(2, 3, 5, 7)$, $(4, 2, 1, 3)$, and $(0, -4, -9, 1)$.

2. Prove that each of the following is a linearly dependent set of vectors.

   *a.* $1, x$, and $2x - 3$;
   *b.* $x + 1$, $2x - 3$, and $3x - 4$;
   *c.* $1, x, x^2$, and $ax^2 + bx + c$;
   *d.* $e^{2x}, e^{-3x}$, and $3e^{2x} - 2e^{-3x}$.

3. Prove that $1, x$, and $x^2$ are linearly independent.

4. Prove that any set of vectors containing the zero vector is a linearly dependent set.

5. Prove that if $\mathbf{v}_1, \mathbf{v}_2, \cdots, \mathbf{v}_m$ are linearly dependent, so are $c_1\mathbf{v}_1, c_2\mathbf{v}_2, \cdots, c_m\mathbf{v}_m$ for arbitrary $c_1, c_2, \cdots, c_m$.

6. Prove that if $\mathbf{v}_1, \mathbf{v}_2, \cdots, \mathbf{v}_m$ are linearly independent, so are $c_1\mathbf{v}_1, c_2\mathbf{v}_2, \cdots, c_m\mathbf{v}_m$ provided no $c_i$ is zero.

## 7.3 MATRIX ALGEBRA

It is apparent from the example of the preceding section that the problem of determining linear dependence or independence of a set of vectors is the problem of determining whether a system of homogeneous linear equations has a nontrivial solution. We have reduced certain such systems to the problem of evaluating determinants. We ultimately do this for any such system. This criterion will be developed using the concept of matrix, which we now introduce.

**Definition 7.31.** *An* $m \times n$ matrix *over the reals is a rectangular array of real numbers with m rows and n columns of the form*

$$\begin{pmatrix} a_{11} & a_{12} & a_{13} & \cdots & a_{1n} \\ a_{21} & a_{22} & a_{23} & \cdots & a_{2n} \\ \vdots & \vdots & \vdots & & \vdots \\ a_{m1} & a_{m2} & a_{m3} & \cdots & a_{mn} \end{pmatrix}.$$

Each $a_{ij}$ in the above matrix is called an *element* of the matrix. The set of elements, in order, in a row (or column) of a matrix is called a *row* (or a *column*) of the matrix. We often use the symbols $A$ or $(a_{ij})$ as an abbreviation for a matrix such as that above; when such notation is used, the number of rows and columns in the matrix must be inferred from the context in which the matrix appears. It should be emphasized that a matrix is an array and does not have a "value." In particular, a *square matrix* (same number of rows and columns) is not a determinant. Thus no "simplification" is possible for the matrix

$$\begin{pmatrix} 1 & 2 \\ 3 & 4 \end{pmatrix}$$

while the determinant

$$\begin{vmatrix} 1 & 2 \\ 3 & 4 \end{vmatrix}$$

has "value" $-2$; that is, the determinant itself is just a somewhat complicated way of writing the real number $-2$. It is true that every square matrix has a determinant associated with it in a natural manner; we call this determinant the *determinant of the matrix*. A matrix need not have more than one row or column. Thus

$$\begin{pmatrix} 1 & 2 & 3 & 7 & -2 \end{pmatrix}$$

and

$$\begin{pmatrix} 1 \\ 2 \\ 3 \end{pmatrix}$$

are matrices, called *row* and *column matrices*, respectively. A real number itself can be considered a one-by-one matrix.

It is possible under certain conditions to define the sum or product of two matrices and the product of a real number and a matrix. We state all three definitions, although in the sequel we are concerned almost exclusively with the product of matrices.

**Definition 7.32.** *Given a real number $c$ and a matrix $A = (a_{ij})$, then $cA$ is the matrix $C = (c_{ij})$ where $c_{ij} = ca_{ij}$.*

As an example, we see that

$$3\begin{pmatrix} 1 & -1 & 4 \\ 2 & 0 & 5 \end{pmatrix} = \begin{pmatrix} 3 & -3 & 12 \\ 6 & 0 & 15 \end{pmatrix}.$$

Note that multiplication here is different from that for determinants. In the matrix case each element is multiplied by the number, while in the determinant case the elements of only one row or one column are so multiplied.

**Definition 7.33.** *Given $m \times n$ matrices $A = (a_{ij})$ and $B = (b_{ij})$, then $A + B$ is the matrix $C = (c_{ij})$ where*

$$c_{ij} = a_{ij} + b_{ij}.$$

Note that the sum of matrices is defined only if the terms have the same number of rows and columns. We see that

$$\begin{pmatrix} 1 & -1 & 4 \\ 2 & 0 & 5 \end{pmatrix} + \begin{pmatrix} 3 & 2 & 1 \\ 1 & 3 & 2 \end{pmatrix} = \begin{pmatrix} 4 & 1 & 5 \\ 3 & 3 & 7 \end{pmatrix}.$$

Addition of matrices is thus simply carried out: we add corresponding elements of the two summand matrices to find an element of the sum matrix. Multiplication of matrices could be defined in an analogous manner, but our applications will be made possible only if we adopt the more intricate multiplication procedure of the following definition.

**Definition 7.34.** *Given an $m \times n$ matrix $A = (a_{ij})$ and an $n \times r$ matrix $B = (b_{jk})$, then $AB$ is the $m \times r$ matrix $C = (c_{ik})$ where*

$$c_{ik} = a_{i1}b_{1k} + a_{i2}b_{2k} + \cdots + a_{in}b_{nk}.$$

We illustrate the product concept with several examples.

**Example 1**

$$\begin{pmatrix} a_{11} & a_{12} \\ a_{21} & a_{22} \end{pmatrix} \begin{pmatrix} b_{11} & b_{12} \\ b_{21} & b_{22} \end{pmatrix} = \begin{pmatrix} a_{11}b_{11} + a_{12}b_{21} & a_{11}b_{12} + a_{12}b_{22} \\ a_{21}b_{11} + a_{22}b_{21} & a_{21}b_{12} + a_{22}b_{22} \end{pmatrix}.$$

**Example 2**

$$\begin{pmatrix} a_{11} & a_{12} \\ a_{21} & a_{22} \end{pmatrix} \begin{pmatrix} x_1 \\ x_2 \end{pmatrix} = \begin{pmatrix} a_{11}x_1 + a_{12}x_2 \\ a_{21}x_1 + a_{22}x_2 \end{pmatrix}.$$

**Example 3**

$$\begin{pmatrix} 1 & 3 \\ 4 & 2 \end{pmatrix} \begin{pmatrix} 3 & 4 \\ 2 & 1 \end{pmatrix} = \begin{pmatrix} 1\cdot 3 + 3\cdot 2 & 1\cdot 4 + 3\cdot 1 \\ 4\cdot 3 + 2\cdot 2 & 4\cdot 4 + 2\cdot 1 \end{pmatrix} = \begin{pmatrix} 9 & 7 \\ 16 & 18 \end{pmatrix}.$$

**Example 4**

$$\begin{pmatrix} 1 & 2 & 1 \\ 0 & 2 & 3 \\ 1 & -1 & -1 \end{pmatrix} \begin{pmatrix} 2 & -1 & 1 \\ 1 & -2 & 0 \\ 2 & 1 & 1 \end{pmatrix} = \begin{pmatrix} 6 & -4 & 2 \\ 8 & -1 & 3 \\ -1 & 0 & 0 \end{pmatrix}.$$

It should be noted that the product is defined only if the first factor has as many columns as the second factor has rows. Thus when $AB$ is defined, $BA$ may not be defined. If $A$ and $B$ are square matrices of the same order, $AB$ and $BA$ are both defined, but $AB$ need not equal $BA$. If we reverse the order of multiplication in Example 4, we find the product matrix

$$\begin{pmatrix} 3 & 1 & -2 \\ 1 & -2 & -5 \\ 3 & 5 & 4 \end{pmatrix},$$

which is distinct from the product matrix of the example.

Matrix multiplication for square matrices is the same in form as for determinant multiplication. Thus, if $A, B$, and $C$ are square matrices with $AB = C$, the determinant of $C$ will be the product of the determinants of $A$ and $B$.

## Exercises 7.3

1. Given

$$A = \begin{pmatrix} 2 & 3 \\ 3 & 5 \end{pmatrix}, \quad B = \begin{pmatrix} 14 & 1 \\ 0 & 3 \end{pmatrix}, \quad C = \begin{pmatrix} 1 & 2 & 3 \\ 4 & 2 & 5 \end{pmatrix},$$

$$D = \begin{pmatrix} 2 & 3 & 3 \\ 1 & 3 & 2 \end{pmatrix}, \quad E = (4 \quad -3), \quad F = (4 \quad -3 \quad 1),$$

$$G = \begin{pmatrix} 2 & 3 & 3 \\ 1 & 3 & 2 \\ 1 & -1 & 2 \end{pmatrix}, \quad H = \begin{pmatrix} 1 & -2 & -3 \\ 2 & 0 & 3 \\ 1 & 1 & -2 \end{pmatrix}, \quad K = \begin{pmatrix} 2 \\ 3 \\ 1 \end{pmatrix},$$

compute, if possible, each of the following:

*a.* $A+B$;
*b.* $B+A$;
*c.* $A+C$;
*d.* $A+E$;
*e.* $C+D$;
*f.* $G+H$;
*g.* $AB$;
*h.* $BA$;
*i.* $AC$;
*j.* $CA$;
*k.* $AE$;
*l.* $EA$;
*m.* $FG$;
*n.* $GF$;
*o.* $GH$;
*p.* $HG$;
*q.* $HK$;
*r.* $KH$;
*s.* $FK$;
*t.* $KF$.

2. Prove that the set of all $m \times n$ matrices is a real vector space under the operations of matrix addition and multiplication by a real number.

## 7.4 RANK OF A MATRIX

In the $m \times n$ matrix

$$A = \begin{pmatrix} a_{11} & a_{12} & \cdot & \cdot & \cdot & a_{1n} \\ a_{21} & a_{22} & \cdot & \cdot & \cdot & a_{2n} \\ \cdot & \cdot & & & & \cdot \\ \cdot & \cdot & & & & \cdot \\ \cdot & \cdot & & & & \cdot \\ a_{m1} & a_{m2} & \cdot & \cdot & \cdot & a_{mn} \end{pmatrix}$$

we note that each row $(a_{k1}, a_{k2}, \cdots, a_{kn})$ is an $n$-tuple and this is a vector in the vector space of $n$-tuples. We call such vectors the *row vectors* of the matrix. In a like manner the columns of the matrix are $m$-tuples and are called the *column vectors* of the matrix. We show that the maximal number of linearly independent row vectors of a matrix can be found by a determinant criterion. This finally will give us a test for the linear dependence or independence of a set of vectors.

**Definition 7.41.** *A matrix has* row rank *$r$ if there exist $r$ linearly independent row vectors in the matrix while any set of more than $r$ row vectors is linearly dependent.*

**Definition 7.42.** *A* submatrix *of a matrix is the matrix itself or a matrix resulting from the deletion of rows and columns of the given matrix.*

**Definition 7.43.** *A matrix has* determinant rank *$s$ if there exists an $s \times s$ submatrix with nonzero determinant, while there is no $t \times t$ submatrix, $t > s$, with nonzero determinant.*

We see that

$$\begin{pmatrix} 1 & 2 & 3 \\ 4 & 3 & 1 \end{pmatrix}$$

has determinant rank two since

$$\begin{vmatrix} 1 & 2 \\ 4 & 3 \end{vmatrix} = -5 \neq 0,$$

while

$$\begin{pmatrix} 1 & 2 & 3 \\ 2 & 4 & 6 \end{pmatrix}$$

has determinant rank one since any $1 \times 1$ submatrix has nonzero determinant and

$$\begin{vmatrix} 1 & 2 \\ 2 & 4 \end{vmatrix} = \begin{vmatrix} 1 & 3 \\ 2 & 6 \end{vmatrix} = \begin{vmatrix} 2 & 3 \\ 4 & 6 \end{vmatrix} = 0.$$

**Theorem 7.41.** *The row rank and determinant rank of a matrix are the same.*

PROOF. Suppose first that every set of $t$ row vectors of the matrix is a linearly dependent set. By Theorem 7.21 we conclude that, given any $t$ rows of the matrix, one of the rows is a linear combination of the others. This implies that every square submatrix which is $t \times t$ or larger will have zero determinant since one row will be a linear combination of the others. We conclude that the determinant rank cannot exceed the row rank.

Suppose now that the determinant rank is $s$ and that the row rank is $r > s$. We may suppose that the first $r$ row vectors of the matrix are linearly independent; thus the row vectors $\mathbf{v}_1, \mathbf{v}_2, \cdots, \mathbf{v}_{s+1}$ of the matrix

$$\begin{pmatrix} a_{11} & a_{12} & \cdots & a_{1n} \\ a_{21} & a_{22} & \cdots & a_{2n} \\ \vdots & \vdots & & \vdots \\ a_{s+1,1} & a_{s+1,2} & \cdots & a_{s+1,n} \end{pmatrix}$$

are linearly independent. We suppose for the moment that $a_{11} \neq 0$ and set

$$\mathbf{w}_1 = \mathbf{v}_1$$
$$\mathbf{w}_i = \mathbf{v}_i - (a_{i1}/a_{11})\mathbf{v}_1, \qquad i > 1.$$

The vectors $\mathbf{w}_1, \mathbf{w}_2, \cdots, \mathbf{w}_{s+1}$ are linearly independent, for

$$\begin{aligned} c_1\mathbf{w}_1 + c_2\mathbf{w}_2 + \cdots + c_{s+1}\mathbf{w}_{s+1} &= (c_1 - c_2a_{21}/a_{11} - \cdots - c_{s+1}a_{s+1,1}/a_{11})\mathbf{v}_1 \\ &\quad + c_2\mathbf{v}_2 + \cdots + c_{s+1}\mathbf{v}_{s+1} \\ &= 0 \end{aligned}$$

will imply

$$c_1 = c_2 = \cdots = c_{s+1} = 0.$$

The matrix of the $\mathbf{w}_i$ vectors is now of the form

$$\begin{pmatrix} a_{11} & a_{12} & \cdots & a_{1n} \\ 0 & a'_{22} & \cdots & a'_{2n} \\ \vdots & \vdots & & \vdots \\ 0 & a'_{s+1,2} & \cdots & a'_{s+1,n} \end{pmatrix}.$$

If all $a_{i1}$ are zero, the corresponding procedure can be carried out using the elements of the first column of the matrix containing a nonzero element (we note that rows can be interchanged if necessary without altering row or determinant rank).

We may now repeat the process to find a new matrix whose rows are linearly independent and for which all elements of the second (or appropriate) column below $a'_{22}$ are zero. We repeat the process and ultimately find a linearly independent set of vectors $\mathbf{u}_1, \mathbf{u}_2, \cdots, \mathbf{u}_{s+1}$ with the property that each $\mathbf{u}_i$ has at least $i-1$ zeros as its first components and that $\mathbf{u}_j$ has more such zero components than $\mathbf{u}_i$ if $j > i$. If we consider the $s+1$ by $s+1$ submatrix of the matrix of the $\mathbf{u}_i$ using all its rows and those columns containing the initial nonzero components of the $\mathbf{u}_i$, we find that it has the determinant

$$\begin{vmatrix} b_{11} & b_{12} & \cdots & b_{1,s+1} \\ 0 & b_{22} & \cdots & b_{2,s+1} \\ 0 & 0 & \cdots & b_{3,s+1} \\ \vdots & \vdots & & \vdots \\ 0 & 0 & \cdots & b_{s+1,s+1} \end{vmatrix} = b_{11}b_{22}\cdots, b_{s+1,s+1} \neq 0.$$

This determinant is also, in view of the manner in which we produced new sets of vectors, the determinant of the corresponding submatrix of the original matrix, whose determinant rank could not have been $s$. We conclude that the row rank cannot exceed the determinant rank.

We have now shown the two ranks equal by showing that neither can exceed the other. This completes the proof of the theorem.

We see now that the determination of linear dependence or linear independence of a set of vector $n$-tuples can be reduced to the question of row rank of the matrix in which they are row vectors, and that this rank can, in theory at least, be determined by evaluating the determinants of all square submatrices of the matrix. Thus the examples following Definition 7.41 show that $(1, 2, 3)$ and $(4, 3, 1)$ are linearly independent, while the vectors $(1, 2, 3)$ and $(2, 4, 6)$ are linearly dependent.

## Exercises 7.4

1. Rework Exercise 7.21 using the results of this section.
2. Determine the rank of each matrix of Exercise 7.31.
3. Prove that $(a_{ij})$ has rank zero if and only if each $a_{ij} = 0$.
4. The column rank of a matrix is the maximal number of linearly inde-

pendent column vectors it contains. Prove that the column rank equals the determinant rank.

## 7.5 BASES FOR VECTOR SPACES

As a consequence of Theorem 7.41 we have

**Theorem 7.51.** *In the vector space of n-tuples any set of $m > n$ vectors is linearly dependent.*

PROOF. Such a set would be the row vectors of an $m \times n$ matrix whose determinant rank can be at most $n$. Since $m > n$, the row rank of the matrix is thus less than $m$.

It is clear that there are sets of $n$ $n$-tuples which are linearly independent, since there are certainly nonzero $n \times n$ determinants. Thus we have

**Theorem 7.52.** *In the vector space of n-tuples there exist sets of n linearly independent vectors.*

The two theorems show that while there exist sets of $n$ linearly independent vectors, any larger set will be linearly dependent in the vector space of $n$-tuples. Thus if we add a vector to a set of $n$ linearly independent ones, it must be a linear combination of the others. (Its scalar multiplier in the dependence equation (7.21) cannot be zero, since this would imply that the original set was linearly dependent.) In particular, it is easy to see that the vector $n$-tuples $(1, 0, 0, \cdots, 0)$, $(0, 1, 0, \cdots, 0) \cdots$, $(0, 0, 0, \cdots, 1)$ are a set of $n$ linearly independent vectors in the space of $n$-tuples; if we add a vector $(x_1, x_2, \cdots, x_n)$ to the set, we see that

$$(x_1, x_2, \cdots, x_n) = x_1(1, 0, 0, \cdots, 0) + x_2(0, 1, 0, \cdots, 0) + \cdots + x_n(0, 0, 0, \cdots, 1).$$

Every vector in the space is thus a linear combination of the original vectors. This suggests

**Definition 7.51.** *A set of linearly independent vectors is a* basis *for a vector space if every vector in the space is a linear combination of the vectors of the set.*

That every set of $n$ linearly independent vectors is a basis for the vector space of $n$-tuples is shown by

**Theorem 7.53.** *In the vector space of n-tuples any set of n linearly independent vectors is a basis.*

PROOF. Suppose that $\mathbf{v}_1, \mathbf{v}_2, \cdots, \mathbf{v}_n$ is such a linearly independent set and that $\mathbf{v}$ is an arbitrary vector in the space. By Theorem 7.51 the vectors $\mathbf{v}_1, \mathbf{v}_2, \cdots, \mathbf{v}_n, \mathbf{v}$ are linearly dependent, whence

$$c_1\mathbf{v}_1 + c_2\mathbf{v}_2 + \cdots + c_n\mathbf{v}_n + c_{n+1}\mathbf{v} = \mathbf{0},$$

not all $c_i$ zero. Moreover, $c_{n+1} \neq 0$, for otherwise the above would be a statement of linear dependence of the set $\mathbf{v}_1, \mathbf{v}_2, \cdots, \mathbf{v}_n$. We thus may solve for $\mathbf{v}$, finding

$$\mathbf{v} = \left(\frac{-c_1}{c_{n+1}}\right)\mathbf{v}_1 + \left(\frac{-c_2}{c_{n+1}}\right)\mathbf{v}_2 + \cdots + \left(\frac{-c_n}{c_{n+1}}\right)\mathbf{v}_n.$$

In view of Definition 7.51 we conclude that $\mathbf{v}_1, \mathbf{v}_2, \cdots, \mathbf{v}_n$ form a basis.

## Exercises 7.5

1. Prove that $(3, 2)$ and $(2, -1)$ form a basis for the space of 2-tuples and express $(1, 3)$ as a linear combination of these vectors.

2. Prove that $(3, 2, 1)$, $(2, -1, 0)$, and $(1, 3, 2)$ form a basis for the space of 3-tuples and express $(1, 1, 1)$ as a linear combination of these vectors.

3. Prove that $1, x$, and $x^2$ form a basis for the vector space of second-degree polynomials and express $3x^2 - 2x + 7$ as a linear combination of these vectors.

4. Prove that $3x^2 + 2x + 1$, $2x^2 - x$, and $x^2 + 3x + 2$ form a basis for the vector space of second-degree polynomials and express $x^2 + x + 1$ as a linear combination of these vectors.

5. Prove that the expression of a vector as a linear combination of the vectors of a basis is unique.

## 7.6 NONSINGULAR MATRICES

We conclude this chapter with a further analysis of square matrices. We first introduce

**Definition 7.61.** *An $n \times n$ matrix is* nonsingular *if it has rank n; otherwise it is* singular.

Nonsingular matrices will be of great importance in the following chapters. Theorem 7.41 gives us a useful criterion for nonsingularity, for it implies at once

**Theorem 7.61.** *A square matrix is nonsingular if and only it its determinant is nonzero.*

We consider another criterion for nonsingularity. We first make two definitions.

**Definition 7.62.** *The $n \times n$ square matrix*

$$I = \begin{pmatrix} 1 & 0 & \cdot & \cdot & \cdot & 0 \\ 0 & 1 & \cdot & \cdot & \cdot & 0 \\ \cdot & \cdot & & & & \cdot \\ \cdot & \cdot & & & & \cdot \\ \cdot & \cdot & & & & \cdot \\ 0 & 0 & \cdot & \cdot & \cdot & 1 \end{pmatrix}$$

*is an* identity *matrix.*

**Definition 7.63.** *If a square matrix $A$ has associated with it a square matrix $A^{-1}$ such that $AA^{-1} = A^{-1}A = I$, then $A^{-1}$ is an* inverse *of $A$.*

The reason for the use of the word "identity" is found in

**Theorem 7.62.** *For any square matrix $A$, $AI = IA = A$.*

The proof is left as an exercise for the reader. We see that $I$ plays the role in matrix multiplication analogous to that played by 1 in the multiplication of real numbers. In a like manner, the inverse of a matrix is analogous to the reciprocal of a real number. Not all matrices have inverses; indeed, only nonsingular ones do, as we prove in

**Theorem 7.63.** *A square matrix has an inverse if and only if it is nonsingular.*

PROOF. Suppose that the matrix $A$ is nonsingular, denote its row vectors by $\mathbf{v}_1, \mathbf{v}_2, \cdots, \mathbf{v}_n$, and denote the row vectors of $I$ by $\mathbf{e}_1, \mathbf{e}_2, \cdots, \mathbf{e}_n$. Since $\mathbf{v}_1, \mathbf{v}_2, \cdots, \mathbf{v}_n$ form a linearly independent set, they are a basis for the vector space of $n$-tuples by Theorem 7.51. Thus we have

$$\begin{aligned} \mathbf{e}_1 &= b_{11}\mathbf{v}_1 + b_{12}\mathbf{v}_2 + \cdots + b_{1n}\mathbf{v}_n \\ \mathbf{e}_2 &= b_{21}\mathbf{v}_1 + b_{22}\mathbf{v}_2 + \cdots + b_{2n}\mathbf{v}_n \\ &\cdot \\ &\cdot \\ &\cdot \\ \mathbf{e}_n &= b_{n1}\mathbf{v}_1 + b_{n2}\mathbf{v}_2 + \cdots + b_{nn}\mathbf{v}_n. \end{aligned} \tag{7.61}$$

Now if we set $A^{-1} = (b_{ij})$, it is evident that $A^{-1}A = I$, for the multiplication of $A$ on the left by $A^{-1}$ combines its rows as shown in equations (7.61). From this it follows that $A^{-1}$ is nonsingular, since its determinant will be the determinant of $I$ divided by the determinant of $A$, and this quotient is not zero. In a like manner, then, we find $(A^{-1})^{-1}$ such that $(A^{-1})^{-1}A^{-1} = I$. Now if

$$A^{-1}A = I,$$

then

$$A^{-1}AA^{-1} = IA^{-1} = A^{-1}$$

and

$$[(A^{-1})^{-1}A^{-1}]AA^{-1} = (A^{-1})^{-1}A^{-1} = I$$
$$= AA^{-1},$$

and $A^{-1}$ is the desired inverse. (The reader can now readily show that $(A^{-1})^{-1} = A$.) Conversely, if $A$ is such that there is an $A^{-1}$ with $A^{-1}A = I$, the product of the determinants of $A$ and $A^{-1}$ will be the determinant of $I$, or 1. Thus the determinant of $A$ is not zero, and by Theorem 7.61 $A$ is nonsingular.

Finally, we use the foregoing theorem to prove

**Theorem 7.64.** *If $A$ is a nonsingular square matrix and if $B = AC$, then $B$ and $C$ have the same rank.*

PROOF. Let the $m \times n$ matrices $B$ and $C$ have ranks $r$ and $s$, respectively. The statement $B = AC$ states that the row vectors of $B$ are linear combinations of the row vectors of $C$ since

$$\mathbf{b}_1 = a_{11}\mathbf{c}_1 + a_{12}\mathbf{c}_2 + \cdots + a_{1m}\mathbf{c}_m$$
$$\mathbf{b}_2 = a_{21}\mathbf{c}_1 + a_{22}\mathbf{c}_2 + \cdots + a_{2m}\mathbf{c}_m$$
$$\vdots$$
$$\mathbf{b}_m = a_{m1}\mathbf{c}_1 + a_{m2}\mathbf{c}_2 + \cdots + a_{mm}\mathbf{c}_m,$$

where $A = (a_{ij})$ and the **b**'s and **c**'s are the row vectors of $B$ and $C$, respectively. If we suppose that $\mathbf{c}_1, \mathbf{c}_2 \cdots, \mathbf{c}_s$ are a maximal set of linearly independent row vectors of $C$, then the row vectors of $B$ are linear combinations of these $s$ vectors. We augment $\mathbf{c}_1, \mathbf{c}_2, \cdots, \mathbf{c}_s$, if necessary, to a basis $\mathbf{c}_1, \mathbf{c}_2, \cdots, \mathbf{c}_s, \cdots, \mathbf{c}'_m$ for the vector space of $m$-tuples. Now we cannot have any $s+1$ row vectors of $B$ linearly independent, for if, say, $\mathbf{b}_1, \mathbf{b}_2, \cdots, \mathbf{b}_{s+1}$ were linearly independent, $\mathbf{b}_1, \mathbf{b}_2, \cdots, \mathbf{b}_{s+1}, \mathbf{c}'_{s+1}, \cdots, \mathbf{c}'_m$ would be a linearly independent set of vectors[1] contradicting Theorem 7.51. We conclude $r \leqslant s$. Now since $A$ is nonsingular, it has an inverse $A^{-1}$, and from $B = AC$ we deduce $C = A^{-1}B$. We use this relation to conclude by a like argument that $s \leqslant r$, whence $s = r$.

[1]If $g_1\mathbf{b}_1 + \cdots + g_{s+1}\mathbf{b}_{s+1} + h_{s+1}\mathbf{c}'_{s+1} + \cdots + h_m\mathbf{c}'_m = \mathbf{0}$, then all $h_i$ are zero, for otherwise we could solve for some $\mathbf{c}_j$ as a linear combination of the **b**'s and **c**'s, hence as a linear combination of all the other **c**'s. This contradicts the linear independence of the **c**'s. The equation then becomes $g_1\mathbf{b}_1 + g_2\mathbf{b}_2 + \cdots + g_{s+1}\mathbf{b}_{s+1} = \mathbf{0}$, whence all the $g_k$ are zero by the linear independence of the **b**'s.

## Exercises 7.6

1. Determine whether each of the following is nonsingular:

*a.* $\begin{pmatrix} 2 & 1 \\ 3 & -1 \end{pmatrix}$

*b.* $\begin{pmatrix} 2 & 1 \\ 6 & 3 \end{pmatrix}$

*c.* $\begin{pmatrix} 2 & 1 & 4 \\ 3 & -1 & 3 \\ 1 & -2 & 1 \end{pmatrix}$

*d.* $\begin{pmatrix} 2 & 1 & 4 \\ 3 & -1 & 3 \\ 1 & -2 & -1 \end{pmatrix}$

2. Prove Theorem 7.62.
3. Prove that if $B = AC$, then the rank of $B$ cannot exceed the rank of $C$.
4. Prove that if $A$ is nonsingular and $AB = AC$, then $B = C$.
5. Prove that the set of all $n \times n$ nonsingular matrices is a group under matrix multiplication.

## REFERENCES

Adler, Claire Fisher, *Modern Geometry, an Integrated First Course*, 2nd ed., McGraw-Hill Book Company, New York, 1967, Chapter 9.

Aitken, A. C., *Determinants and Matrices*, 9th ed., Oliver and Boyd, Edinburgh, 1958, Chapters 1 to 3.

Beaumont, Ross, *Linear Algebra*, Harcourt, Brace & World, New York, 1965, Chapters 2 to 4.

Birkhoff, Garrett, and Saunders MacLane, *A Survey of Modern Algebra*, 3rd ed., The Macmillan Company, New York, 1965, Chapters 7 to 9.

Curtis, Charles W., *Linear Algebra: An Introductory Approach*, 2nd ed., Allyn & Bacon, Boston, 1968.

Finkbeiner, Daniel T., II, *Introduction to Matrices and Linear Transformations*, 2nd ed., W. H. Freeman and Company, San Francisco, 1966, Chapters 2 to 4, 6.

Fuller, Leonard E., *Linear Algebra with Applications*, Dickenson Publishing Company, Belmont, Calif., 1966, Chapters 1 to 4.

Graustein, William C., *Introduction to Higher Geometry*, The Macmillan Company, New York, 1930, Chapter 1.

Hodge, W. V. D., and D. Pedoe, *Methods of Algebraic Geometry*, vol. I, Cambridge University Press, Cambridge, 1947, Chapter 2.

Hoffman, Kenneth, and Ray Kunze, *Linear Algebra*, Prentice-Hall, Englewood Cliffs, N.J., 1961, Chapters 1 to 3.

Hohn, Franz, E., *Elementary Matrix Algebra*, 2nd ed., The Macmillan Company, New York, 1964, Chapters 1, 3 to 5.

Marcus, Marvin, and Henryk Minc, *Elementary Linear Algebra*, The Macmillan Company, New York, 1968, Chapters 1, 2.

Marcus, Marvin, and Henryk Minc, *Introduction to Linear Algebra*, The Macmillan Company, New York, 1965, Chapters 1, 2.

Murdoch, D. C., *Linear Algebra for Undergraduates*, John Wiley and Sons, New York, 1957, Chapters 1 to 3.

Perlis, Sam, *Theory of Matrices*, Addison-Wesley Publishing Company, Reading, Mass., 1952, Chapters 1 to 4.

Paige, Lowell J., and J. Dean Swift, *Elements of Linear Algebra*, Ginn and Company, Boston, 1961, Chapters 2 to 3, 6 to 7.

Rosenbaum, Robert A., *Introduction to Projective Geometry and Modern Algebra*, Addison-Wesley Publishing Company, Reading, Mass., 1963, Chapters 3, 7.

Stoll, Robert R., *Linear Algebra and Matrix Theory*, McGraw-Hill Book Company, New York, 1952, Chapters 1 to 3.

Stoll, Robert R., and Edward T. Wong, *Linear Algebra*, Academic Press, New York, 1968, Chapters 1, 2, 4, 5.

Vinograde, Bernard, *Linear and Matrix Algebra*, Raytheon Education Company, Boston, 1967.

Weiss, Marie J., and Roy Dubisch, *Higher Algebra for the Undergraduate*, 2nd ed., John Wiley and Sons, New York, 1962, Chapters 6 to 7.

Zelinsky, Daniel, *A First Course in Linear Algebra*, Academic Press, New York, 1968, Chapters 1, 3, 5.

## CHAPTER EIGHT

# FUNDAMENTALS OF ANALYTIC PROJECTIVE GEOMETRY

Although points are not associated with unique vectors, since many vectors are coordinates for a given point, the vector theory of the previous chapter can still be carried over into projective geometry. In this chapter we reexamine and extend the material of Chapter 5 using analytic methods based on the algebraic theories of Chapter 7.

## 8.1 DEPENDENT AND INDEPENDENT POINTS

We consider in this section the geometric interpretation of linear dependence and independence. To carry over the vector methods, we must introduce something corresponding to scalar multiplication and vector addition in geometry.

**Definition 8.11.** *If the vector n-tuples* $\mathbf{v}_1, \mathbf{v}_2, \cdots, \mathbf{v}_m$ *are coordinates of the points* $P_1, P_2, \cdots, P_m$*, respectively, then*

$$c_1P_1 + c_2P_2 + \cdots + c_mP_m$$

*will be the point with coordinates given by the vector*

$$c_1\mathbf{v}_1 + c_2\mathbf{v}_2 + \cdots + c_m\mathbf{v}_m$$

*if this is not the zero vector. If the vector sum is* $\mathbf{0}$, *we write*

$$c_1P_1 + c_2P_2 + \cdots + c_mP_m = \theta.$$

Thus if $P_1$ has coordinates $(x_1, x_2, \cdots, x_n)$ and $P_2$ has coordinates $(y_1, y_2, \cdots, y_n)$, $c_1P_1 + c_2P_2$ is the point with coordinates $(c_1x_1 + c_2y_1, c_1x_2 + c_2y_2, \cdots, c_1x_n + c_2y_n)$. This definition applies equally well to points in the plane with triples of numbers for coordinates, to points on a line with pairs of numbers for coordinates, and to points in space with 4-tuples for coordinates; that is, $n$ may be 2, 3, or 4. The point represented by such an expression depends on the coordinates chosen for the given points. Suppose that $P_1$ has coordinates (2, 1, 1) and $P_2$ has coordinates (3, 2, 2). Then $P_1 + 2P_2$ is the point with coordinates $(2, 1, 1) + 2(3, 2, 2) = (8, 5, 5)$. On the other hand if we take (4, 2, 2) as coordinates of $P_1$, then $P_1 + 2P_2$ has coordinates $(4, 2, 2) + 2(3, 2, 2) = (10, 6, 6)$. The resulting points are distinct, for their Euclidean coordinates are $(\frac{8}{5}, 1)$ and $(\frac{5}{3}, 1)$, respectively.

A similar definition applies to lines, and by duality we define

**Definition 8.12.** *If the vector n-tuples* $\mathbf{v}_1, \mathbf{v}_2, \cdots, \mathbf{v}_m$ *are coordinates of the lines* $l_1, l_2, \cdots, l_m$, *respectively, then*

$$c_1l_1 + c_2l_2 + \cdots + c_ml_m$$

*will be the line with coordinates given by the vector*

$$c_1\mathbf{v}_1 + c_2\mathbf{v}_2 + \cdots + c_m\mathbf{v}_m$$

*if this is not the zero vector. If the vector sum is* $\mathbf{0}$, *we write*

$$c_1l_1 + c_2l_2 + \cdots + c_ml_m = \theta.$$

We can also carry over to the geometric context the idea of linear combination, and we say $P$ is a *linear combination* of the points $P_1, P_2, \cdots, P_m$ if

$$P = c_1P_1 + c_2P_2 + \cdots + c_mP_m.$$

A similar agreement is made for a linear combination of lines.

In spite of the ambiguity involved in the use of linear combinations of points brought about by the possibility of different vectors representing the same point, the concepts of linear dependence and linear independence can be carried over to the geometry context. We make the obvious

**Definition 8.13.** *A set of points is* independent *or* dependent *according as the set of vectors serving as coordinates of the points is a linearly independent or a linearly dependent set.*

It is crucial here that the conclusion of linear independence or dependence of the vectors involved must be independent of the vectors chosen to represent the points, for otherwise the definition makes no sense. This is indeed the case as can easily be verified (see Exercises 7.25 and 7.26). To preserve duality we introduce

**Definition 8.14.** *A set of lines is* independent *or* dependent *according as the set of vectors serving as coordinates of the lines is a linearly independent or a linearly dependent set.*

We note that Theorem 7.21 applies; that is, if a set of points is dependent, one of the points is a linear combination of the others.

The main purpose of this section is an investigation of the geometric interpretations of the concepts of dependence and independence of points or lines. We shall see that dependence is associated with incidence properties; for example, dependence of three points will imply their collinearity. The result will be that analytic investigations of such incidence relations can be reduced to questions of linear dependence of vectors, which in turn can be reduced to questions of matrix rank and determinant value. We first consider the dependence of a pair of points. The conclusion is obvious in this case, but we include the result for the sake of completeness.

**Theorem 8.11.** *Two points are dependent if and only if they coincide.*

The theorem follows easily from the fact that the different sets of natural homogeneous coordinates of the same point are proportional. The details of the proof are left to the reader. By duality we have

**Theorem 8.12.** *Two lines are dependent if and only if they coincide.*

**Corollary 8.121.** *Distinct pairs of points or lines are independent.*

A more interesting and less trivial result is obtained if we consider the case of three points.

**Theorem 8.13.** *Three points are dependent if and only if they are collinear.*

In this theorem we do not assume the points are distinct, nor do we restrict ourselves to the projective plane. The theorem holds whether the coordinates are pairs, triples, or 4-tuples (see Exercise 8.19 for the space case).

PROOF. We suppose first that the points are collinear. If the points are on the projective line with pairs as coordinates, the matrix whose row vectors are the coordinates is of the form

$$\begin{pmatrix} x_1 & x_2 \\ x_1' & x_2' \\ x_1'' & x_2'' \end{pmatrix}.$$

This matrix can have rank at most two by Theorem 7.41. Hence its row vectors are linearly dependent and the points are dependent. It is clear that the converse is true for points on the projective line.

Suppose now that the points are in the projective plane. They will lie on the line $[u_1, u_2, u_3]$ if and only if the system

$$\begin{aligned} u_1x_1 + u_2x_2 + u_3x_3 &= 0 \\ u_1x_1' + u_2x_2' + u_3x_3' &= 0 \\ u_1x_1'' + u_2x_2'' + u_3x_3'' &= 0 \end{aligned}$$

has a nontrivial solution. This will be so if and only if

$$\begin{vmatrix} x_1 & x_2 & x_3 \\ x_1' & x_2' & x_3' \\ x_1'' & x_2'' & x_3'' \end{vmatrix} = 0.$$

Finally, this determinant condition is fulfilled if and only if the points are dependent.

**Corollary 8.131.** *Any three points on the projective line are dependent.*

We note that the remaining question in the proof of Theorem 6.43 is now resolved, for Theorems 7.41 and 8.13 guarantee that at least one of the determinants in the conclusion of the theorem is not zero.

By duality we have

**Theorem 8.14.** *Three lines are dependent if and only if they are concurrent.*

Finally we have

**Theorem 8.15.** *Any four points in the projective plane are dependent.*

The proof is left to the reader.

The previous set of theorems has established the fact that questions of collinearity and concurrence reduce to questions of dependence of the points or lines involved. The determination of the dependence can in turn be referred back to the problem of determination of the rank of the matrix whose rows are the coordinates of the elements involved. Finally by Theorem 7.41 the rank question can be settled by the evaluation of one or more determinants. We illustrate the approach with an example.

**Example.** In the projective plane are the points $(1, 2, 3)$, $(2, 1, 2)$, and $(-1, -2, 1)$ collinear? If the points were collinear, they would be dependent and the matrix whose rows are their coordinates,

$$\begin{pmatrix} 1 & 2 & 3 \\ 2 & 1 & 2 \\ -1 & -2 & 1 \end{pmatrix},$$

would have rank at most two. By Theorem 7.41 the determinant of the matrix would be zero. We find

$$\begin{vmatrix} 1 & 2 & 3 \\ 2 & 1 & 2 \\ -1 & -2 & 1 \end{vmatrix} = -12 \neq 0,$$

showing the matrix has rank three and that the points are not collinear.

We noted at the beginning of this section that a linear combination of points does not determine a unique point. The resulting ambiguity can sometimes be used to advantage, and we shall make use on occasion of the following theorem and its successors.

**Theorem 8.16.** *If $P_1, P_2, \cdots, P_m$ are independent while $P_1, P_2, \cdots, P_m, P_{m+1}$ are dependent, then $P_{m+1}$ can be written as a linear combination of $P_1, P_2, \cdots, P_m$ and the nonzero scalars in the resulting expansion can all be taken to be one.*

PROOF. We have

$$c_1P_1 + c_2P_2 + \cdots + c_mP_m + c_{m+1}P_{m+1} = \theta$$

where $c_{m+1} \neq 0$, for otherwise the above would be a statement of the dependence of $P_1, P_2, \cdots, P_m$. If we set $a_i = -c_i/c_{m+1}$, $i = 1, 2, \cdots, m$, we find

$$P_{m+1} = a_1P_1 + a_2P_2 + \cdots + a_mP_m.$$

Now if $a_i \neq 0$, then the vector associated with $a_iP_i$ is a vector which is also a set of coordinates for $P_i$, whence for the appropriate choice of coordinates we could replace each such $a_iP_i$ by $P_i$.

**Theorem 8.17.** *If $P_1, P_2, P_3$, and $P_4$ are such that no three are collinear, for a suitable choice of coordinates*

$$P_1 + P_2 + P_3 + P_4 = \theta.$$

The proof is left to the reader.

**Theorem 8.18.** *If $P_1$ and $P_2$ are distinct, $P_3$ is on the line $P_1P_2$ if and only if it is a linear combination of $P_1$ and $P_2$.*

PROOF. Since $P_1$ and $P_2$ are distinct, they are independent by Theorem 8.11. The reasoning in the proof of Theorem 8.16 shows that $P_1, P_2$, and $P_3$ are dependent if and only if $P_3$ is a linear combination of $P_1$ and $P_2$. The result now follows from Theorem 8.13.

**Corollary 8.181.** *If $P_1, P_2$, and $P_3$ are collinear, if $P_1$ and $P_2$ are distinct, and if $P_2$ and $P_3$ are distinct, coordinates may be chosen so that*

$$P_3 = P_1 + cP_2.$$

**Corollary 8.182.** *If $P_1$, $P_2$, and $P_3$ are collinear and distinct, coordinates may be chosen so that*

$$P_3 = P_1 + P_2.$$

**Corollary 8.183.** *If $P_1$, $P_2$, and $P_3$ are collinear, if $P_1$ and $P_2$ are distinct, if $P_2$ and $P_3$ are distinct, and if $P_1$ and $P_2$ have coordinates fixed, then coordinates of $P_3$ may be chosen so that*

$$P_3 = P_1 + cP_2.$$

## Exercises 8.1

1. Determine whether each of the following sets of points is dependent or independent:
   *a.* (2, 7, 9), (3, 1, 6);
   *b.* (2, 7, 9), (3, 1, 6), and (1, 1, 2);
   *c.* (2, 7, 9), (3, 1, 6), (1, 1, 2), and (1, 1, 2);
   *d.* (2, 7, 9), (3, 1, 6), and (−1, −13, −12).

2. If the three points of Exercise 1*d* are called $P$, $Q$, and $R$ in that order, choose coordinates for these points in such a way that $P = Q + R$.

3. Apply Corollaries 8.181 and 8.183 to the following points $P_1$, $P_2$, and $P_3$, stated in that order:
   *a.* (2, 7, 9), (3, 1, 6), and (0, 19, 15);
   *b.* (2, 7, 9), (3, 1, 6), and (4, 14, 18).

4. Apply Theorem 8.17 to the points (2, 7, 9), (3, 1, 6), (1, 1, 2), and (1, −6, 3).

5. Prove Theorem 8.11.

6. Prove Theorem 8.15.

7. Prove Theorem 8.17.

8. Write the duals of the results of this section starting with Theorem 8.15.

9.* This section has been concerned with the concept of dependence of points in the projective plane. Extend the results to space as follows:
   *a.* Write the space dual of Definition 8.11;
   *b.* Write the space dual of Definition 8.13;
   *c.* Prove that Theorem 8.13 is valid in space;
   *d.* Prove that four points are dependent if and only if they are coplanar;
   *e.* Prove that any five points are dependent;
   *f.* Write a space analogue of Theorem 8.18 and prove it;
   *g.* Write space duals of parts *d*, *e*, and *f* of this exercise.

10.* We have not considered line coordinates for lines in space. The following suggests how this might be done. The coordinates are known as Plücker line coordinates, after their creator (1829), and have important generalization in higher dimensional projective geometry. Since two points determine a line, associated with a line are matrices of the form

$$\begin{pmatrix} x_1 & x_2 & x_3 & x_4 \\ x_1' & x_2' & x_3' & x_4' \end{pmatrix}$$

whose rows are coordinates of points on the line. From the matrix we find twelve determinants of the form

$$p_{ij} = \begin{vmatrix} x_i & x_j \\ x_i' & x_j' \end{vmatrix}, i, j = 1, 2, 3, 4; i \neq j.$$

*a.* Show that $p_{ij} = -p_{ji}$.

*b.* Show that choice of two different points to determine the line leads to a set of $p_{ij}'$ proportional to the above.

*c.* Show that not all $p_{ij}$ are zero if distinct points are used in forming the matrix.

*d.* Show that $p_{12}p_{34} + p_{13}p_{42} + p_{14}p_{23} = 0$. (It is this relation that creates the difficulty. The six $p_{ij}$ in this relation are the Plücker line coordinates. Note that in view of the relation not every 6-tuple is a set of Plücker line coordinates for a line in space. [Ref: Struik, *Analytic and Projective Geometry*, pp. 177 ff.])

11. A *subspace* of a vector space is a subset of the vector space which is itself a vector space under the same operations. Prove that there is a one-to-one correspondence between the points in the real projective plane and the one-dimensional subspaces of the vector space of all 3-tuples of real numbers.

12.* Prove there is a one-to-one correspondence between the points in real projective space and the one-domensional subspaces of the vector space of all 4-tuples of real numbers.

## 8.2 APPLICATIONS OF THE DEPENDENCE RELATION

Let us now reconsider three results we have previously stated and see how they can be proved analytically. The first two have very brief, elegant analytic proofs; the third is more involved but still uses only the dependence concept. We consider the theorem of Desargues first.

**Theorem 8.21 (Desargues).** *Triangles perspective from a point are perspective from a line, and conversely.*

PROOF. We suppose first that the triangles are perspective from a point $P$. We observe that the theorem is trivial if a pair of corresponding vertices coincide, for then the corresponding sides meet in only two distinct points, which certainly determine an axis of perspectivity [see Figure 8.1 (*a*)]; accordingly we assume that corresponding vertices are distinct. If this is so, the point $P$ is a linear combination of each of these pairs since it lies on a line with them. We thus have

$$P = aA + a'A' = bB + b'B' = cC + c'C'.$$

From these we obtain the relations

$$\begin{aligned} aA + a'A' &= bB + b'B' \\ bB + b'B' &= cC + c'C' \\ cC + c'C' &= aA + a'A', \end{aligned}$$

which in turn can be written in the form

$$aA - bB = b'B' - a'A' \tag{8.21}$$
$$bB - cC = c'C' - b'B' \tag{8.22}$$
$$cC - aA = a'A' - c'C'. \tag{8.23}$$

Now either of the expressions in equation (8.21) represents a point, since $A$ and $B$, or $A'$ and $B'$, are distinct. The left-hand member states the point is on the line $AB$ while the right-hand member states it is on the line $A'B'$. We conclude that the point in question is the point of intersection $C''$ of the two lines. In a like manner we conclude that the

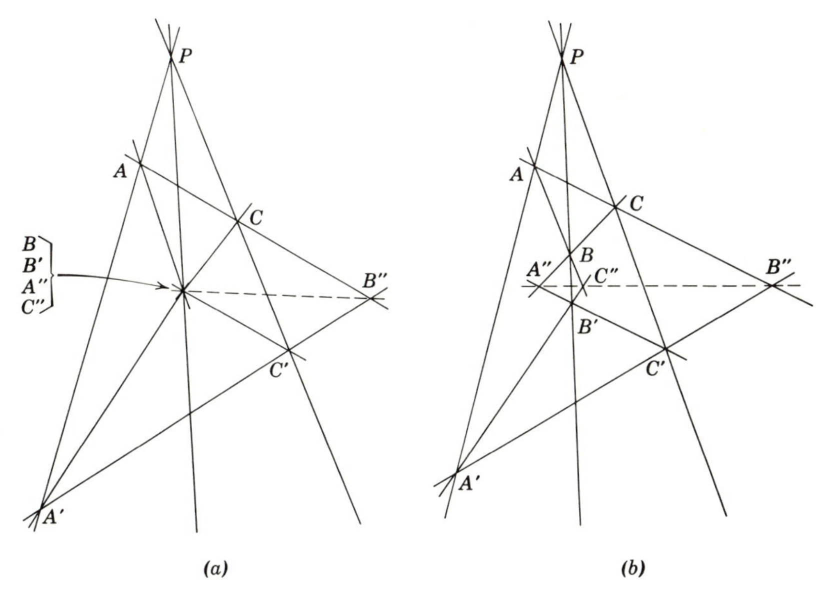


**Figure 8.1**

members of equations (8.22) and (8.23) are $A''$ and $B''$, respectively. We note that the three equations sum to $\theta$. This means that

$$C'' + A'' + B'' = \theta,$$

showing that these points are dependent, hence collinear. The converse proof can be carried out by essentially reversing the steps.[1] The details are left to the reader.

In Chapter 5 when we defined a complete quadrangle and a complete quadrilateral, we stated that the resulting diagonal triangles were proper triangles, since their vertices could not be collinear. We did not prove the statement at that time, but we can now give a simple analytic proof.

**Theorem 8.22.** *The diagonal points of a complete quadrangle are non-collinear.*

PROOF. Let $P_1$, $P_2$, $P_3$, and $P_4$ be the vertices of the quadrangle and let $D_1$, $D_2$, and $D_3$ be the diagonal points (Figure 8.2). We know that no three of the vertices are dependent, since no three of them may be collinear. By Theorem 8.17 we may choose coordinates of the vertices such that

$$P_1 + P_2 + P_3 + P_4 = \theta.$$

From this we find that

$$\begin{aligned} P_1 + P_2 &= -P_3 - P_4 \\ P_2 + P_3 &= -P_4 - P_1 \\ P_3 + P_1 &= -P_2 - P_4. \end{aligned}$$

[1]Note that this proof is either a proof for the plane case or the space case, depending on whether the coordinates of the points are plane or space coordinates. We have not had to prove the two cases separately as we did in Chapter 5.

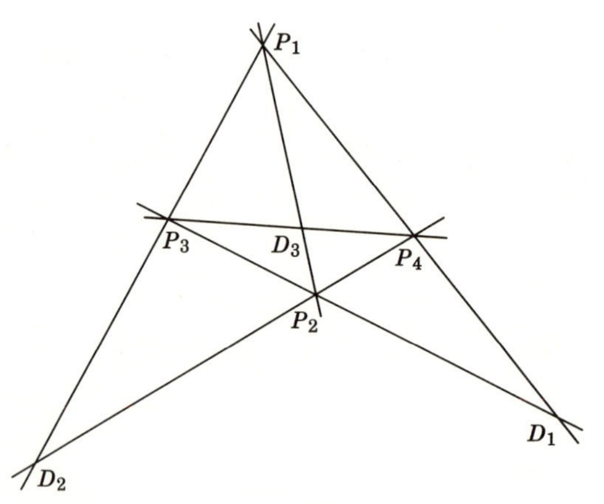


**Figure 8.2**

Now the first equation has members representing a point on $P_1P_2$ and $P_3P_4$ simultaneously. This point must be $D_3$. Similar arguments apply to the other equations, and we conclude that

$$D_3 = P_1 + P_2$$
$$D_1 = P_2 + P_3$$
$$D_2 = P_3 + P_1.$$

We must show that $D_1$, $D_2$, and $D_3$ are independent; that is, if

$$c_1D_1 + c_2D_2 + c_3D_3 = \theta,$$

then $c_1 = c_2 = c_3 = 0$. We have

$$c_1D_1 + c_2D_2 + c_3D_3 = (c_2 + c_3)P_1 + (c_1 + c_3)P_2 + (c_1 + c_2)P_3.$$

Since $P_1$, $P_2$, and $P_3$ are independent, we must have

$$c_2 + c_3 = 0$$
$$c_1 + c_3 = 0$$
$$c_1 + c_2 = 0.$$

This system has only the solution $c_1 = c_2 = c_3 = 0$, since its determinant of coefficients

$$\begin{vmatrix} 0 & 1 & 1 \\ 1 & 0 & 1 \\ 1 & 1 & 0 \end{vmatrix} = 2 \neq 0.$$

Finally, we give an analytic proof of the theorem of Pappus. The computation is involved in this case, but the arguments are based only on the dependence concept.

**Theorem 8.23 (Pappus).** *If distinct points $A$, $B$, and $C$ lie on a line and if distinct points $A'$, $B'$, and $C'$ lie on a second distinct (coplanar) line, the points of intersection of $AB'$ and $A'B$, of $AC'$ and $A'C$, and of $BC'$ and $B'C$ are collinear.*

PROOF. Since the lines $AC$ and $A'C'$ meet in only one point, we may assume that $B'$ is not that point, whence $A$, $B'$, and $C$ are noncollinear and therefore independent. The coordinates of these points can thus be taken to be a basis in the vector space of triples, and we can thus write each point as a linear combination of the points $A$, $B'$, and $C$. Since $Q$ is not collinear with any pair of these points, we choose coordinates of $A$, $B'$, and $C$ such that

$$Q = A + B' + C. \tag{8.24}$$

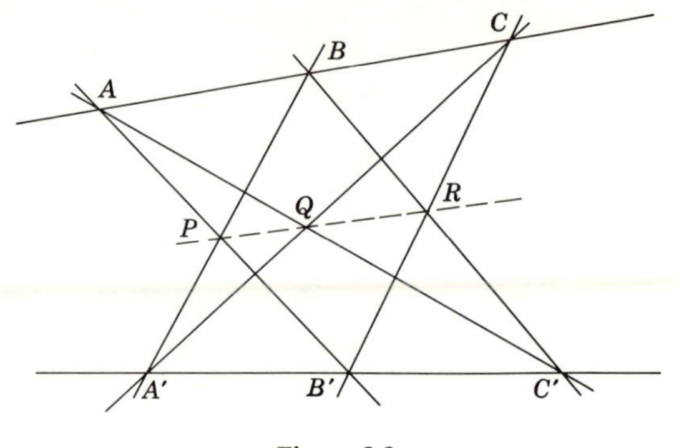


**Figure 8.3**

In the following we make repeated use of Corollary 8.183, for $A, B'$, and $C$ now have their coordinates fixed. Since $B$ is on $AC$, we may choose its coordinates such that

$$B = rA + C. \tag{8.25}$$

Noting that $A'$ is on $QC$ and $C'$ is on $A'B'$, we choose coordinates for these two points such that

$$A' = Q + t'C = A + B' + tC \tag{8.26}$$

(where $t = t' + 1$) and

$$C' = A' + s'B' = A + sB + tC \tag{8.27}$$

(where $s = s' + 1$). Now $A$, $C'$, and $Q$ are collinear. It is apparent from equations (8.24) and (8.27) that any one of these points can be a linear combination of the others only if $s = t$; we thus replace equation (8.27) by

$$C' = A + tB' + tC. \tag{8.28}$$

Now since $R$ is on $BC'$, we may choose its coordinates such that

$$\begin{aligned} R &= pC' + B \\ &= p(A + tB' + tC) + (rA + C) \\ &= (p + r)A + ptB' + (pt + 1)C. \end{aligned}$$

Since $R$ is also on $B'C$, we must have $p = -r$ and

$$R = -rtB' + (-rt + 1)C. \tag{8.29}$$

Finally, since $P$ is on $A'B$, we may choose its coordinates so that

$$\begin{aligned} P &= A' + mB \\ &= (A + B' + tC) + m(rA + C) \\ &= (1 + mr)A + B' + (t + m)C. \end{aligned}$$

Since $P$ is also on $AB'$ we have $m = -t$ and

$$P = (1 - tr)A + B'. \qquad (8.210)$$

From equations (8.24), (8.29), and (8.210) we conclude that

$$P - (1 - tr)Q + R = (tr - rt)B' + (tr - rt)C = \theta$$

showing that $P$, $Q$, and $R$ are dependent, hence collinear.

## Exercises 8.2

1. Prove the converse portion of Theorem 8.21.
2. Prove that the three points in which three nonconcurrent sides of a complete quadrangle, one through each diagonal point, meet the opposite sides of the diagonal triangle are collinear.

## 8.3 CHANGE OF COORDINATES; GENERAL HOMOGENEOUS COORDINATES

Once we had added the ideal points and proved the basic incidence relations in the synthetic treatment of projective geometry, ideal elements no longer occupied a special position, and in the later development there was no need to determine whether an element was ideal or ordinary. This is not yet so in the analytic treatment, for ideal elements are characterized by a last coordinate of zero. Natural homogeneous coordinates are very closely tied to the Euclidean coordinates of Euclidean analytic geometry. We desire to remove this restriction and consider more general coordinate systems. Central to such a discussion is the idea of a basis for a vector space as developed in Section 7.5.

In the vector space of triples we know that (1, 0, 0), (0, 1, 0), and (0, 0, 1) form a basis. Let these triples be coordinates of the points $P_1$, $P_2$, and $P_3$, respectively. Since any vector triple is a linear combination of these three vectors, any point in the plane is a linear combination of the points $P_1$, $P_2$, and $P_3$. In particular, if $P$ has coordinates $(x_1, x_2, x_3)$, we see that

$$P = x_1P_1 + x_2P_2 + x_3P_3;$$

that is, the scalar multipliers of $P_1$, $P_2$, and $P_3$ are coordinates of the point $P$. For the above choice of coordinates of $P_1$, $P_2$, and $P_3$ there is not a unique set of multipliers leading to a given point $P$, but any two such sets are in proportion. It is true, however, that a different choice of coordinates for $P_1$, $P_2$, and $P_3$, say $(2, 0, 0)$, $(0, 3, 0)$, and $(0, 0, 1)$, will lead to sets of coefficients that are as easily determined. Thus for the point $(4, -3, 2)$ we find

$$(4, -3, 2) = 4(1, 0, 0) - 3(0, 1, 0) + 2(0, 0, 1),$$

but

$$(4, -3, 2) = 2(2, 0, 0) - 1(0, 3, 0) + 2(0, 0, 1).$$

Let us consider the point $P_4$ with coordinates $(1, 1, 1)$. Clearly for our original coordinate choice

$$P_4 = P_1 + P_2 + P_3.$$

If we assign coordinates to $P_1$, $P_2$, and $P_3$ in such a way as to preserve this relation, the coordinates are of the form $(k, 0, 0)$, $(0, k, 0)$, and $(0, 0, k)$, respectively, and the coefficients in the expression for any point will indeed be coordinates of the point. The configuration consisting of the three points serving as a "basis" together with this point $(1, 1, 1)$ is basic to the natural coordinate system. We make the following definition.

**Definition 8.31.** *The triangle with vertices* $(1, 0, 0)$, $(0, 1, 0)$, *and* $(0, 0, 1)$ *is the* reference triangle *for the coordinate system. The point* $(1, 1, 1)$ *is the* unit point. *The reference triangle together with the unit point is the* reference system *for the coordinate system.*

We observe that the characteristic property of the unit point is that its coordinates must all be equal.

We consider now a generalization of this discussion. Suppose that we take three independent points $P_1$, $P_2$, and $P_3$ in the plane. Given any other point $P$ in the plane, we know by Theorem 8.16 that we may write

$$P = x_1'P_1 + x_2'P_2 + x_3'P_3, \tag{8.31}$$

where $(x_1', x_2', x_3')$ are in general not natural homogeneous coordinates of the point $P$. Let us consider how many of the properties of homogeneous coordinates these numbers have.

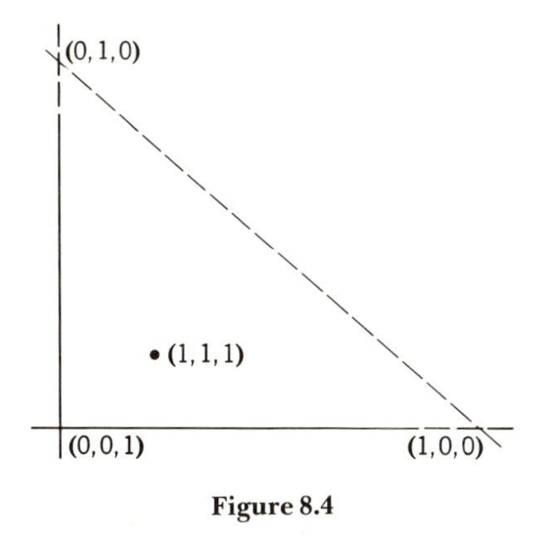


**Figure 8.4**

For a fixed choice of natural homogeneous coordinates for $P_1, P_2$, and $P_3$ the linear combination of these points with coefficients $kx_1'$, $kx_2'$, and $kx_3'$, $k \neq 0$, also represents the same point $P$. This follows since we have, if this new combination represents a point $P'$, $kP - P' = \theta$, from which the identity of $P$ and $P'$ follows from Theorem 8.11. On the other hand, if the coefficients in the expression of a point $P'$ are not in proportion to those of $P$, no such relation of dependence can exist between $P$ and $P'$, and again by Theorem 8.11 the points are distinct. We see that each point is associated with a complete set of proportional nonzero triples, and we therefore replace equation (8.31) by

$$kP = x_1'P_1 + x_2'P_2 + x_3'P_3,\ k \neq 0. \tag{8.32}$$

No point can have all its coefficients zero in such an expansion since such a linear combination produces the zero vector, which does not represent any point.

Let the points $P_1$, $P_2$, and $P_3$ have fixed natural homogeneous coordinates $(a_{11}, a_{21}, a_{31})$, $(a_{12}, a_{22}, a_{32})$, and $(a_{13}, a_{23}, a_{33})$, respectively, and let $P$ have natural homogeneous coordinates $(x_1, x_2, x_3)$. Then it follows from equation (8.32) that

$$\begin{aligned} kx_1 &= a_{11}x_1' + a_{12}x_2' + a_{13}x_3' \\ kx_2 &= a_{21}x_1' + a_{22}x_2' + a_{23}x_3' \\ kx_3 &= a_{31}x_1' + a_{32}x_2' + a_{33}x_3' \end{aligned} \tag{8.33}$$

give the $x_i$ in terms of the $x_j'$. Since $P_1$, $P_2$, and $P_3$ are independent, the matrix $(a_{ij})$ has rank three,[2] hence is nonsingular and has a nonzero determinant. This means that system (8.33) can be solved to give linear

[2]Since its column rank is three. See Exercise 7.44.

relations for the $x'_j$ in terms of the $x_i$. If we let $A$ represent the matrix $(a_{ij})$, $X$ represent the column matrix $\begin{pmatrix} x_1 \\ x_2 \\ x_3 \end{pmatrix}$, and $X'$ represent the matrix $\begin{pmatrix} x'_1 \\ x'_2 \\ x'_3 \end{pmatrix}$, then (8.33) can be written in the compact form $kX = AX'$, where $A$ is nonsingular. The solution for the $x'_j$ becomes $k'X' = A^{-1}X$, where $k' = 1/k \neq 0$.

If we regard system (8.33) as a change of coordinates in the projective plane, lines will continue to have first-degree homogeneous equations after the change; that is, the line

$$u_1x_1 + u_2x_2 + u_3x_3 = 0$$

has new equations

$$u_1(a_{11}x'_1 + a_{12}x'_2 + a_{13}x'_3) + u_2(a_{21}x'_1 + a_{22}x'_2 + a_{23}x'_3) + u_3(a_{31}x'_1 + a_{32}x'_2 + a_{33}x'_3) = 0,$$

which is linear and homogeneous in $x'_1$, $x'_2$, and $x'_3$.

Suppose the points $P, Q, R, \cdots$, have natural homogeneous coordinates $(x_1, x_2, x_3)$, $(y_1, y_2, y_3)$, $(z_1, z_2, z_3)$, $\cdots$, and are associated with the sets of coefficients $(x'_1, x'_2, x'_3)$, $(y'_1, y'_2, y'_3)$, $(z'_1, z'_2, z'_3)$ $\cdots$. If we set

$$B = \begin{pmatrix} k_1x_1 & k_2y_1 & k_3z_1 \cdots \\ k_1x_2 & k_2y_2 & k_3z_2 \cdots \\ k_1x_3 & k_3y_3 & k_3z_3 \cdots \end{pmatrix}$$

and

$$C = \begin{pmatrix} x'_1 & y'_1 & z'_1 \cdots \\ x'_2 & y'_2 & z'_2 \cdots \\ x'_3 & y'_3 & z'_3 \cdots \end{pmatrix},$$

then in view of (8.33) $B = AC$ where $A$ is nonsingular. It follows from Theorem 7.64 that $B$ and $C$ have the same rank. This means that the same dependence and independence relations hold for the sets of coefficients as hold for the natural homogeneous coordinates and for the points themselves. We conclude that all results of Section 8.1 on dependent and independent points will apply in terms of the new coefficient triples being considered here.

These considerations show that the sets of coefficients have the properties of natural homogeneous coordinates needed to preserve all previous results of this chapter, and we shortly define them to be homogeneous coordinates of the points. The change to the new coordinates is given by (8.33) or its matrix form $kX = AX'$. The points $P_1$, $P_2$, and $P_3$ have new coordinates (1, 0, 0), (0, 1, 0), and (0, 0, 1), respectively, and are the vertices of the reference system for the new coordinate system. Now the reference triangle need not have one ideal side. The three points do not determine new coordinates uniquely, for a change in the old natural homogeneous coordinates of the points may change the resulting new coordinates.

Given any point $P_4$ not on a line with any pair of the points $P_1$, $P_2$, and $P_3$, for an appropriate choice of coordinates we have

$$kP_4 = P_1 + P_2 + P_3,$$

and $P_4$ is the unit point in the new reference system. The new coordinates are now uniquely chosen, for any change of natural homogeneous coordinates that preserves $P_4$'s property of being the unit point must change the coordinates of $P_1$, $P_2$, and $P_3$ by the same proportionality factor. This will not alter the form of (8.33).

We now have generalized the idea of homogeneous coordinates and are able to change coordinates or to assign coordinates without giving a preferred position to the ideal points of the plane. We summarize our results in the following definition and pair of theorems.

**Definition 8.32.** *If $P_1$, $P_2$, and $P_3$ are independent points, the point $P$ has (general)* homogeneous coordinates *$(x_1, x_2, x_3)$ if $P = x_1P_1 + x_2P_2 + x_3P_3$ for a fixed set of natural homogeneous coordinates of $P_1$, $P_2$, and $P_3$.*

**Theorem 8.31.** *Given four points, no three of which are collinear, the coordinate system determined by these points taken as the points of a reference system in a given order is uniquely determined.*

**Theorem 8.32.** *A projective change of coordinates is given by $kX = AX'$, $k \neq 0$, where $X$ is the column matrix of the old coordinates, $X'$ is the column*

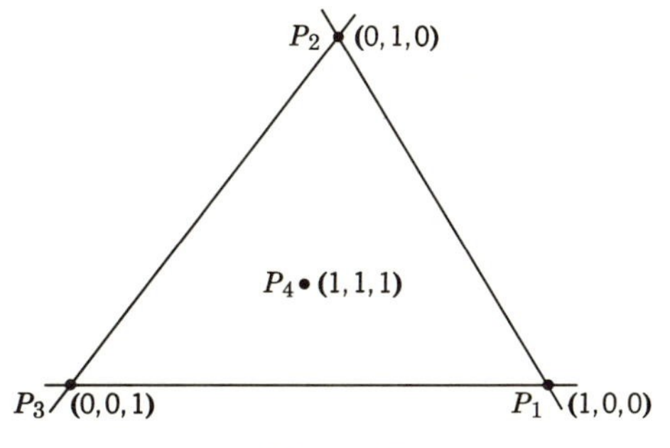


**Figure 8.5**

*matrix of the new coordinates, and $A$ is a nonsingular matrix, its columns being the old coordinates of the vertices of the new reference triangle.*

It should be noted that Theorem 8.32 applies even to a projective change of coordinates in which neither is the natural system, for the reasoning used in its derivation makes no use of any peculiar property of the natural system. Moreover, all previous results of this chapter, derived for natural coordinate systems, remain valid for general homogeneous coordinate systems.

Suppose that we change coordinates from $(x_1, x_2, x_3)$ to $(x_1', x_2', x_3')$ then carry out a second change from $(x_1', x_2', x_3')$ to $(x_1'', x_2'', x_3'')$. The pair of changes is the same as the single change from $(x_1, x_2, x_3)$ to $(x_1'', x_2'', x_3'')$. We seek the relation between these changes. The first change can be written as $kX = AX'$ or in detail as equations (8.33). The second can be written as $k'X' = BX''$ or

$$\begin{aligned} k'x_1' &= b_{11}x_1'' + b_{12}x_2'' + b_{13}x_3'' \\ k'x_2' &= b_{21}x_1'' + b_{22}x_2'' + b_{23}x_3'' \\ k'x_3' &= b_{31}x_1'' + b_{32}x_2'' + b_{33}x_3''. \end{aligned} \tag{8.34}$$

If we substitute from equations (8.34) into equations (8.33) and set $k'' = kk'$, we find that

$$k''x_i = (a_{i1}b_{11} + a_{i2}b_{21} + a_{i3}b_{31})x_1'' + (a_{i1}b_{12} + a_{i2}b_{22} + a_{i3}b_{32})x_2'' + (a_{i1}b_{13} + a_{i2}b_{23} + a_{i3}b_{33})x_3'',$$

$i = 1, 2, 3$. From this we see that if the single change of coordinates is given by $k''X = CX''$, then

$$c_{ij} = a_{i1}b_{1j} + a_{i2}b_{2j} + a_{i3}b_{3j}$$

and $C = AB$; that is, we multiply matrices associated with the consecutive coordinate changes to find the matrix associated with the equivalent single change. We note that if $A$ and $B$ are nonsingular, so also is $C$ in view of Theorem 7.61. Here we have the justification for the definition of product of matrices made in Section 7.3. A simpler multiplication rule could have been used, but it would not be useful in the present application. We have proved

**Theorem 8.33.** *If successive projective changes of coordinates are given by $kX = AX'$ and $k'X' = BX''$, $A$ and $B$ being nonsingular, the single change equivalent to the pair of successive changes is given by $k''X = CX''$ where $C = AB$ and is nonsingular.*

Let us illustrate the foregoing with an example. The points with natural homogeneous coordinates $(1, 1, 0)$, $(1, 0, 1)$, $(0, 1, 1)$, and $(3, -2, -1)$

have the property that no three of them are collinear, as can be verified by finding all the determinants formed with any three of these sets for rows. We seek a new coordinate system in which the first three of these are the vertices of the reference triangle and the last is the unit point. We find

$$(3,-2,-1)=(1,1,0)+2(1,0,1)-3(0,1,1).$$

Thus suitable coordinates for the vertices are $(1,1,0)$, $(2,0,2)$, and $(0,-3,-3)$ since

$$(3,-2,-1)=(1,1,0)+(2,0,2)+(0,-3,-3).$$

The matrix associated with the change can be taken as

$$\begin{pmatrix} 1 & 2 & 0 \\ 1 & 0 & -3 \\ 0 & 2 & -3 \end{pmatrix};$$

that is, it is the matrix whose columns are the suitable coordinates of the vertices of the new reference triangle. Old coordinates are given in terms of the new by

$$\begin{aligned} kx_1 &= x_1' + 2x_2' \\ kx_2 &= x_1' \qquad\quad - 3x_3' \\ kx_3 &= \qquad 2x_2' - 3x_3' \end{aligned} \tag{8.35}$$

since

$$k\begin{pmatrix} x_1 \\ x_2 \\ x_3 \end{pmatrix}=\begin{pmatrix} 1 & 2 & 0 \\ 1 & 0 & -3 \\ 0 & 2 & -3 \end{pmatrix}\begin{pmatrix} x_1' \\ x_2' \\ x_3' \end{pmatrix}.$$

The line $x_1+3x_2-2x_3=0$ will have equation

$$\begin{aligned} (x_1'+2x_2')+3(x_1'-3x_3')-2(2x_2'-3x_3') \\ = 4x_1'-2x_2'-3x_3'=0. \end{aligned}$$

If a relation for the new coordinates in terms of the old is desired, we solve equations (8.35) for the $x_j'$ finding

$$\left(\frac{1}{k}\right)x_1' = (\tfrac{1}{2})x_1+(\tfrac{1}{2})x_2-(\tfrac{1}{2})x_3$$

$$\left(\frac{1}{k}\right)x_2' = (\tfrac{1}{4})x_1-(\tfrac{1}{4})x_2+(\tfrac{1}{4})x_3$$

$$\left(\frac{1}{k}\right)x_3' = (\tfrac{1}{6})x_1-(\tfrac{1}{6})x_2-(\tfrac{1}{6})x_3.$$

Thus the point with old coordinates $(2,-3,1)$ or $(24,-36,12)$ will have new coordinates

$$\begin{aligned} x_1' &= (\tfrac{1}{2})(24) + (\tfrac{1}{2})(-36) - (\tfrac{1}{2})(12) = -12 \\ x_2' &= (\tfrac{1}{4})(24) - (\tfrac{1}{4})(-36) + (\tfrac{1}{4})(12) = 18 \\ x_3' &= (\tfrac{1}{6})(24) - (\tfrac{1}{6})(-36) - (\tfrac{1}{6})(12) = 8 \end{aligned}$$

or $(-6, 9, 4)$.

Suppose we have a second change given by

$$k' \begin{pmatrix} x_1' \\ x_2' \\ x_3' \end{pmatrix} = \begin{pmatrix} 1 & 2 & 7 \\ 2 & 1 & 2 \\ 1 & 1 & 1 \end{pmatrix} \begin{pmatrix} x_1'' \\ x_2'' \\ x_3'' \end{pmatrix}$$

where

$$\begin{vmatrix} 1 & 2 & 7 \\ 2 & 1 & 2 \\ 1 & 1 & 1 \end{vmatrix} = 6 \neq 0.$$

The single change from the original system is then

$$k'' \begin{pmatrix} x_1 \\ x_2 \\ x_3 \end{pmatrix} = \begin{pmatrix} 1 & 2 & 0 \\ 1 & 0 & -3 \\ 0 & 2 & -3 \end{pmatrix} \begin{pmatrix} 1 & 2 & 7 \\ 2 & 1 & 2 \\ 1 & 1 & 1 \end{pmatrix} \begin{pmatrix} x_1'' \\ x_2'' \\ x_3'' \end{pmatrix} = \begin{pmatrix} 5 & 4 & 11 \\ -2 & -1 & 4 \\ 1 & -1 & 1 \end{pmatrix} \begin{pmatrix} x_1'' \\ x_2'' \\ x_3'' \end{pmatrix}$$

or

$$\begin{aligned} k''x_1 &= 5x_1'' + 4x_2'' + 11x_3'' \\ k''x_2 &= -2x_1'' - x_2'' + 4x_3'' \\ k''x_3 &= x_1'' - x_2'' + x_3''. \end{aligned}$$

Our development has been in terms of plane analytic projective geometry. The results specialize readily to the projective line, where we have the following:

**Definition 8.33.** *The points* $(1,0)$ *and* $(0,1)$ *are the* reference set *for the coordinate system on the projective line. The point* $(1,1)$ *is the* unit point. *The reference set together with the unit point is the* reference system *on the line.*

**Definition 8.34.** *Given distinct points* $P_1$ *and* $P_2$, *the point* $P$ *on the line determined by these points has (general)* homogeneous coordinates $(x_1, x_2)$ *on this line if* $P = x_1P_1 + x_2P_2$ *for a fixed set of natural homogeneous coordinates of* $P_1$ *and* $P_2$.

**Theorem 8.34.** *Given three distinct points on the projective line, the coordinate system determined by these points taken as the points of a reference system in a given order is uniquely determined.*

In the above specializations it should be noted that Theorems 8.32 and

8.33 remain valid; that is, in a change of coordinates $kX = AX'$ is valid whether $X$ and $X'$ have two or three elements with $A$ being $2 \times 2$ or $3 \times 3$, respectively.

We conclude this section with a theorem about the relation between homogeneous coordinates in the projective plane and on the projective line which we shall find useful in the next section.

**Theorem 8.35.** *The two nonzero coordinates on a side of the reference triangle in the projective plane are homogeneous coordinates on that line.*

PROOF. Suppose we consider the side determined by $P_1$ and $P_2$ with respective coordinates $(1, 0, 0)$ and $(0, 1, 0)$. Now any point on this line is the linear combination $x_1P_1 + x_2P_2$. The assertion now follows at once from Definition 8.34.

## Exercises 8.3

1. Find the matrix of the change of coordinates in which the vertices of the reference triangle are $(1, 1, 0)$, $(1, 1, 2)$, and $(0, 1, 4)$ and the unit point is $(5, 9, 22)$. Find the new coordinates of the point with old coordinates $(2, -3, 4)$.

2. Find the matrix of the change of coordinates in which the vertices of the reference triangle are $(1, 2, 1)$, $(2, 1, 2)$, and $(1, -1, 2)$ and the unit point is $(1, -1, 3)$. Find the new coordinates of the point with old coordinates $(1, 1, 1)$.

3. Find the matrix of the change of coordinates in which the vertices of the reference set are $(1, 2)$ and $(2, 3)$ and the unit point is $(-1, 0)$. Find the new coordinates of the point with old coordinates $(3, -2)$.

4. Find the matrix of the change of coordinates in which the vertices of the reference set at $(2, -1)$ and $(4, -3)$ and the unit point is $(3, 2)$. Find the new coordinates of the point with old coordinates $(5, 7)$.

5. Any change of coordinates $kX = AX'$ causes the equation of every line to be

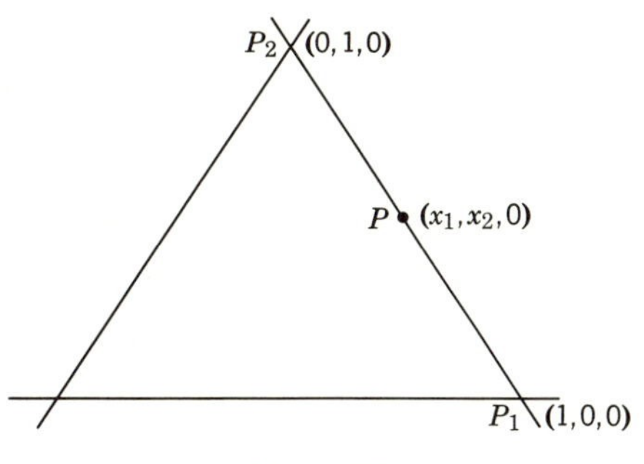


**Figure 8.6**

changed. If a line has old coordinates $[u_1, u_2, u_3]$, find its new coordinates $[u_1', u_2', u_3']$ in terms of the old line coordinates and the elements of the matrix $A$. Write the result in matrix form.

6.* Generalize all definitions and theorems of this section to projective space.

7.* Find the matrix of the change of coordinates in which the vertices of the reference tetrahedron are (1, 1, 0, 0), (1, 0, 1, 0), (1, 0, 0, 1), and (0, 0, 1, 1) and the unit point is (4, 1, 4, 3).

## 8.4 PROJECTIVE TRANSFORMATIONS BETWEEN LINES

In the preceding section we saw that the relations

$$\begin{aligned} kx_1 &= a_{11}x_1' + a_{12}x_2' \\ kx_2 &= a_{21}x_1' + a_{22}x_2', \end{aligned}$$

or, more simply, $kX = AX'$, are associated with a change of coordinates on the projective line. As interpreted there $(x_1, x_2)$ and $(x_1', x_2')$ were coordinates of the same point on the projective line, these coordinates being associated with two different reference systems and the above relations telling how to change from one set of coordinates to the other.

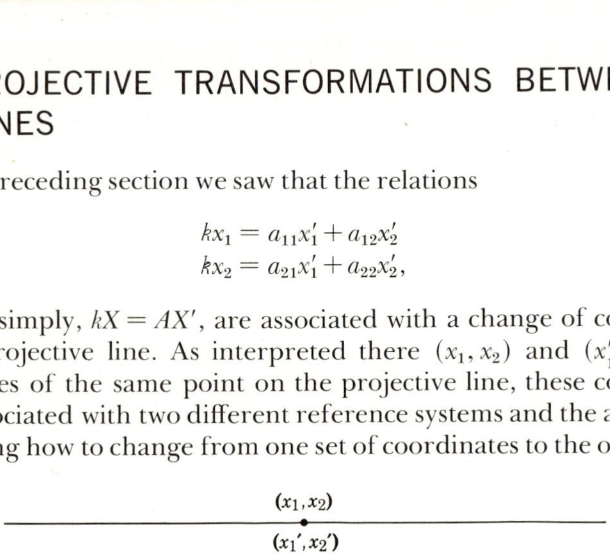

$(x_1, x_2)$

$(x_1', x_2')$

**Figure 8.7**

There is a second possible interpretation for these relations. We can assume that the coordinate system does not change. Then $(x_1, x_2)$ and $(x_1', x_2')$ are coordinates of two points which will in general be distinct. If we call these points $P$ and $P'$, respectively, the relations above associate the unique point $P$ with the point $P'$. This association is exactly that which we used to define a function in Section 5.3. Thus the relations can be interpreted as the analytic representation of some function $T$ for which $T(P') = P$. We observe that the points $P$ and $P'$ need not be on the same line in this interpretation, for the coordinates $(x_1, x_2)$ and $(x_1', x_2')$ could be coordinates on distinct projective lines. We consider the

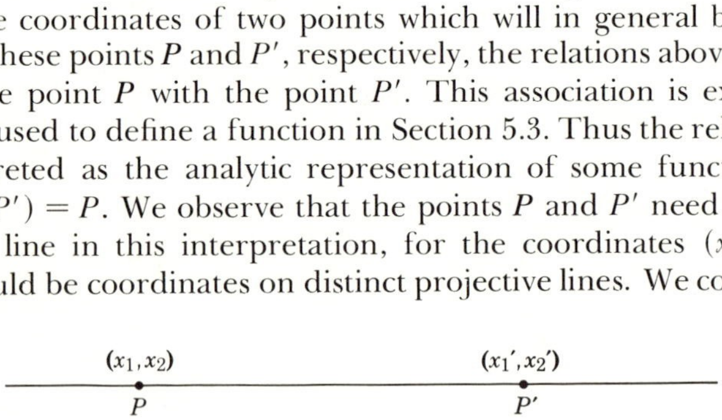

$(x_1, x_2)$ $(x_1', x_2')$

$P$ $P'$

**Figure 8.8**

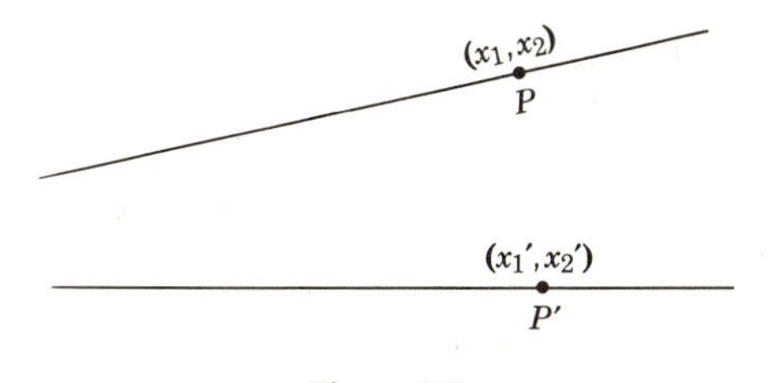


**Figure 8.9**

nature of the function $T$ given by $kX = AX'$ where $A$ is nonsingular. We shall see that the set of all such functions is the set of all projective transformations. If we establish this, the results of the preceding section can be translated into this new interpretation to give results on projective transformations. To establish the connection with projectivities we first prove

**Theorem 8.41.** *A central perspectivity has analytic representation $kX = AX'$ where $k \neq 0$ and $A$ is nonsingular.*

PROOF. We may as well assume the lines are distinct, for a perspectivity of a line onto itself is the identity and has representation $kX = IX'$. We choose homogeneous coordinates so that the two lines are the sides $[1, 0, 0]$ and $[0, 1, 0]$ of the reference triangle. By Theorem 8.35 we may take the nonzero coordinates on the two related lines as homogeneous coordinates on these lines regarded as projective lines. Since the center of perspectivity must lie on neither line, its coordinates will be of the form $(a_1, a_2, a_3)$ where $a_1 a_2 \neq 0$. If the points $(x_1, 0, x_2)$ and $(0, x_1', x_2')$ correspond, they must be collinear with the center of perspectivity, and

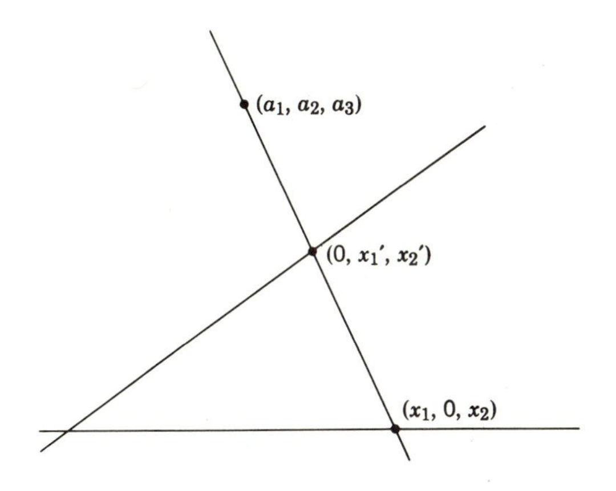


**Figure 8.10**

we have

$$\begin{vmatrix} x_1 & 0 & x_2 \\ 0 & x_1' & x_2' \\ a_1 & a_2 & a_3 \end{vmatrix} = x_1(a_3x_1' - a_2x_2') + x_2(-a_1x_1') = 0$$

or

$$kx_1 = -a_1x_1'$$
$$kx_2 = a_3x_1' - a_2x_2',$$

where

$$\begin{vmatrix} -a_1 & 0 \\ a_3 & -a_2 \end{vmatrix} = a_1a_2 \neq 0.$$

Finally, we note that the relation holds for other choices of coordinates on the two lines, for if we change coordinates by $k_2X = B\bar{X}$ and $k_3X' = C\bar{X}'$, the perspectivity will be given by $k_4\bar{X} = B^{-1}AC\bar{X}'$, where $B^{-1}AC$ is nonsingular by Theorem 7.61.

The extension of this theorem to projectivities is simple, for we know that a projectivity is a product of central perspectivities. Each perspectivity is given by a relation of the form $kX = AX'$ by Theorem 8.41 and by Theorem 8.33 a succession of such relations is of the same form. Thus we have

**Theorem 8.42.** *A projective transformation has analytic representation $kX = AX'$ where $k \neq 0$ and $A$ is nonsingular.*

To justify the identification of projective transformations and the set of transformations defined by $kX = AX'$, we must still prove the converse of Theorem 8.42.

**Theorem 8.43.** *The transformation of points of a line onto points of a line with analytic representation $kX = AX'$, $k \neq 0$ and $A$ nonsingular, is a projective transformation.*

PROOF. Let $T$ represent the transformation defined by $kX = AX'$, and let $P', Q'$, and $R'$ have coordinates $(1, 0)$, $(0, 1)$, and $(1, 1)$, respectively. If we apply Theorem 8.34 to the points $T(P')$, $T(Q')$, and $T(R')$, we conclude that the matrix condition for the change of coordinates to make these points the points of a reference system is $kX = AX'$. In view of Theorem 8.42 we see that the projective transformation $T'$ mapping $P'$, $Q'$, and $R'$ onto $T(P')$, $T(Q')$, and $T(R')$, respectively, must have representation $kX = AX'$, whence $T \doteq T'$.

We see that if interpreted in terms of transformations rather than change in coordinates, Theorem 8.34 is just

**Theorem 8.44 (Fundamental Theorem of Projective Geometry).** *There exists one and only one projective transformation mapping three distinct points on a line onto three distinct points on a line in a given order.*

We note that the corresponding pairs of points can be taken to be the points of the reference systems on the two lines if the lines are distinct; that is, we may assume the projectivity maps $(1,0)$ onto $(1,0)$, $(0,1)$ onto $(0,1)$, and $(1,1)$ onto $(1,1)$. Clearly the analytic representation for the projectivity in this case is $kX = IX'$. Thus we have

**Theorem 8.45.** *For a suitable choice of coordinates on two distinct lines a projectivity between the lines has the representation* $kX = IX'$.

This theorem shows that in any problem involving such a projectivity between distinct lines we might as well use the simple representation $kX = IX'$ if we are concerned only with the points on the lines themselves. The theorem should not be confused, however, with Theorem 8.35, which enables us to imbed a line in the plane using the nonzero coordinates on the line as coordinates of points of the line. The suitable coordinate system of Theorem 8.45 may not be an acceptable one for imbedding both lines as sides of the reference triangle. Note that in the case where the lines are sides of the reference triangle the relation $kX = IX'$ implies that the intersection of the lines corresponds to itself. Thus the projectivity would have to be a perspectivity.

## Exercises 8.4

1. Prove that if $T$ is a projectivity, $T(aA + bB) = aT(A) + bT(B)$.

2. Find the analytic representation of the projectivity carrying the points $(1,2)$, $(3,-1)$, and $(4,2)$ into the points $(-4,7)$, $(2,-3)$, and $(1,1)$, respectively.

3. Find the analytic representation of the projectivity carrying the points $(3,4)$, $(-1,3)$, and $(5,2)$ into the points $(-1,3)$, $(3,1)$, and $(2,5)$, respectively.

4. Find the analytic representation of the projectivity carrying the points $(3,4)$, $(-1,3)$, and $(5,2)$ into the points $(-1,3)$, $(3,1)$, and $(1,-3)$, respectively.

## 8.5 CROSS RATIO

The concept of cross ratio is one of the oldest of the concepts now known to be part of projective geometry; indeed, it is believed that the theory was known to Pappus. Cross ratio is frequently introduced using metric concepts from Euclidean geometry and making appropriate extensions to allow for the presence of ideal elements. We prefer here to introduce the concept analytically in

**Definition 8.51.** *If the points $A$, $B$, $C$, and $D$, at least three of which are*

*distinct, on a projective line have coordinates* $(a_1, a_2)$, $(b_1, b_2)$, $(c_1, c_2)$, *and* $(d_1, d_2)$, *respectively, the real number*

$$R(A,B,C,D) = \frac{\begin{vmatrix} a_1 & a_2 \\ c_1 & c_2 \end{vmatrix} \begin{vmatrix} b_1 & b_2 \\ d_1 & d_2 \end{vmatrix}}{\begin{vmatrix} b_1 & b_2 \\ c_1 & c_2 \end{vmatrix} \begin{vmatrix} a_1 & a_2 \\ d_1 & d_2 \end{vmatrix}}$$

*is the* cross ratio *of the four points in the order A, B, C, D, if it exists. If the number does not exist, the cross ratio is said to be infinite.*

This definition uses a particular coordinate system on the line. If the definition is to be reasonable, we must prove

**Theorem 8.51.** *Cross ratio is independent of the coordinate system used.*

PROOF. Suppose that we change coordinates by $kX = AX'$ with the points having the new coordinates $(a_1', a_2')$, $(b_1', b_2')$, $(c_1', c_2')$, and $(d_1', d_2')$, respectively. Then we have

$$k_a \begin{pmatrix} a_1 \\ a_2 \end{pmatrix} = A \begin{pmatrix} a_1' \\ a_2' \end{pmatrix}$$

and

$$k_c \begin{pmatrix} c_1 \\ c_2 \end{pmatrix} = A \begin{pmatrix} c_1' \\ c_2' \end{pmatrix}.$$

These can now be combined to give the matrix relation

$$\begin{pmatrix} k_a a_1 & k_c c_1 \\ k_a a_2 & k_c c_2 \end{pmatrix} = A \begin{pmatrix} a_1' & c_1' \\ a_2' & c_2' \end{pmatrix},$$

which in turn, interchanging rows and columns, we use to obtain the determinant relation

$$k_a k_c \begin{vmatrix} a_1 & a_2 \\ c_1 & c_2 \end{vmatrix} = |A| \begin{vmatrix} a_1' & a_2' \\ c_1' & c_2' \end{vmatrix}$$

where $|A|$ denotes the determinant of the matrix $A$. The invariance under a coordinate change now follows, since

$$\frac{\begin{vmatrix} a_1 & a_2 \\ c_1 & c_2 \end{vmatrix} \begin{vmatrix} b_1 & b_2 \\ d_1 & d_2 \end{vmatrix}}{\begin{vmatrix} b_1 & b_2 \\ c_1 & c_2 \end{vmatrix} \begin{vmatrix} a_1 & a_2 \\ d_1 & d_2 \end{vmatrix}} = \frac{\dfrac{|A|}{k_a k_c} \begin{vmatrix} a_1' & a_2' \\ c_1' & c_2' \end{vmatrix} \dfrac{|A|}{k_b k_d} \begin{vmatrix} b_1' & b_2' \\ d_1' & d_2' \end{vmatrix}}{\dfrac{|A|}{k_b k_c} \begin{vmatrix} b_1' & b_2' \\ c_1' & c_2' \end{vmatrix} \dfrac{|A|}{k_a k_d} \begin{vmatrix} a_1' & a_2' \\ d_1' & d_2' \end{vmatrix}}$$

$$= \frac{\begin{vmatrix} a_1' & a_2' \\ c_1' & c_2' \end{vmatrix} \begin{vmatrix} b_1' & b_2' \\ d_1' & d_2' \end{vmatrix}}{\begin{vmatrix} b_1' & b_2' \\ c_1' & c_2' \end{vmatrix} \begin{vmatrix} a_1' & a_2' \\ d_1' & d_2' \end{vmatrix}}.$$

The argument using the coordinate change $kX = AX'$ is just as much an argument considering a projective transformation with this representation. We have accordingly also proved

**Theorem 8.52.** *Cross ratio is preserved under a projective transformation.*

The formulas hold for any choice of coordinates. Given points $A$, $B$, $C$, and $D$, it is sometimes convenient to study their cross ratio with coordinates chosen to simplify the computation. In particular we note that if $A$ has coordinates $(1,0)$, $B$ has coordinates $(0,1)$, $C$ has coordinates $(1,1)$, and $D$ has coordinates $(x_1, x_2)$, then $R(A,B,C,D) = x_1/x_2$. It is apparent that for this $A$, $B$, and $C$ the ratio of coordinates of $D$ is determined when a value for the cross ratio is specified. Thus we have

**Theorem 8.53.** *For fixed distinct collinear points $A$, $B$, and $C$ and a fixed number $t$ (which may be infinite), there is one and only one point $D$ for which $R(A,B,C,D) = t$.*

The same judicious choice of coordinates for $A$, $B$, $C$, and $D$ leads to a simple proof of the following theorem, which shows that cross ratio is in some sense a generalization of the concept of a harmonic sequence.

**Theorem 8.54.** *If $H(A,B;C,D)$, then $R(A,B,C,D) = -1$ and conversely.*

PROOF. We choose coordinates and carry out the harmonic point construction as shown in Figure 8.11. We can have $H(A,B;C,D)$ if and only if $D$, $(0,1,1)$, and $(1,1,0)$ are collinear. This is the case if and only if

$$\begin{vmatrix} 0 & 1 & 1 \\ 1 & 1 & 0 \\ x_1 & 0 & x_2 \end{vmatrix} = 0$$

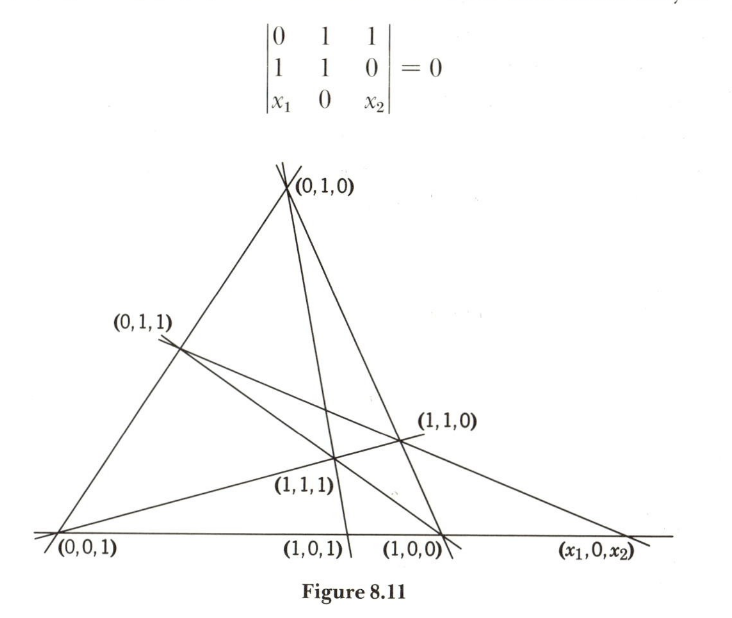


**Figure 8.11**

or

$$-x_1 - x_2 = 0$$

or

$$\frac{x_1}{x_2} = -1.$$

But this ratio is $R(A, B, C, D)$.

The following theorem will be of use later and is easily proved using the definition of cross ratio.

**Theorem 8.55.** *If $P, Q, A, B$, and $C$ are distinct collinear points, then*

$$R(P, Q, A, B)R(P, Q, B, C) = R(P, Q, A, C).$$

The proof is left to the reader.

If cross ratio is defined for four collinear points, we should preserve duality by defining it for four concurrent lines. This we do and observe a relation between the dual concepts in Section 8.8.

## Exercises 8.5

1. Find the cross ratio of the following points (in the given order):
   *a.* $(2, -5), (3, 4), (-3, -1), (2, 3)$;
   *b.* $(1, 5), (-1, -1), (3, 4), (7, -2)$;
   *c.* $(1, 5), (-1, -1), (3, 4), (2, 2)$;
   *d.* $(1, 5), (-1, -1), (2, 10), (2, 2)$;
   *e.* $(1, 5), (-1, -1), (0, 2), (3, 7)$.

2. Suppose that $A$, $B$, $C$, and $D$ have coordinates $(0, 1)$, $(1, 1)$, $(x, 1)$, and $(y, 1)$, respectively, and that $H(A, B; C, D)$. Find the relation between $x$ and $y$. Show that 1 is the harmonic mean of $x$ and $y$. (The harmonic mean of two numbers is the reciprocal of the arithmetic mean of their reciprocals.)

3. If $C = A + rB$ and $D = A + sB$, prove $R(A, B, C, D) = r/s$.

4. Prove Theorem 8.55.

5#. Show that if $A$, $B$, $C$, and $D$ are ordinary and if $AC$ denotes the directed length of the segment from $A$ to $C$, etc., then

$$R(A, B, C, D) = \frac{AC}{BC} \Big/ \frac{AD}{BD}.$$

6. If $R(A, B, C, D) = t$, show that the cross-ratio values obtained by considering all rearrangements of the order of the points are $t$, $1/t$, $1-t$, $1/(1-t)$, $(t-1)/t$, and $t/(t-1)$.

7. Use Theorems 8.52 and 8.53 to obtain an alternate proof of the Fundamental Theorem.

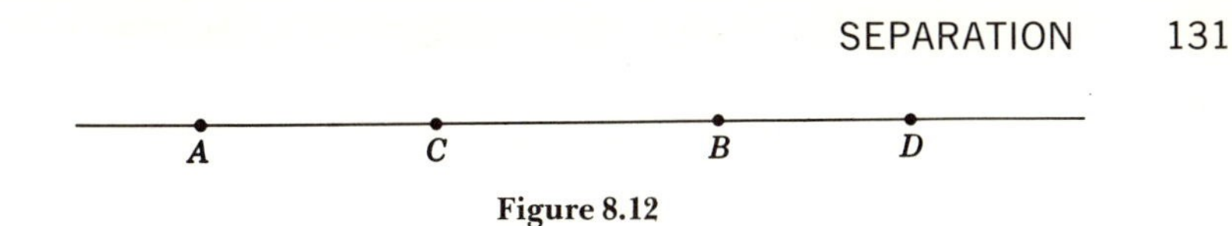


Figure 8.12

## 8.6 SEPARATION

If we consider four collinear points, we have an intuitive feeling on the Euclidean line for what it means to say that one pair of them interlocks with or separates the other pair. Thus in Figure 8.12 we would say that $A$ and $B$ separate $C$ and $D$ or that $C$ and $D$ separate $A$ and $B$. We would not claim, however, that $A$ and $D$ separate $B$ and $C$. This concept of separation is related to order in the Euclidean plane, but can be introduced in the projective plane where betweenness cannot. We have noted earlier that the projective line is better portrayed as a closed curve, as shown in Figure 8.13. In this figure we can agree to the same intuitive feelings about separation that we did in Figure 8.12. A clue to an appropriate definition is to be found in the result of Exercise 8.55 as applied to Figure 8.12. We thus make the following definition.

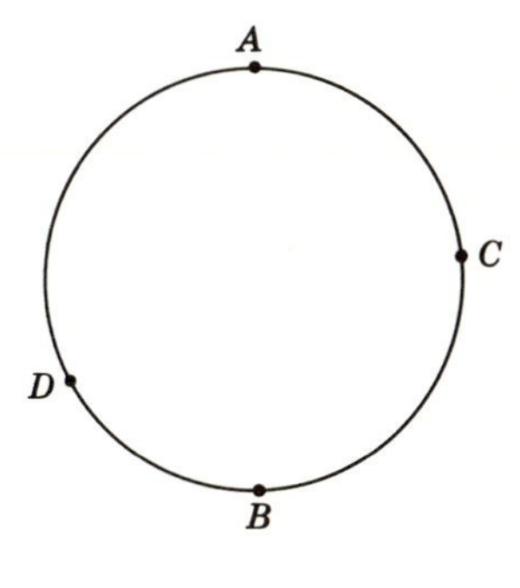


Figure 8.13

**Definition 8.61.** *Given four distinct collinear points $A$, $B$, $C$, and $D$, we say that $A$ and $B$* separate *$C$ and $D$, denoted by $AB\|CD$, if $R(A,B,C,D) < 0$.*

As an immediate consequence of the definition and Theorem 8.54 we have

**Theorem 8.61.** *If $H(A,B;C,D)$, then $AB\|CD$.*

We also see directly from Theorem 8.52 that we have

**Theorem 8.62.** *If $T$ is a projectivity and $AB\|CD$, then $T(A)T(B)\|T(C)T(D)$.*

In the proof of Theorem 8.53 we noted that if coordinates are suitably chosen,

$$R(A,B,C,D) = t \tag{8.61}$$

where $t = x_1/x_2$ and $D$ has coordinates $(x_1, x_2)$. If we choose these coordinates, we can easily verify that

$$R(C,D,A,B) = t, \tag{8.62}$$
$$R(A,B,D,C) = 1/t, \tag{8.63}$$
$$R(A,C,B,D) = 1-t, \tag{8.64}$$

and

$$R(A, D, B, C) = (t-1)/t. \tag{8.65}$$

The following theorems constitute the basis for the properties of separation and follow at once from the definition and the five equations. The reader is urged to illustrate them with a figure similar to Figure 8.13.

**Theorem 8.63.** *If $AB \| CD$, then $CD \| AB$.*

**Theorem 8.64.** *If $AB \| CD$, then $AB \| DC$.*

**Theorem 8.65.** *If $AB \| CD$, then $AC \nparallel BD$.*

PROOF. If $R(A, B, C, D) = t < 0$, then by equation (8.64) $R(A, C, B, D) = 1 - t > 0$.

**Theorem 8.66.** *If A, B, C, and D are distinct collinear points, then one and only one of the following holds: $AB \| CD$, $AC \| BD$, $AD \| BC$.*

PROOF. In view of equations (8.61), (8.64), and (8.65), we must consider the signs of $t$, $1-t$, and $(t-1)/t$. If $t < 0$, only the first of these is negative. If $0 < t < 1$, only the third is negative. If $t > 1$, only the second is negative. If $t = 0$ or $t = 1$, the points are not distinct.

**Theorem 8.67.** *If $AB \| CD$ and $AC \| BE$, then $AB \| DE$.*

PROOF. If $AB \| CD$, then by Theorem 8.64 $AB \| DC$ and $R(A, B, D, C) < 0$. If $AC \| BE$, then by Theorem 8 65 $AB \nparallel CE$ and $R(A, B, C, E) > 0$. Since

$$R(A, B, D, C)\, R(A, B, C, E) = R(A, B, D, E)$$

by Theorem 8.55, we must have $R(A, B, D, E) < 0$ and $AB \| DE$.

## Exercises 8.6

1. Prove Theorem 8.63.
2. Prove Theorem 8.64.
3. Prove that $AB \| CD$ implies $BA \| CD$.

4.# Prove that if the points $A$, $B$, $C$, and $D$ have coordinates $a$, $b$, $c$, and $d$, respectively, then $AB \| CD$ if and only if $a < c < b < d$ or $a < d < b < c$ or $a > c > b > d$ or $a > d > b > c$.

## 8.7 PROJECTIVE TRANSFORMATIONS OF A LINE ONTO ITSELF; INVOLUTIONS

Let us now restrict ourselves to a consideration of projective transformations of the points of a line onto the points of the same line. In a transformation of a set onto itself we can investigate the existence and number of *fixed points* of the transformation; that is, points which are mapped onto themselves by the transformation. The fixed point question is quickly settled for a projectivity of a line onto itself. First, if there are three or more fixed points, the transformation must be the identity in view of the Fundamental Theorem, since the identity would have these fixed points and is the unique projectivity having them. We conclude that a nonidentity projective transformation of a line onto itself has at most two distinct fixed points. It is customary and useful to classify projectivities according to the number of fixed points they have as follows:

**Definition 8.71.** *A nonidentity projectivity of a line onto itself is* elliptic, parabolic, *or* hyperbolic *as it has no, one, or two fixed points.*

We seek an analytic criterion to tell whether a projectivity is elliptic, parabolic, or hyperbolic.

**Theorem 8.71.** *A projectivity of a line onto itself given by $kX = AX'$ where*

$$A = \begin{pmatrix} a_{11} & a_{12} \\ a_{21} & a_{22} \end{pmatrix} \neq I$$

*is elliptic, parabolic, or hyperbolic as $(a_{22} - a_{11})^2 + 4a_{12}a_{21}$ is negative, zero, or positive.*

PROOF. If a point has homogeneous coordinates $(x_1, x_2)$ where $x_2 \neq 0$, we say that $x = x_1/x_2$ if the *nonhomogeneous* coordinate of the point. This coordinate is defined for all but one of the points of a line and is just the Euclidean coordinate if the homogeneous coordinates are the natural ones. A projective transformation

$$kx_1 = a_{11}x_1' + a_{12}x_2'$$
$$kx_2 = a_{21}x_1' + a_{22}x_2'$$

is given in nonhomogeneous coordinates by

$$x = \frac{a_{11}x' + a_{12}}{a_{21}x' + a_{22}}$$

where $a_{11}a_{22} - a_{12}a_{21} \neq 0$ to insure $A$ is nonsingular. Now if a point with nonhomogeneous coordinate $x$ is a fixed point, we must have

$$x = \frac{a_{11}x + a_{12}}{a_{21}x + a_{22}}$$

or

$$a_{21}x^2 + (a_{22} - a_{11})x - a_{12} = 0.$$

This equation has no, one, or two real solutions, and the transformation is elliptic, parabolic, or hyperbolic, as the discriminant of the equation, $(a_{22}-a_{11})^2+4a_{12}a_{21}$, is negative, zero, or positive. If the point with $x_2=0$ ($x$ undefined) is fixed, the discriminant condition will still be consistent. If the transformation is hyperbolic, we have $a_{21}=0$, $a_{22}-a_{11}\neq 0$ and the discriminant will be positive. If the point with $x_2=0$ is the only fixed point, we have $a_{21}=a_{22}-a_{11}=0$, and the discriminant will be zero.

We observe that when we have a set of transformations of a line onto itself, the product of two such transformations is defined. In particular if $T$ is such a transformation, $T^2=TT$, $T^3=TTT$, etc., are all defined. For such a $T$ it may be the case that for some $n$ we have $T^n=I$, the identity. We make

**Definition 8.72.** *A transformation of a set onto itself has* period n *if n is the least positive integer for which* $T^n=I$.

Of special interest here are projective transformations of period two. These were first investigated by Desargues, much of his work being rediscovered by de la Hire. We introduce

**Definition 8.73.** *A projective transformation of a line onto itself of period two is an* involution.

We seek an analytic condition that a projectivity of a line onto itself be an involution. If $A$ is the matrix of an involution, we must have $A^2=kI$, $k\neq 0$. We find by direct computation that

$$A^2=\begin{pmatrix} a_{11}^2+a_{12}a_{21} & a_{11}a_{12}+a_{12}a_{22} \\ a_{21}a_{11}+a_{22}a_{21} & a_{12}a_{21}+a_{22}^2 \end{pmatrix}.$$

Now if $a_{11}^2+a_{12}a_{21}$ and $a_{12}a_{21}+a_{22}^2$ are to be equal, we must have $a_{11}^2=a_{22}^2$. If we had $a_{11}=a_{22}$, the requirement that the other two matrix elements be zero would lead to the conclusion that $a_{12}=a_{21}=0$. In this case $A$ would represent the identity, which has period one and is not an involut on. We conclude $a_{11}=-a_{22}$ if we have an involution. Conversely if $a_{11}=-a_{22}$, it is clear that $A^2$ is of the desired form and we have an involution. We have proved

**Theorem 8.72.** *A projectivity of a line onto itself with associated matrix* $(a_{ij})$ *is an involution if and only if* $a_{11}=-a_{22}$.

We can now use this criterion to investigate the fixed points of an involution. We find

**Theorem 8.73.** *An involution is elliptic or hyperbolic.*

PROOF. If $A$ is the matrix of the involution, it must be nonsingular.

Using Theorem 8.72, we see that this requires

$$a_{11}a_{22} - a_{12}a_{21} = -a_{11}^2 - a_{12}a_{21} \neq 0.$$

From Theorems 8.71 and 8.72 we see that if the involution were parabolic, we would have

$$(a_{22} - a_{11})^2 + 4a_{12}a_{21} = 4(a_{11}^2 + a_{12}a_{21}) = 0.$$

This contradicts the first statement. Hence the involution cannot be parabolic.

The Fundamental Theorem asserts that a projectivity is determined when three distinct sets of corresponding points are specified. If we require that the projectivity be an involution, we have imposed an additional condition, and it is not reasonable to assume that three sets of corresponding points can be specified. If this were the case, every projectivity of a line onto itself would be an involution! It is true that an involution is determined when two sets of corresponding points are specified, as is shown by the next theorem.

**Theorem 8.74.** *An involution is uniquely determined when two distinct pairs of corresponding points are given.*

PROOF. Let $A$ and $B$, and $C$ and $D$, be the distinct pairs of points. By the Fundamental Theorem we know that there is a unique projective transformation $T$ for which

$$T(A, B, C) = B, A, D.$$

By Theorem 8.52 we must have

$$\begin{aligned} R(A, B, C, D) &= R(B, A, D, T[D]) \\ &= R(A, B, T[D], D). \end{aligned}$$

By an obvious generalization of Theorem 8.53 we conclude that $T[D] = C$, and $T$ has the desired effect on the given pairs of points. Now if $P$ is any other point on the line,

$$\begin{aligned} R(A, B, P, T[P]) &= R(B, A, T[P], T^2[P]) \\ &= R(A, B, T^2[P], T[P]). \end{aligned}$$

Hence $T^2[P] = P$, and $T$ is an involution.

We know that if an involution has fixed points, it has exactly two of them. These fixed points play an interesting role in the theory of involutions. The following is one in which the involution and its fixed points are shown to be closely related to the harmonic sequence concept.

**Theorem 8.75.** *If $A$ and $B$ are fixed points of an involution, distinct points $C$ and $D$ correspond under the involution if and only if $H(A, B; C, D)$.*

PROOF. We suppose first that $T$ is an involution and $T(A, B, C) = A, B, D$. If $R(A, B, C, D) = t$, it is clear from Definition 8.51 that $R(A, B, D, C) = 1/t$. By Theorem 8.52 we must have $R(A, B, C, D) = R(A, B, D, C)$. This requires $t = 1/t$ or $t^2 = 1$. If $t = 1$, we claim that $C$ and $D$ are not distinct, since $R(A, B, C, C) = 1$ and Theorem 8.53 then requires that $C = D$. We conclude that $t = -1$ and by Theorem 8.54 $H(A, B; C, D)$. Conversely if $H(A, B; C, D)$, then $R(A, B, C, D) = -1$ and $R(A, B, D, C) = 1/(-1) = -1 = R(A, B, C, D)$. The projectivity $T$ for which $T(A, B, C) = A, B, D$ is then the desired involution; it is an involution, since $T^2(A, B, C) = A, B, C$, which implies $T^2 = I$ by the Fundamental Theorem. Note that we must have $T(D) = C$ to preserve cross ratio under $T$.

Another cross-ratio requirement for an involution is the following

**Theorem 8.76.** *The pairs $A$ and $A'$, $B$ and $B'$, and $C$ and $C'$ correspond under an involution if and only if*

$$R(A, B, C, C') = R(A', B', C', C).$$

PROOF. If the pairs correspond under an involution $T$, the cross ratio equality follows from Theorem 8.52, since $T(A, B, C, C') = A', B', C', C$. Conversely, suppose that

$$R(A, B, C, C') = R(A', B', C', C).$$

By Theorem 8.74 the pairs $B$ and $B'$, and $C$ and $C'$ determine an involution. Let $A''$ correspond to $A$ under this involution. By the first part of this theorem

$$R(A, B, C, C') = R(A'', B', C', C).$$

Hence

$$R(A', B', C', C) = R(A'', B', C', C)$$

or

$$R(B', C', C, A') = R(B', C', C, A'').$$

By Theorem 8.53 we conclude that $A'$ and $A''$ coincide.

We conclude this section by using the preceding theorem to prove a classical theorem known to Pappus on the involution property of a complete quadrangle.

**Theorem 8.77.** *Any line cuts the pairs of opposite sides of a complete quadrangle in pairs of points which correspond under an involution.*

PROOF. Let the pairs of points be $A$ and $A'$, $B$ and $B'$, and $C$ and $C'$ as indicated in Figure 8.14. If $T$ is the perspectivity from the line of these

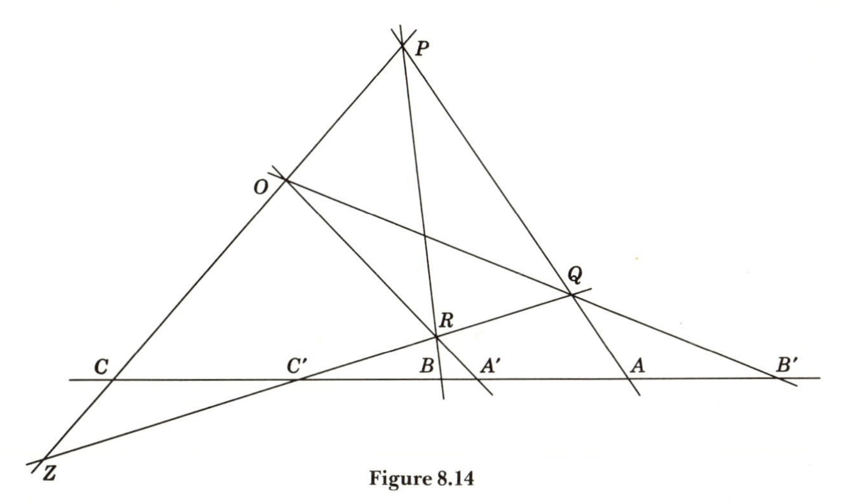


**Figure 8.14**

points to the line $QR$ with center $P$, we have $T(A, B, C, C') = Q, R, Z, C'$. Now if $S$ is the perspectivity from $QR$ back to the original line with center $O$, we have $S(Q, R, Z, C') = B', A', C, C'$. We thus have a projectivity of the line onto itself given by $ST$ where $[ST](A, B, C, C') = B', A', C, C'$. If we follow $ST$ by an involution under which $B'$ and $A'$, and $C$ and $C'$ correspond, we get a transformation $U$ for which $U(A, B, C, C') = A', B', C', C$. By Theorem 8.52 we have

$$R(A, B, C, C') = R(A', B', C', C).$$

Hence by Theorem 8.76 the pairs correspond under an involution.

## Exercises 8.7

1. Determine whether each of the following is the matrix of an elliptic, parabolic, or hyperbolic transformation:

*a.* $\begin{pmatrix} 4 & 1 \\ -1 & 2 \end{pmatrix}$;

*b.* $\begin{pmatrix} 4 & 1 \\ 1 & 2 \end{pmatrix}$;

*c.* $\begin{pmatrix} 1 & -4 \\ 2 & 1 \end{pmatrix}$.

2. Determine the matrix of the involution for each of the following pairs of corresponding points:

*a.* $(2, 1), (3, 2)$ and $(-1, 4), (1, 1)$;

*b.* $(3, 5), (4, 1)$ and $(-2, 7), (2, 3)$;

*c*. $(4,-5)$, $(1,2)$ and $(5,-4)$, $(2,1)$;

*d*. $(2,1)$, $(3,2)$ and $(-1,4)$, $(-4,4)$;

*e*. $(2,1)$, $(2,1)$ and $(-1,4)$, $(-1,4)$.

3. Verify the nonsingularity of the matrix $(a_{ij})$ of Theorem 8.74.

4. Prove that if $T$ is a hyperbolic projectivity of a line onto itself with fixed points $M$ and $N$, then $R(M,N,A,T[A])$ is a constant; that is, it is independent of $A$.

5. Suppose $T$ and $S$ are projectivities of a line onto itself. Each of these may be elliptic, parabolic, or hyperbolic. What can be concluded about the classification of $ST$ as elliptic, parabolic, or hyperbolic?

6.# If the ideal point of a line is not a fixed point for an involution, it corresponds to an ordinary point, called the central point of the involution. Prove that the product of the distances of two corresponding points to the central point is a constant.

7. Prove that a projectivity of a line onto itself has period three if and only if $a_{11}{}^2+a_{22}{}^2=-a_{11}a_{22}-a_{12}a_{21}$.

8. Prove that there are no periodic parabolic projectivities.

9. Prove that every parabolic or hyperbolic projectivity can be represented as the product of two perspectivities.

10. Prove that any projectivity of a line onto itself can be represented as the product of two involutions.

## 8.8 OTHER PROJECTIVE TRANSFORMATIONS

In our considerations of projective transformations we have thus far restricted ourselves to transformations between points of coplanar lines, although the reader was invited in Exercise 5.49 to consider the plane dual of this type of transformation. The analytic representation of a projective transformation was found to be the matrix relation $kX=AX'$. This representation certainly suggests possible generalizations of the idea of projective transformation, for there is nothing in this relation to require that the projectively related lines be coplanar. Indeed, even the definition of projectivity as a product of central perspectivities need not have required this. There is seemingly no restriction that requires $X$ and $X'$ to represent points on projective lines. They presumably could represent points in the plane or points in space, in which cases the transformation would be between planes of points or of a space of points onto itself. Finally, the matrices $X$ and $X'$ might well be associated with line or plane coordinates rather than with point coordinates. In this section we briefly outline extensions of the idea of projective transformation suggested by the potential generality of the relation $kX = AX'$.

In our previous discussion the matrices $X$ and $X'$ were column matrices whose two elements were homogeneous coordinates of points on a line. We note that the duals of the set of points on a line are a *pencil of lines* (plane dual) and a *pencil of planes* (space dual). By Theorem 8.14 we know that the lines of the pencil are of the form $c_1l_1 + c_2l_2$, where $l_1$ and $l_2$ are distinct lines of the pencil. By duality with Definition 8.34 we would define $(c_1, c_2)$ to be *homogeneous coordinates for the pencil of lines.*

Thus, to obtain homogeneous coordinates for the pencil of lines through the point $(0, 1, -1)$, we first seek two distinct lines through the point. The lines $[2, 2, 2]$ and $[2, 1, 1]$ are such since $(0, 1, -1)$ satisfies

$$2x_1 + 2x_2 + 2x_3 = 0$$

and

$$2x_1 + x_2 + x_3 = 0.$$

Now any line of the pencil must be of the form

$$c_1[2, 2, 2] + c_2[2, 1, 1]$$

and its homogeneous coordinates, *as a member of the pencil,* will be $(c_1, c_2)$. Thus the line

$$3[2, 2, 2] - 2[2, 1, 1] = [2, 4, 4]$$

has homogeneous coordinates $(3, -2)$ as a member of the pencil. The unit line $(1, 1)$ is

$$[2, 2, 2] + [2, 1, 1] = [4, 3, 3].$$

Similarly the pencil through the point $(10, -1, -4)$ can be written in the form

$$c_1'[-1, 2, -3] + c_2'[1, 2, 2]$$

with homogeneous coordinates $(c_1', c_2')$. The relation

$$k\begin{pmatrix} c_1 \\ c_2 \end{pmatrix} = \begin{pmatrix} 1 & 0 \\ 0 & 1 \end{pmatrix}\begin{pmatrix} c_1' \\ c_2' \end{pmatrix}$$

then determines a transformation of one pencil onto the other in which the line $[2, 2, 2]$ is mapped onto the line $[-1, 2, -3]$, the line $[2, 1, 1]$ is mapped onto the line $[1, 2, 2]$, and the unit line $[4, 3, 3]$ is mapped onto the unit line $[0, 4, -1]$.

In a similar manner the space dual of Theorem 8.13 asserts that the planes in a pencil are of the form $c_1\pi_1 + c_2\pi_2$, where $\pi_1$ and $\pi_2$ are distinct planes of the pencil, and we take $(c_1, c_2)$ to be *homogeneous coordinates in the pencil of planes*. The relation $kX = AX'$ could define a transformation between any pair of sets each of which is either a line of points, a pencil of lines, or a pencil of planes. The matrices $X$ and $X'$ need not represent the same types of configurations; for example, $kX = AX'$ could be interpreted as a transformation of the line of points with coordinates $(x_1', x_2')$ onto the pencil of lines with coordinates $(x_1, x_2)$, a point being mapped onto a line.

Suppose now that $X$ and $X'$ contain three elements. They could then represent coordinates for the points in a plane or for the lines in a plane. By reasoning similar to that above we could conclude they could also represent homogeneous coordinates in a *bundle of planes* or a *bundle of lines* (set of planes or lines in space through a common point), these being the space duals of a plane of points and a plane of lines. Now $kX = AX'$ could define a transformation between any pair of sets, each of which is a plane of points, a plane of lines, a bundle of planes, or a bundle of lines.

Finally, we note that if $X$ and $X'$ are 4-tuples, they represent coordinates for the points or planes in space. In this case $kX = AX'$ could represent a transformation between a pair of sets, each of which is the space of points or the space of planes.

The foregoing discussion leads us to the following definitions:

**Definition 8.81.** *A line of points, a pencil of lines, and a pencil of planes are* one-dimensional primitive forms. *A plane of points, a plane of lines, a bundle of planes, and a bundle of lines are* two-dimensional primitive forms. *The space of points and the space of planes are* three-dimensional primitive forms.

**Definition 8.82.** *A* projective transformation *is a mapping of the elements of an n-dimensional primitive form onto the elements of an n-dimensional primitive form* ($n = 1, 2,$ *or* $3$) *in which homogeneous coordinates are related by* $kX = AX'$, *A nonsingular.*

A primitive form of dimension greater than one contains primitive forms of lower dimension. Thus a plane of points contains lines of points. It is readily verified that a projective transformation between higher dimensional primitive forms maps these lower dimensional primitive forms onto primitive forms of the same dimension. If $T$ maps the plane of points $\pi$ onto the plane of points $\pi'$, every line of points in $\pi$ is mapped onto a line of points in $\pi'$. If $S$ maps the plane of points $\pi$ onto the plane of lines $\pi'$, it maps every line of points on $\pi$ onto a pencil of lines on $\pi'$.

We note that the argument of Theorem 8.44 is still applicable in the more general use of the term "projective transformation," and that we can generalize this theorem to

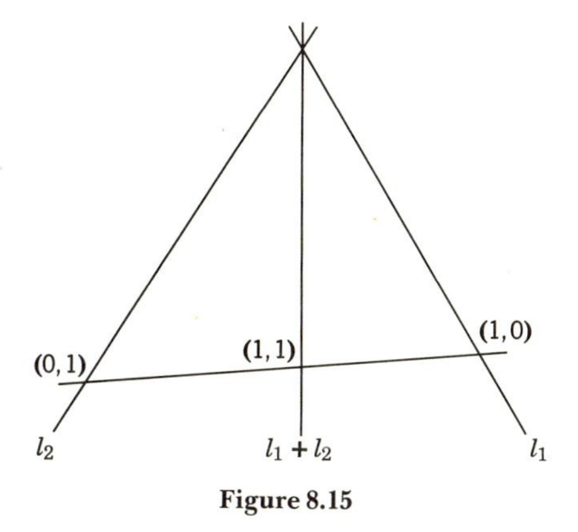


**Figure 8.15**

**Theorem 8.81 (Fundamental Theorem of Projective Geometry).** *There exists one and only one projective transformation mapping $n+2$ elements of an n-dimensional primitive form, any $n+1$ of which are independent, onto $n+2$ distinct elements of an n-dimensional primitive form, any $n+1$ of which are independent, in a given order.*

We consider an example of a more general projective transformation which we use in the next chapter. The relation between the lines of a pencil and the points on a line in the plane of the pencil in which each point on the line corresponds to the line of the pencil on which it lies is a projectivity. Suppose we have assigned homogeneous coordinates on the line. If we let $l_1$ contain the point $(1, 0)$, $l_2$ the point $(0, 1)$, and $l_1 + l_2$ the point $(1, 1)$, the homogeneous coordinates $(c_1, c_2)$ in the pencil associated with the representation $c_1 l_1 + c_2 l_2$ will be such that the point $(c_1, c_2)$ lies on the line $(c_1, c_2)$ of the pencil. The matrix of the transformation is the identity matrix.

The plane duals of the cross-ratio results of Section 8.5 should now be

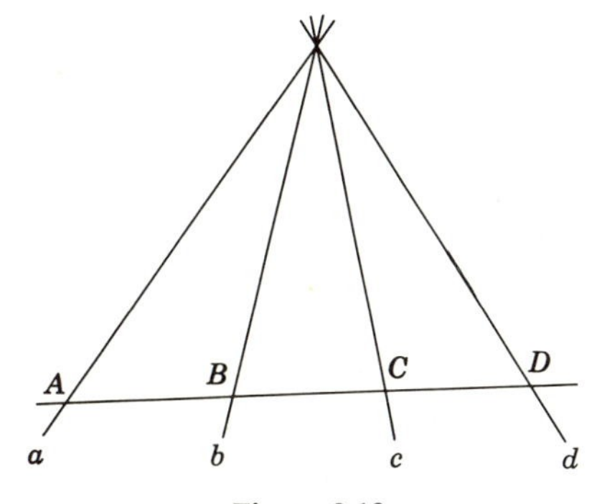


**Figure 8.16**

apparent in terms of the homogeneous coordinates of the lines of a pencil of lines. The above transformation indicates the following relation between the cross ratio of points and of lines.

**Theorem 8.82.** *If concurrent lines $a$, $b$, $c$, and $d$ meet a line not through the point of concurrence in points $A$, $B$, $C$, and $D$, respectively,* $R(A,B,C,D) = R(a,b,c,d)$.

## Exercises 8.8

1. Given a line of points and a pencil of planes whose axis does not meet the given line, show that the correspondence associating each point of the line with the plane of the pencil on which it lies is a projectivity.

2. Given a plane of lines and a bundle of planes whose center is not on the given plane, show that the correspondence associating each line of the plane with the plane of the bundle in which it lies is a projectivity.

3. Given a plane of points and a bundle of lines whose center is not on the given plane, show that the correspondence associating each point of the plane with the line of the bundle on which it lies is a projectivity.

## REFERENCES

Adler, Claire Fisher, *Modern Geometry, An Integrated First Course*, 2nd ed., McGraw-Hill Book Company, New York, 1967, Chapters 4, 5, 7 to 10.

Artzy, Rafael, *Linear Geometry*, Addison-Wesley Publishing Company, Reading, Mass., 1965, Chapter 3.

Busemann, Herbert, and Paul J. Kelly, *Projective Geometry and Projective Metrics*, Academic Press, New York, 1953, Chapter 1.

Coxeter, H. S. M., *Non-Euclidean Geometry*, 5th ed., University of Toronto Press, Toronto, 1965, Chapter 4.

Gans, David, *Transformations and Geometries*, Appleton-Century-Crofts, New York, 1969, Chapters 5, 6, 8, 9.

Graustein, William C., *Introduction to Higher Geometry*, The Macmillan Company, New York, 1945, Chapters 3, 5, 19.

Gruenberg, K. W., and A. J. Weir, *Linear Geometry*, D. Van Nostrand Company, Princeton, 1967, Chapter 2.

Klein, Felix, translated by E. R. Hedrick and C. A. Noble, *Elementary Mathematics from an Advanced Standpoint*, vol. 2, Dover Publications, New York, 1939, Part Two, II.

Levy, Harry, *Projective and Related Geometries*, The Macmillan Company, New York, 1961, Chapters 2, 3.

Maxwell, E. A., *The Methods of Plane Projective Geometry Based on the Use of General Homogeneous Coordinates*, Cambridge University Press, Cambridge, 1960, Chapters 1 to 3.

Modenov, P. S., and A. S. Parkhomenko, translated by Michael B. P. Slater, *Geometric Transformations*, vol. 2: *Projective Transformations*, Academic Press, New York, 1965, Chapter 1.

O'Hara, C. W., and D. R. Ward, *An Introduction to Projective Geometry*, Oxford University Press, New York, 1937, Chapters 7, 8.

Rainich, G. Y., and S. M. Dowdy, *Geometry for Teachers*, John Wiley and Sons, New York, 1968, Chapter 3.

Rosenbaum, Robert A., *Introduction to Projective Geometry and Modern Algebra*, Addison-Wesley Publishing Company, Reading, Mass., 1963, Chapters 5, 6.

Seidenberg, A., *Lectures in Projective Geometry*, D. Van Nostrand Company, Princeton, 1962, Chapter 8.

Springer, C. E., *Geometry and Analysis of Projective Spaces*, W. H. Freeman and Company, San Francisco, 1965, Chapters 1, 2, 4 to 7.

Struik, Dirk J., *Lectures on Analytic and Projective Geometry*, Addison-Wesley Publishing Company, Reading, Mass., 1953, Chapters 1 to 3.

Tuller, Annita, *A Modern Introduction to Geometries*, D. Van Nostrand Company, Princeton, 1967, Chapter 5.

# CHAPTER NINE

# CONICS

In Euclidean geometry the loci studied after lines and planes are usually the conic sections in plane geometry and the quadric surfaces in solid geometry. This is true even in secondary school geometry where we usually confine ourselves to the particular conics or quadrics known as circles and spheres, respectively. From the analytic point of view the conics follow naturally after the study of lines since they are the loci having second-degree equations. In projective geometry we have considered loci whose equations are first degree, and it is natural to turn next to loci defined by second-degree equations. These are indeed the conics and quadric surfaces. The conics are often defined by synthetic methods which are probably new to the reader. We introduce the conics in a synthetic manner and in Section 9.4 show their relation to second-degree equations.[1]

## 9.1 SYNTHETIC INTRODUCTION TO THE CONICS

We have seen in Section 8.8 that projectivities can be generalized to correspondences between sets other than the points of two coplanar lines. In particular we can have a projectivity between two coplanar pencils of

[1]The focus-directrix definition of a conic section or the constant-sum-of-the-distances-to-the-foci definition of an ellipse are Euclidean. The Euclidean synthetic development dates back to the Greeks, but the projective approach to be considered here is due to Steiner (1796 to 1863).

lines. We recall that the case of a projectivity between two such pencils in which corresponding lines meet in collinear points is that of an *axial perspectivity* (the plane dual of central perspectivity) with the line being called the *axis* of the perspectivity (see Exercise 5.49). Keeping this in mind, we make

**Definition 9.11.** *A* point conic *is the set of points of intersection of corresponding lines of two projectively related coplanar pencils of lines. If the pencils are distinct and the projectivity is not an axial perspectivity, the point conic is* nondegenerate; *otherwise the point conic is* degenerate.

Figure 9.1 shows a nondegenerate point conic determined by the projectivity between the two pencils in which $l_1$, $l_2$, and $l_3$ correspond to $l'_1$, $l'_2$, and $l'_3$, respectively.

If $P$ and $P'$ are centers of distinct pencils determining a point conic, the line $PP'$ is a member of each pencil and must meet its corresponding lines in each of the pencils at the points $P$ and $P'$. If the points $P$ and $P'$ coincide, it is clear that every pair of corresponding lines meet at least at this point. Thus we have proved

**Theorem 9.11.** *The center or centers of the pencils determining a point conic are on the point conic.*

Let us consider the degenerate conics and suppose first that the two pencils coincide. If the perspectivity is the identity, the intersections consist of all points on all lines of the pencil; thus the conic contains all points of the plane [Figure 9.2(*a*)]. If the projectivity is not the identity,

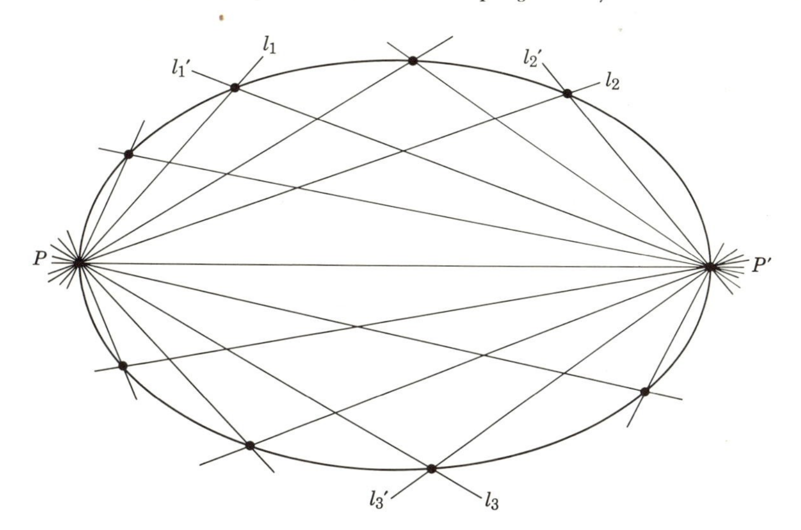


**Figure 9.1**

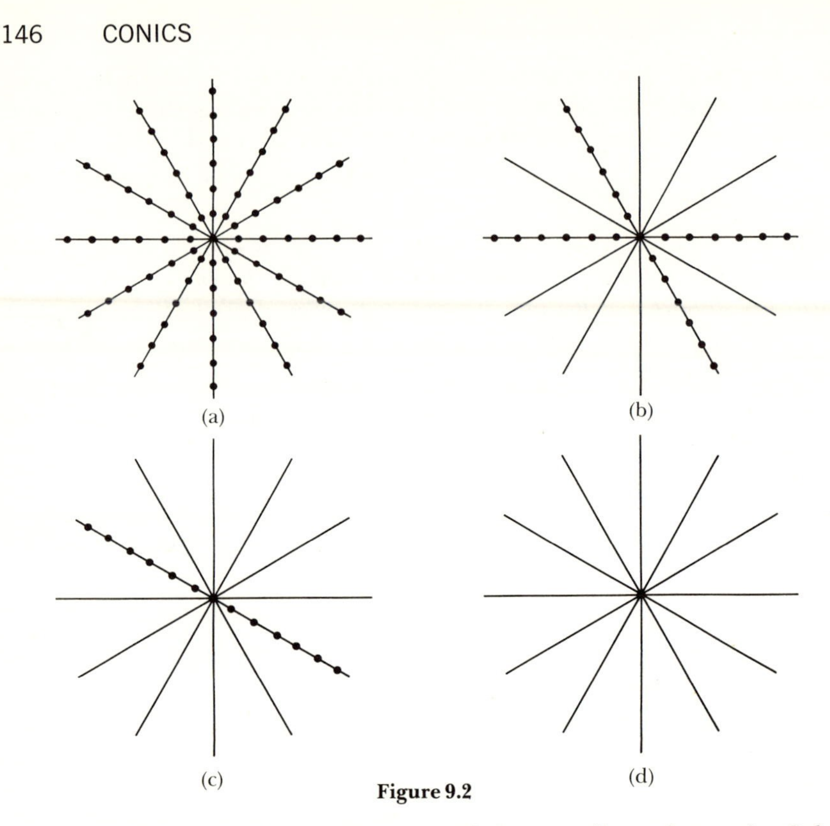


**Figure 9.2**

we certainly have at least the center of the pencil on the conic. Other points can be only those on self-corresponding lines under the projectivity, of which there will be two, one, or none. Thus we may have two lines [Figure 9.2(*b*)], one line [Figure 9.2(*c*)], or a point [Figure 9.2(*d*)]. Suppose now that we have distinct pencils and an axial perspectivity. In this case the points will be the points on the axis of perspectivity and the points of the line joining the centers of the pencils, since this line must correspond to itself under an axial perspectivity (Figure 9.3). We thus have

**Theorem 9.12.** *A degenerate point conic is a plane of points, the set of points on a pair of lines, the points on a line, or a point.*

It is clear from Definition 9.11 that a conic is a plane figure; in this chapter we thus consider plane projective geometry exclusively. We need to carry over plane duality by introducing a dual to Definition 9.11.

**Definition 9.12.** *A* line conic *is the set of lines joining corresponding points of two projectively related coplanar lines of points. If the lines are distinct and*

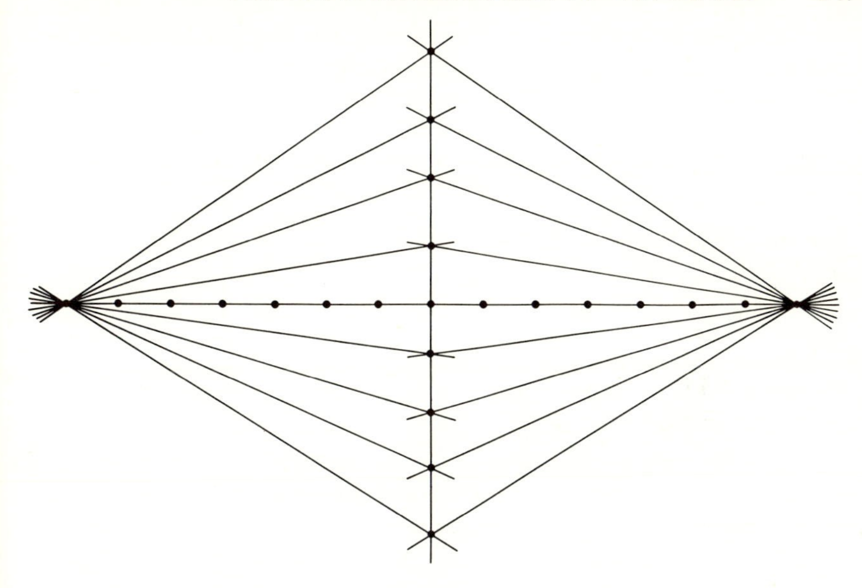

**Figure 9.3**

*the projectivity is not a central perspectivity, the line conic is* nondegenerate; *otherwise it is* degenerate.

Figure 9.4 shows a nondegenerate line conic determined by the projectivity in which $P_1$, $P_2$, and $P_3$ correspond to $P_1'$, $P_2'$, and $P_3'$, respectively. It should be emphasized that a line conic is a set of lines, not the

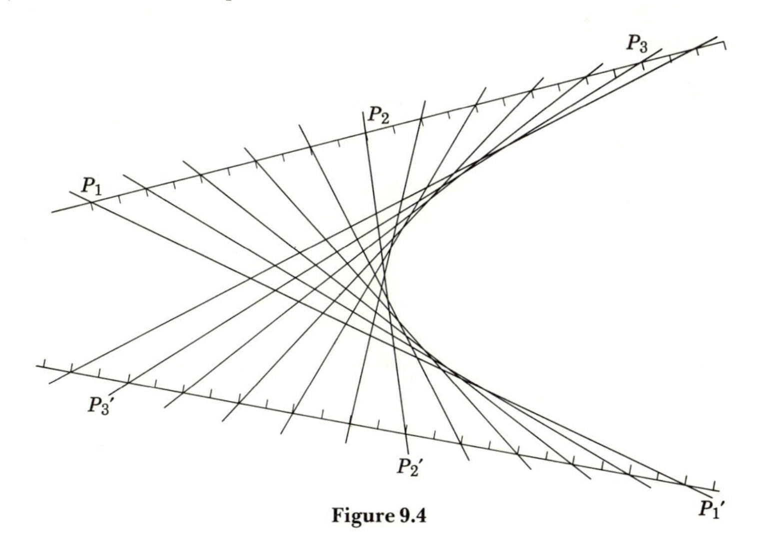


**Figure 9.4**

"curve" these lines seem to outline. This curve seems to be a point conic, and it is a reasonable conjecture that the elements of a line conic are the tangents of a point conic. We prove in Section 9.3 that this is indeed the case.

With Definition 9.12 made we can now write the duals of Theorems 9.11 and 9.12. We have

**Theorem 9.13.** *The projectively related lines used to determine a line conic are on the line conic.*

Thus in Figure 9.4 the lines $P_1P_3$ and $P_1'P_3'$ are elements of the line conic.

**Theorem 9.14.** *A degenerate line conic is a plane of lines, a pair of pencils of lines, a pencil of lines, or a line.*

It is by no means clear at this point that conic as defined here is the familiar ellipse, parabola, or hyperbola (extended from the Euclidean to the projective plane), although the reader will find, if he draws point conics, that they always seem to be such. That this is true will become more reasonable when we obtain an analytic representation of a conic. In Section 11.2 we show the precise relation between the Euclidean conics and the projective conics as defined here.

We can prove one theorem at this point indicating the "second-degree" nature of a conic in the following sense.

**Theorem 9.15.** *A line meets a nondegenerate point conic in at most two distinct points.*

PROOF. If a line met the conic in three distinct points, these points would be the points of intersection of three distinct pairs of lines in the two pencils. These pairs determine a unique projectivity by the Fundamental Theorem, but they determine an axial perspectivity with the given line as axis. This contradicts the assumption of nondegeneracy.

**Corollary 9.151.** *If all points of a line are not on a point conic, the line meets the point conic in at most two distinct points.*

This is merely a restatement of the theorem using the enumeration of Theorem 9.12.

By duality we have

**Theorem 9.16.** *A point lies on at most two distinct lines of a nondegenerate line conic.*

## Exercises 9.1

1. Show the types of projective transformation leading to each of the degenerate line conics of Theorem 9.14.

2. Prove that if five distinct points on a conic are such that no three are collinear, the conic is nondegenerate.

3. *a.* Show that the two pencils used to define a nondegenerate point conic induce a projectivity $T$ of any line of the plane (except the line joining their centers) onto itself as follows: if $l$ and $l'$ correspond in the two pencils and meet the line at $P$ and $P'$, respectively, then $T(P) = P'$.

*b.* Show this transformation is hyperbolic, parabolic, or elliptic as the line meets the conic in two, one, or no points.

## 9.2 PASCAL'S THEOREM

Pascal's theorem was first proved by Pascal in 1640 by methods considerably different from those we use here.[2] We first prove a very special version of the theorem, then use this to prove the theorem itself and two other important relations about conics.

**Theorem 9.21.** *Given distinct points $A$, $B$, $C$, $D$, $E$, and $F$, no three of which are collinear, let $P$ be the intersection of $AE$ and $CF$, $Q$ the intersection of $AD$ and $CB$, and $R$ the intersection of $BE$ and $DF$. Then $C$, $D$, $E$, and $F$ are on a nondegenerate point conic determined by projectively related pencils centered at $A$ and $B$ if and only if $P$, $Q$, and $R$ are collinear.*

PROOF. We suppose $C$, $D$, $E$, and $F$ are distinct points, all distinct from $A$ and $B$, on a nondegenerate conic determined by pencils centered at $A$ and $B$. Let $BC$ meet $FD$ at $L$, $AC$ meet $FD$ at $M$, and $AE$ meet $FD$ at $N$.

[2]Pascal proved the theorem by proving it for a circle, then showing that the result remained valid under a projective transformation of a plane onto a plane which mapped the circle onto an arbitrary conic.

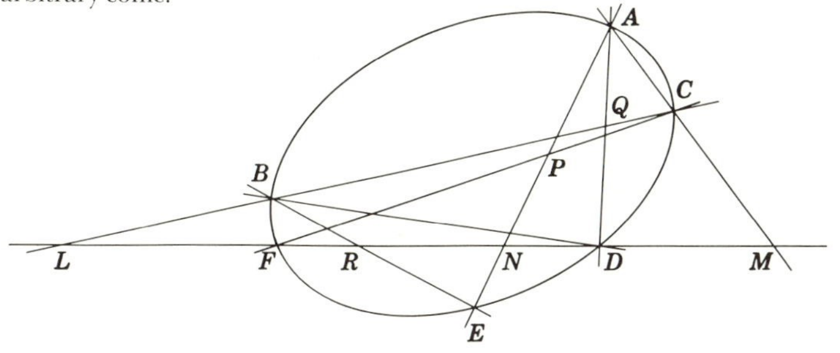


**Figure 9.5**

We know that

$$R(AC, AD, AE, AF) = R(BC, BD, BE, BF) \tag{9.21}$$

since the pencils centered at $A$ and $B$ are projectively related and since the lines involved are distinct and well determined by hypothesis. By Theorem 8.82 we have

$$R(AC, AD, AE, AF) = R(M, D, N, F) \tag{9.22}$$

and

$$R(BC, BD, BE, BF) = R(L, D, R, F), \tag{9.23}$$

whence

$$R(M, D, N, F) = R(L, D, R, F). \tag{9.24}$$

Now if $T$ is the projectivity of $DF$ onto itself for which $T(M, D, F) = L, D, F$, we must have $T(N) = R$ to preserve cross ratio. The points $M, D$, and $F$ are carried onto the points $L, D$, and $F$ by successive central perspectivities, the first from $DF$ to $CF$ with center $A$, and the second from $CF$ back to $DF$ with center $Q$ (the hypothesis that no three of the points be collinear implies that $CF$ and $DF$ are distinct and that neither $A$ nor $Q$ is on either of these lines). By the Fundamental Theorem the product of these perspectivities must be $T$, from which these perspectivities must carry $N$ to $P$, then $P$ back to $R$. This requires that $P$ and $R$ be on a line with $Q$.

Conversely, if $P$, $Q$, and $R$ are collinear, this product of perspectivities leads to a projectivity $T$ with $T(M, D, N, F) = L, D, R, F$. We thus know that equations (9.22) to (9.24) hold, from which equation (9.21) also holds. We now apply the same reasoning to the pencils at $A$ and $B$ that we previously applied to the points on $DF$. In the projectivity $S$ for which $S(AC, AD, AE) = BC, BD, BE$ we must have $S(AF) = BF$ in view of equation (9.21) and the Fundamental Theorem. Hence $C, D$, $E$, and $F$ are on a conic determined by pencils centered at $A$ and $B$. The nondegeneracy of this conic follows from the assumption of noncollinearity of any three of the points (see Exercise 9.12). This completes the proof of the theorem.

In the statement of Theorem 9.21 there is no change in the determination of the points $P$, $Q$, and $R$ if we interchange $A$ and $C$, $B$ and $D$, and $E$ and $F$. Thus if $F$ is on the conic determined by pencils at $A$ and $B$ and passing through the points $C$, $D$, and $E$, the points $P$, $Q$, and $R$ are collinear. This in turn will imply by the theorem that $F$ is on the conic with centers of pencils at $C$ and $D$ and passing through the points $A$, $B$, and $E$. Since $C$ and $D$ could be any points on the conic determined by the pencils at $A$ and $B$, a point conic can be determined by any pair of points on it and a set of projectively related pencils through these points.

In this argument we have assumed $A$, $B$, $C$, and $D$ are all distinct. Actually this need not be so as far as our final conclusion is concerned. If we wish to change from pencils at $A$ and $B$ to pencils at $A$ and $B'$, we may change from pencils at $A$ and $B$ to pencils at $C$ and $D$, then from pencils at $C$ and $D$ to pencils at $A$ and $B'$. We have proved

**Theorem 9.22 (Steiner's theorem).** *A nondegenerate point conic can be determined as the points of intersection of two projectively related pencils with centers at arbitrary distinct points on the conic.*

Theorem 9.22 in turn leads to a simple proof of the "five-point" theorem on conics which the reader may have encountered in his analytic geometry study of the conic sections.

**Theorem 9.23.** *There exists one and only one nondegenerate conic passing through five distinct points, no three of which are collinear.*

PROOF. Let the points be $A$, $B$, $C$, $D$, and $E$. By Theorem 9.22 we may determine any conic through $A$ and $B$ using pencils centered at these points. If no three of the points are collinear, we may be sure that $AC$, $AD$, and $AE$ are distinct and that $BC$, $BD$, and $BE$ are distinct. By the Fundamental Theorem there is one and only one projectivity $T$ between the pencils for which $T(AC, AD, AE) = BC, BD, BE$; hence there is one and only one point conic passing through the five points, the conic being nondegenerate in view of the noncollinearity assumption.

Before we state Pascal's theorem, we must reconsider the concept of polygon. The complete quadrangle is clearly not the same as the quadrilateral of Euclidean geometry. We can define the equivalent of the Euclidean quadrilateral in the projective plane although sides will have to be lines rather than line segments. The polygons of Euclidean geometry generalize to the *simple* polygons of projective geometry, which we now define.

**Definition 9.21.** A simple $n$-sided polygon *is the configuration consisting*

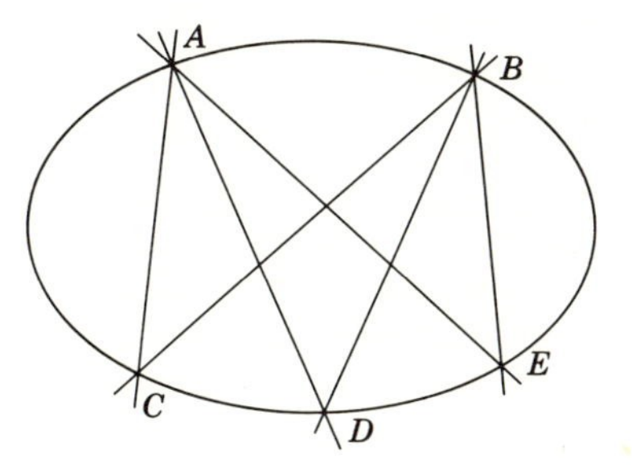


**Figure 9.6**

*of n points, $P_1, P_2, \cdots, P_n$, no three consecutive ones of which are collinear, and the n lines $P_1P_2, P_2P_3, \cdots, P_nP_1$.*

In a simple polygon the order in which the vertices are given is pertinent, for the sides are the lines joining consecutive vertices (the order is cyclic; that is, $P_2$ follows $P_1$, $P_3$ follows $P_2, \cdots, P_1$ follows $P_n$); in a complete polygon all pairs of vertices are used to determine sides. It is apparent that a simple polygon is a self-dual figure.

Let us now reexamine Theorem 9.21 in the light of Theorem 9.22. In view of the latter the points $A$ and $B$ in the statement of Theorem 9.21 do not occupy preferred positions, and the theorem can be interpreted as a statement about six points on a conic. These points determine a simple hexagon inscribed in the conic; in fact they determine many simple hexagons inscribed in the conic depending on the order in which the vertices are given. If we consider hexagon $ADFCBE$, shown in Figure 9.7, which is Figure 9.5 with the extraneous lines omitted, we see that the points $P$, $Q$, and $R$ are the intersections of the opposite sides of the hexagon. We can now state Theorem 9.21 in the form which is known as Pascal's theorem.

**Theorem 9.24 (Pascal's theorem).** *If a simple hexagon is inscribed in a nondegenerate point conic, the points of intersection of its opposite sides are collinear. If a simple hexagon has no three of its vertices collinear and if the points of intersection of its opposite sides are collinear, it is inscribed in a nondegenerate point conic.*

In Figure 9.8 are shown two other simple hexagons determined by the points on the conic of Figure 9.5.[3] It should be remarked that the theorem of Pappus is Pascal's theorem for a degenerate conic consisting of the points on two lines.

The dual of Pascal's theorem was first proved in 1806 by Brianchon without use of duality and carries its discoverer's name.

**Theorem 9.25 (Brianchon's theorem).** *If a simple hexagon has elements of a nondegenerate line conic for its sides, the lines joining opposite vertices are concurrent. If a simple hexagon has no three of its sides concurrent and if the*

[3]The lines $PQR$ of Figure 9.8 are called *Pascal lines*, and the points $P$, $Q$, and $R$ are called *Pascal points*. Given six points on a conic, there are determined 60 simple hexagons, 45 Pascal points, and 60 Pascal lines. Each Pascal line actually contains four Pascal points. The Pascal lines meet by threes in 20 *Steiner points*, which lie by fours on 15 *Steiner-Plücker lines*. The Pascal lines also meet by threes in 60 *Kirkman points*, which lie by threes on 20 *Cayley-Salmon lines*, which in turn pass by fours through 15 points. The reader will find it interesting and frustrating to draw the entire configuration for a given six points on the conic and keep the entire figure on a single sheet of paper.

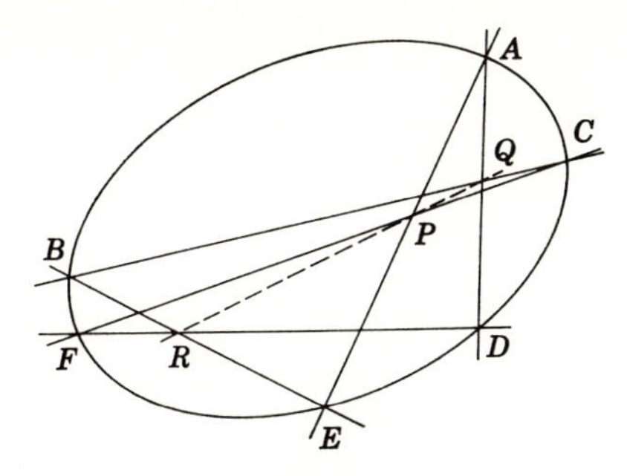


**Figure 9.7**

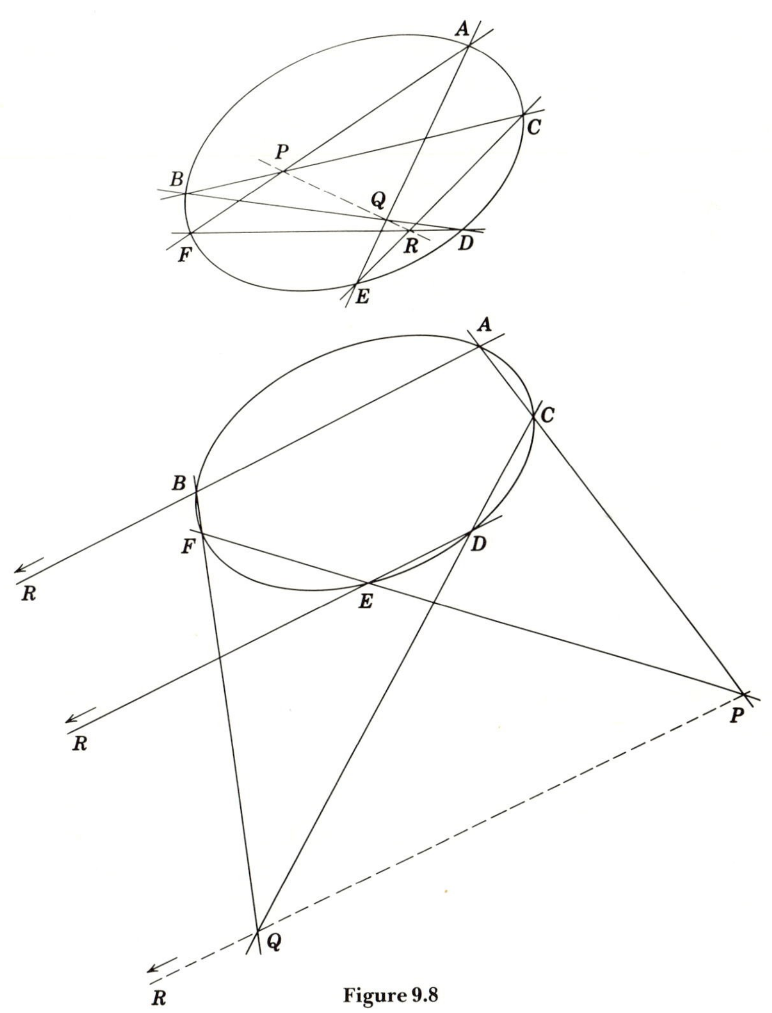


**Figure 9.8**

*lines joining opposite vertices are concurrent, the sides of the hexagon are elements of a nondegenerate line conic.*

## Exercises 9.2

1. State the duals of Theorems 9.21 and 9.22.

2. What conics are determined by five collinear points? By five points, four of which are collinear? By five points, three of which are collinear?

3. Prove that if four points $A$, $B$, $C$, and $D$ are on a non-degenerate conic and if the lines joining them to a point $P$ on the conic are harmonically related, the relation holds for all positions of $P$ on the conic.

4. Prove that if a line meets a nondegenerate conic in two points, these points correspond under an involution together with the points in which any inscribed complete quadrangle meets the given line.

## 9.3 TANGENTS AND POINTS OF CONTACT

We now investigate the relation between point and line conics. Our goal is proof of the fact that a line conic does indeed consist of the set of tangents to a point conic. The definition given in calculus in terms of a limiting process for a tangent is not appropriate in a synthetic study of conics. Since any line not on a conic meets it in at most two points, the familiar elementary geometry definition of tangent to a circle can be generalized and used here. We have

**Definition 9.31.** *A line is a* tangent line *to a nondegenerate point conic if it meets the conic in exactly one point.*

The existence and uniqueness of tangent lines is shown in

**Theorem 9.31.** *There exists one and only one tangent line at each point of a nondegenerate point conic.*

PROOF. Let $P$ be an arbitrary point on the conic. By Steiner's theorem we may generate the conic using two pencils, one of which is centered at $P$. We claim the desired tangent is the line $l$ of the pencil passing through $P$ corresponding to the line $l'$ of the other pencil which passes through $P$. It is clear that no other line through $P$ can be a tangent since it will meet the conic at a second point where it meets its corresponding line of the other pencil. The line $l$ cannot meet the conic at a second point because it would then coincide with the member of the pencil through $P$ determining the other point on the conic and would corre-

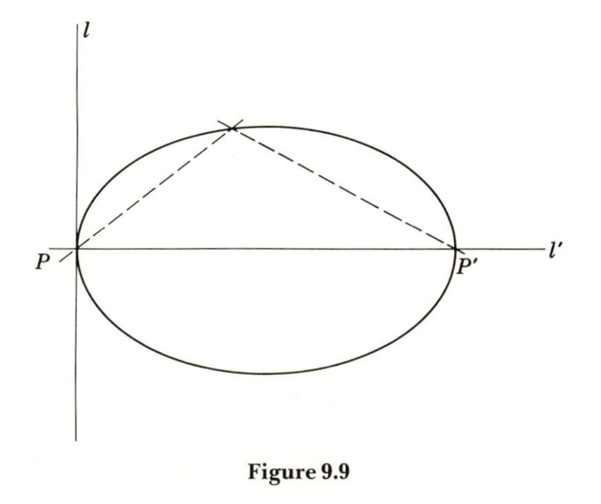


**Figure 9.9**

spond to two lines of the second pencil, contradicting the one-to-one property of a projectivity.

We must carry forward duality. The dual of tangent line is defined in

**Definition 9.32.** *A point is a* point of contact *of a nondegenerate line conic if it lies on exactly one line of the conic.*

These points of contact are the points in a line conic figure which seem to trace out a point conic; indeed, the dual of the conjecture that a line conic is the set of tangents of a point conic is the conjecture that a point conic is the set of points of contact of a line conic. By duality with Theorem 9.31 we have

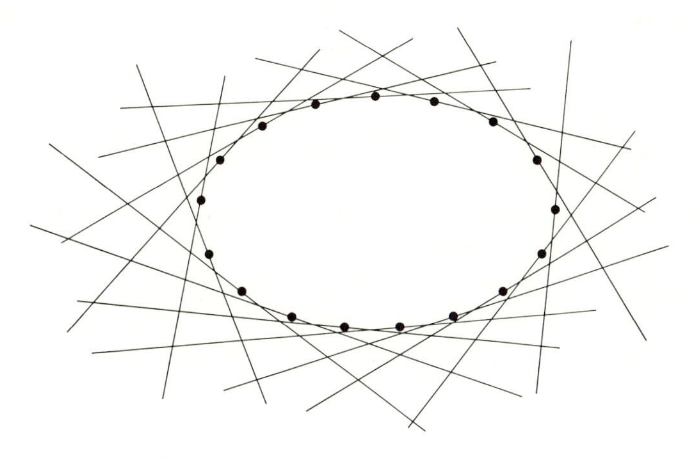

**Figure 9.10**

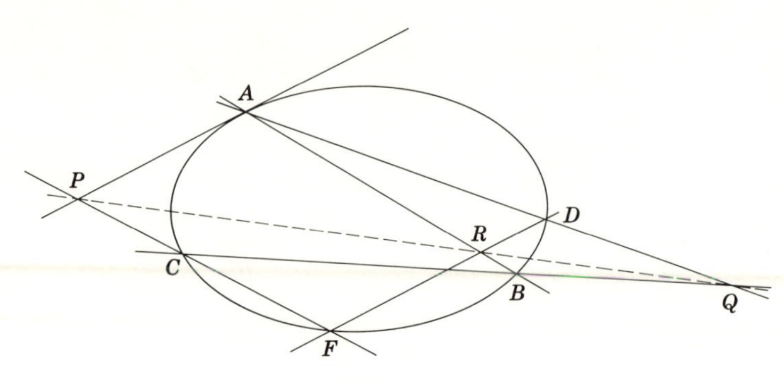


**Figure 9.11**

**Theorem 9.32.** *There exists one and only one point of contact on each line of a nondegenerate line conic.*

In the proof of Theorem 9.21 it was assumed that the points $A$, $B$, $C$, $D$, $E$, and $F$ were all distinct. The theorem is still valid if some of these points coincide. The lines $AC$, $AD$, $BC$, $BD$, etc., of equation (9.21) are lines understood to meet the conic at these points and only these points. If $A$ and $E$ coincide, we must replace $AE$ in equation (9.21) by the tangent line to the conic at $A$, but the argument in the proof remains valid. In this case the simple hexagon becomes a simple pentagon, and Pascal's theorem becomes

**Theorem 9.33.** *If a simple pentagon is inscribed in a nondegenerate point conic, the tangent at a vertex meets the opposite side in a point collinear with the points of intersection of the other two pairs of nonadjacent sides.*

**Theorem 9.34.** *If the sides of a simple pentagon are lines of a nondegenerate line conic, the line joining a point of contact on one side to the opposite vertex is concurrent with the lines joining the other two pairs of nonconsecutive vertices.*

Theorem 9.34 is the dual of Theorem 9.33.

Further specialization is possible. If in Theorem 9.21 $A$ and $E$ coincide and $B$ and $C$ coincide, the simple hexagon becomes a simple quadrangle, $AE$ becomes the tangent at $A$, and $BC$ becomes the tangent at $B$, but the proof is still valid. We thus have

**Theorem 9.35.** *If $A$, $B$, $D$, and $F$ are vertices of a simple quadrangle inscribed in a nondegenerate point conic, the tangent at $A$ and side $BF$, the tangent at $B$ and side $AD$, and the sides $AB$ and $DF$ meet in collinear points.*

By duality we have

**Theorem 9.36.** *If $a$, $b$, $d$, and $f$ are sides of a simple quadrangle which are elements of a nondegenerate line conic, the point of contact of $a$ and the inter-*

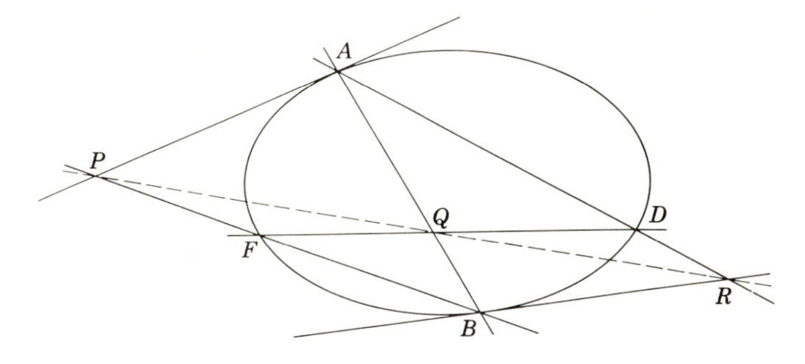


**Figure 9.12**

*section of b and f, the point of contact of b and the intersection of a and d, and the intersections of a and b and of d and f determine concurrent lines.*

We can also very readily get another quadrangle theorem if we let $A$ and $D$ coincide and $B$ and $C$ coincide in Theorem 9.21.

**Theorem 9.37.** *If $A$, $B$, $E$, and $F$ are the vertices of a simple quadrangle inscribed in a nondegenerate point conic, the tangents at $A$ and $B$, the tangents at $E$ and $F$, and the sides $AE$ and $BF$ meet in collinear points.*

PROOF. Let the points of intersection be $P$, $Q$, and $R$ as shown in Figure 9.13 and let $AF$ meet $BE$ at $S$. Theorem 9.21 here states, under the assumption of coincidence of $A$ and $D$ and of $B$ and $C$, that $P$, $Q$, and $S$ are collinear. If we interchange the role of $A$ and $F$ and of $B$ and $E$, we can conclude that $Q$, $R$, and $S$ are collinear by the same reasoning. The two lines so found are identical, having two common points, from which $P$, $Q$, and $R$ are collinear.

The dual of this theorem is

**Theorem 9.38.** *If $a$, $b$, $e$, and $f$ are sides of a simple quadrangle which are elements of a nondegenerate line conic, the points of contact of $a$ and $b$, the points of contact of $e$ and $f$, and the points of intersection of $a$ and $e$ and of $b$ and $f$ determine concurrent lines.*

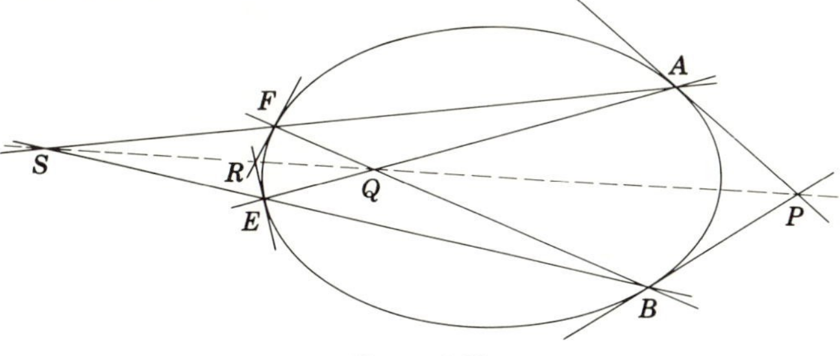


**Figure 9.13**

Theorems 9.35 and 9.37 are the main ones in the above set, for they are necessary in the proof of the main theorem of this section, which we may now consider.

**Theorem 9.39.** *The tangents to a nondegenerate point conic form a nondegenerate line conic.*

PROOF. Let $A$, $B$, $C$, and $P$ be points on a point conic, and let the tangents at these points meet in the points $M, N, R, S$, and $U$, as shown in Figure 9.14. Let $V$ be the intersection of $AB$ and $CP$, and let $W$ be the intersection of $AP$ and $BC$. We think of $A$, $B$, and $C$ as being fixed distinct points on the conic and $P$ as being an arbitrary point. The former points determine a projectivity $T$ between the tangent lines at $B$ and $C$, where $T(U, M, B) = C, N, U$, and the lines joining corresponding points are the tangents at $C, A$, and $B$. If we can show $T(S) = R$, the line joining $S$ and $P$ will be tangent at $P$, and since $P$ is arbitrary, we will have shown that the tangents are a line conic. If we apply Theorem 9.37 to the quadrangle $ACBP$, we conclude that $M$, $W$, and $R$ are collinear. Similar reasoning applied to the quadrangle $ABCP$ shows that $S$, $W$, and $N$ are collinear. If we apply Theorem 9.35 to the quadrangle $ABCP$, we see that $V$, $N$, and $W$ are collinear. We conclude that $M$, $W$, and $R$ are collinear and that $V$, $N$, $W$, and $S$ are collinear. We now seek the axis of homology (Theorem 5.52 and Definition 5.51) of the projectivity $T$. Since $UN$ and $MC$ meet at $C$, while $MU$ and $BN$ meet at $B$, we conclude $BC$ is this axis. Now $SN$ and $MT(S)$ must meet on $BC$. Since $SN$ meets $BC$ at $W$, $T(S)$ must be the intersection of $MW$ with the tangent at $C$. This point is $R$. We conclude that $T(S) = R$. This completes the proof of the theorem.

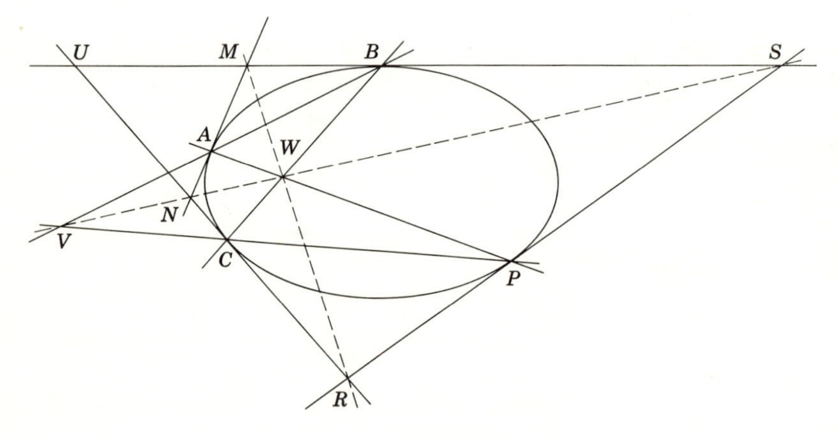


**Figure 9.14**

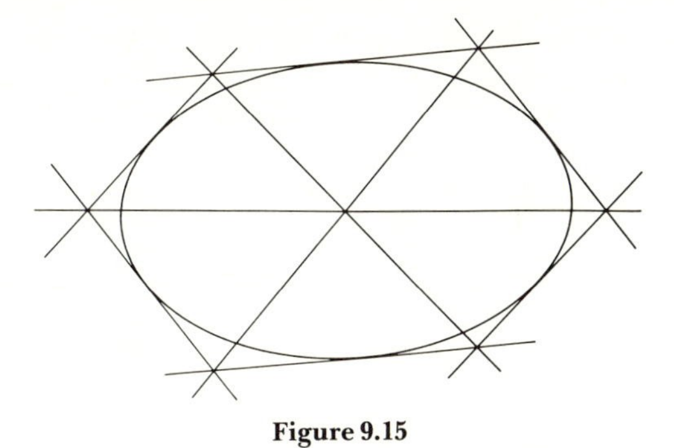

**Figure 9.15**

By duality we have also proved

**Theorem 9.310.** *The points of contact of nondegenerate line conic form a nondegenerate point conic.*

In view of the relation between line conic and point conic we can now restate Brianchon's theorem as

**Theorem 9.311 (Brianchon's theorem).** *The lines joining the opposite vertices of a simple hexagon circumscribed about a nondegenerate point conic are concurrent.*

Similar restatements in terms of figures circumscribed about a point conic can also be made for Theorems 9.34, 9.36, and 9.38.

## Exercises 9.3

1. Determine whether the converses of Theorems 9.33 to 9.38 are valid.

2. Restate Theorems 9.34, 9.36, and 9.38 in terms of figures circumscribed about a point conic.

3. Prove that if a triangle is inscribed in a point conic, the tangents to the conic at the vertices meet the opposite sides of the triangle in collinear points.

4. Prove that if a triangle is inscribed in a conic while a second triangle is circumscribed about the conic, having its sides tangent to the conic at the vertices of the first triangle, the two triangles are Desarguean.

5. Prove that if two triangles are circumscribed about a conic, they are also inscribed in a conic.

## 9.4 ANALYTIC REPRESENTATION OF A CONIC

Before proving that every conic has an equation of second degree, let us first consider a special example. In Section 8.8 we constructed a projectivity between the two pencils with centers at $(0,1,-1)$ and $(10,-1,-4)$, where coordinates of the pencils were $c_1[2,2,2]+c_2[2,1,1]$ and $c_1'[-1,2,-3]+c_2'[1,2,2]$, respectively; that is, lines of the first pencil have coordinates of the form

$$[2c_1+2c_2, 2c_1+c_2, 2c_1+c_2]$$

and lines of the second pencil have coordinates of the form

$$[-c_1'+c_2', 2c_1'+2c_2', -3c_1'+2c_2'].$$

We found that $(c_1, c_2)$ and $(c_1', c_2')$ could be taken as homogeneous coordinates in the two pencils, and the projectivity we developed had defining relations $kc_1=c_1'$, $kc_2=c_2'$. Let us now seek the equation of the point conic determined by the pair of pencils and the given projectivity between them. If a point $(x_1, x_2, x_3)$ is to lie on corresponding lines of the pencil for a particular $c_1$ and $c_2$, we must have

$$(2c_1+2c_2)x_1+(2c_1+c_2)x_2+(2c_1+c_2)x_3=0$$

and

$$(-c_1'+c_2')x_1+(2c_1'+2c_2')x_2+(-3c_1'+2c_2')x_3=0.$$

From these relations we find that

$$\frac{c_1}{c_2}=-\frac{2x_1+x_2+x_3}{2x_1+2x_2+2x_3}$$

and

$$\frac{c_1'}{c_2'}=-\frac{x_1+2x_2+2x_3}{-x_1+2x_2-3x_3}.$$

Now in view of the projectivity given, these ratios must be the same. We thus must have

$$(2x_1+x_2+x_3)(-x_1+2x_2-3x_3)=(2x_1+2x_2+2x_3)(x_1+2x_2+2x_3),$$

or

$$\begin{aligned}-2x_1^2+4x_1x_2-6x_1x_3+2x_2^2-x_2x_1-3x_2x_3-3x_3^2-x_3x_1+2x_3x_2\\ =2x_1^2+4x_1x_2+4x_1x_3+4x_2^2+2x_2x_1\\ +4x_2x_3+4x_3^2+2x_3x_1+4x_3x_2.\end{aligned}$$

We can write this in the form

$$4x_1^2+0x_1x_2+10x_1x_3+2x_2^2+3x_2x_1+7x_2x_3+7x_3^2+3x_3x_1+2x_3x_2=0.$$

Further simplification of this equation is possible, but we prefer to leave it in the above form for the moment.

We certainly have obtained a second-degree equation for the above point conic. Let us now generalize our analysis to an arbitrary conic.

We suppose the line coordinates in the two pencils are of the form

$$c_1[u_1, u_2, u_3] + c_2[v_1, v_2, v_3]$$

and

$$c_1'[u_1', u_2', u_3'] + c_2'[v_1', v_2', v_3']$$

and that the projectivity is given by $kc_1 = c_1'$, $kc_2 = c_2'$. Again, if a point $(x_1, x_2, x_3)$ is to lie on corresponding members of the pencils, we must have

$$(c_1u_1 + c_2v_1)x_1 + (c_1u_2 + c_2v_2)x_2 + (c_1u_3 + c_2v_3)x_3 = 0$$

and

$$(c_1'u_1' + c_2'v_1')x_1 + (c_1'u_2' + c_2'v_2')x_2 + (c_1'u_3' + c_2'v_3')x_3 = 0.$$

As in the special example we must have

$$\frac{c_1}{c_2} = \frac{c_1'}{c_2'} = -\frac{v_1x_1 + v_2x_2 + v_3x_3}{u_1x_1 + u_2x_2 + u_3x_3}$$
$$= -\frac{v_1'x_1 + v_2'x_2 + v_3'x_3}{u_1'x_1 + u_2'x_2 + u_3'x_3},$$

from which it follows that

$$b_{11}x_1^2 + b_{12}x_1x_2 + b_{13}x_1x_3 + b_{22}x_2^2 + b_{21}x_2x_1 + b_{23}x_2x_3$$
$$+ b_{33}x_3^2 + b_{31}x_3x_1 + b_{32}x_3x_2 = 0,$$

where

$$b_{ij} = u_iv_j' - u_j'v_i, \quad i, j = 1, 2, 3.$$

We conclude that a point conic does indeed have a homogeneous second-degree equation. In the equation above it need not be the case that $b_{ij} = b_{ji}$ when $i \neq j$. Of course, we could combine the terms $b_{ij}x_ix_j$ and $b_{ji}x_jx_i$. Instead, let us set

$$a_{ij} = a_{ji} = \frac{(b_{ij} + b_{ji})}{2}.$$

Then we can replace $b_{ij}x_ix_j + b_{ji}x_jx_i$ by $a_{ij}x_ix_j + a_{ji}x_jx_i$ where $a_{ij} = a_{ji}$. Thus we could write our example in the form

$$4x_1^2 + (\tfrac{3}{2})x_1x_2 + (\tfrac{13}{2})x_1x_3 + 2x_2^2 + (\tfrac{9}{2})x_2x_3$$
$$+ (\tfrac{3}{2})x_2x_1 + (\tfrac{13}{2})x_3x_1 + 7x_3^2 + (\tfrac{9}{2})x_3x_2 = 0.$$

In general we have shown that a point conic has equation of the form

$$\sum_{i,j=1}^{3} a_{ij}x_ix_j = 0,$$

where $a_{ij} = a_{ji}$.

The advantage of writing the equation in the symmetric form is that it permits us to introduce simple matrix notation in the equation. Let $A$ denote the matrix $(a_{ij})$, $X$ denote the column matrix of coordinates as before, and $X^*$ denote the row matrix[4] $(x_1\ x_2\ x_3)$. It is then readily verified that the expression $\Sigma a_{ij}x_ix_j$ is merely the one-by-one matrix $X^*AX$. Thus in matrix form a point conic has equation

$$X^*AX = (0).$$

In particular, the example has matrix equation

$$(x_1\ x_2\ x_3)\begin{pmatrix} 4 & \frac{3}{2} & \frac{13}{2} \\ \frac{3}{2} & 2 & \frac{9}{2} \\ \frac{13}{2} & \frac{9}{2} & 7 \end{pmatrix}\begin{pmatrix} x_1 \\ x_2 \\ x_3 \end{pmatrix} = (0).$$

The matrix $A$ in the matrix form of the equation of a conic is said to be *symmetric* in view of the fact that $a_{ij} = a_{ji}$. We summarize our results in

**Theorem 9.41.** *A point conic has equation of the form*

$$\sum_{i,j=1}^{3} a_{ij}x_ix_j = 0$$

*where* $a_{ij} = a_{ji}$, *or in matrix form*

$$X^*AX = (0),$$

*where* $A$ *is a symmetric matrix.*

In Euclidean analytic geometry it was possible to reduce any conic to a standard and simpler form by an appropriate change of coordinates; for example, the ellipse

$$52x^2 - 72xy + 73y^2 + 400x - 950y + 2725 = 0$$

can be reduced to the simpler form

$$4x''^2 + 16y''^2 = 64$$

by a rotation of axes through $\sin^{-1}(\frac{3}{5})$ and a translation of axes to a new origin at $(5, 5)$. We wish to determine whether a similar procedure is possible in the projective study of conics.

Let us consider the conic

$$2x_1{}^2 + 8x_1x_2 + 4x_1x_3 + 7x_2{}^2 - 10x_2x_3 - 3x_3{}^2 = 0.$$

We choose here not to keep the conic written in symmetric form. Our first object is to change coordinates in such a way as to eliminate the

[4]The matrix formed by interchanging the rows and columns of a given matrix is said to be its *transpose*. We denote the transpose of a matrix $C$ by $C^*$; here $X^*$ is the transpose of $X$.

$x_1x_2$ term, thereby obtaining a simpler equation for the conic. To eliminate such a term, we need only complete the square; the above equation can be written in the form

$$2(x_1{}^2+4x_1x_2+4x_2{}^2)+4x_1x_3-8x_2{}^2+7x_2{}^2-10x_2x_3-3x_3{}^2=0,$$

or

$$2(x_1+2x_2)^2+4(x_1+2x_2)x_3-x_2{}^2-18x_2x_3-3x_3{}^2=0.$$

Let us now change coordinates and introduce new ones $(x_1', x_2', x_3')$ where

$$\begin{aligned} kx_1' &= x_1+2x_2 \\ kx_2' &= \quad\;\; x_2 \\ kx_3' &= \qquad\quad x_3. \end{aligned}$$

We note that this change of coordinates is a permissible one since the matrix of the change is nonsingular, having determinant

$$\begin{vmatrix} 1 & 2 & 0 \\ 0 & 1 & 0 \\ 0 & 0 & 1 \end{vmatrix} = 1 \neq 0.$$

In terms of the new coordinates the conic has equation

$$2x_1'^2+4x_1'x_3'-x_2'^2-18x_2'x_3'-3x_3'^2=0,$$

and we have eliminated the $x_1x_2$ term. To eliminate the $x_1'x_3'$ term, we complete the square again, finding

$$2(x_1'^2+2x_1'x_3'+x_3'^2)-2x_3'^2x_2'^2-18x_2'x_3'-3x_3'^2=0.$$

If we set

$$\begin{aligned} k'x_1'' &= x_1' \qquad + x_3' \\ k'x_2'' &= \quad\; x_2' \\ k'x_3'' &= \qquad\quad\; x_3' \end{aligned}$$

(the change again has nonsingular matrix), the equation becomes

$$2x_1''^2-x_2''^2-18x_2''x_3''-5x_3''^2=0.$$

Finally the $x_2''x_3''$ term can be eliminated in the same way, reducing the equation of the conic to

$$2x_1'''^2-x_2'''^2+76x_3'''^2=0.$$

The method would not work on the conic whose equation is

$$4x_1x_2+3x_3{}^2=0,$$

but here the change of coordinates

$$\begin{aligned} kx_1 &= x_1' - x_2' \\ kx_2 &= x_1' + x_2' \\ kx_3 &= \qquad\quad x_3', \end{aligned}$$

which is allowable, since

$$\begin{vmatrix} 1 & -1 & 0 \\ 1 & 1 & 0 \\ 0 & 0 & 1 \end{vmatrix} = 2 \neq 0 \tag{9.41}$$

changes the conic to the form

$$4x_1'^2 - 4x_2'^2 + 3x_3'^2 = 0.$$

Such changes will in general reduce a conic equation to the form

$$a_{11}x_1^2 + a_{22}x_2^2 + a_{33}x_3^2 = 0.$$

The coordinate changes will always be permissible, for any single square completion will have an associated determinant with 1*s* on the main diagonal and only one nonzero element off this diagonal; the value of such a determinant is one. The change necessary for the case where squared terms are missing always leads to a determinant like that of (9.41). If the individual changes are associated with the nonsingular matrices $A, B, C, \cdots$, the overall change will be associated with the matrix $ABC \cdots$, which will be nonsingular.

A further simplification is still possible. We reduced the first conic to the form

$$2x_1^2 - 1x_2^2 + 76x_3^2 = 0.$$

Let us now introduce the change of coordinates

$$\begin{aligned} kx_1' &= \sqrt{2}x_1 \\ kx_2' &= \sqrt{1}x_2 \\ kx_3' &= \sqrt{76}x_3. \end{aligned}$$

The determinant associated with the change is $\sqrt{152} \neq 0$, and the equation reduces to

$$x_1'^2 - x_2'^2 + x_3'^2 = 0.$$

Such a change is always possible. Given an equation

$$a_{11}x_1^2 + a_{22}x_2^2 + a_{33}x_3^2 = 0,$$

set $kx_i' = b_i x_i$, where $b_i = \sqrt{|a_{ii}|}$ if $a_{ii} \neq 0$ and $b_i = 1$ if $a_{ii} = 0$. The resulting equation will be of the form

$$c_1x_1'^2 + c_2x_2'^2 + c_3x_3'^2 = 0,$$

where each $c_i$ is $1, -1$, or 0. We have proved

**Theorem 9.42.** *For a suitable choice of coordinates a point conic has equation*

$$c_1x_1^2 + c_2x_2^2 + c_3x_3^2 = 0,$$

*where each* $c_i$ *is* $1, -1$, *or* 0.

Let us now reconsider this result from the matrix point of view. The argument applies to any second-degree homogeneous equation, whether or not it represents a conic. Changes of coordinates are given in matrix form by a relation $kX = BX'$ where $B$ is nonsingular. In terms of the row vectors $X^*$ and $X'^*$ associated with $X$ and $X'$, the coordinate change is given by $kX^* = X'^*B^*$, where $B^*$ is the transpose of $B$. We thus find by substitution that

$$X^*AX = \left(\frac{1}{k^2}\right)X'^*B^*ABX',$$

from which it follows that the equation of the conic in terms of the new coordinates can be taken as $X'^*CX'$ where $C = B^*AB$. Thus Theorem 9.42 has the matrix form of

**Theorem 9.43.** *Given a symmetric matrix* $A$, *there exists a nonsingular matrix* $B$ *such that* $B^*AB$ *is a matrix with zeros off the main diagonal and only* 1s, $-$1s, *and* 0s *on the main diagonal.*

Theorem 9.42 enables us to write any conic as one of just six second-degree equations for a suitable choice of coordinates. The six cases are shown in Figure 9.16. The first case is not a conic, but the remaining five are. If we wish to state a converse of Theorem 9.41, we must either allow the first case to be an exception or define it to be a conic. We shall follow the latter course. The agreement of nondegeneracy in the following definition is to enable us to state Theorem 9.45.

**Definition 9.41.** *An equation of the form* $\Sigma a_{ij}x_ix_j = 0$ *satisfied by the coordinates of no point is said to determine an* imaginary conic. *An imaginary conic is nondegenerate.*

With this agreement, we have by inspection of Figure 9.16.

**Theorem 9.44.** *An equation of the form* $\Sigma a_{ij}x_ix_j = 0$ *is the equation of a conic.*

In view of Theorem 7.64, if the matrices $A$ and $C$ are related by $C = B^*AB$ where $B$ (hence also $B^*$) is nonsingular, the ranks of $C$ and $A$ are the same. Thus the rank of the matrix associated with the equation of a conic is invariant under a projective change of coordinates. The ranks of the six cases of Figure 9.16 are clearly 3, 3, 2, 2, 1, and 0 in that order.

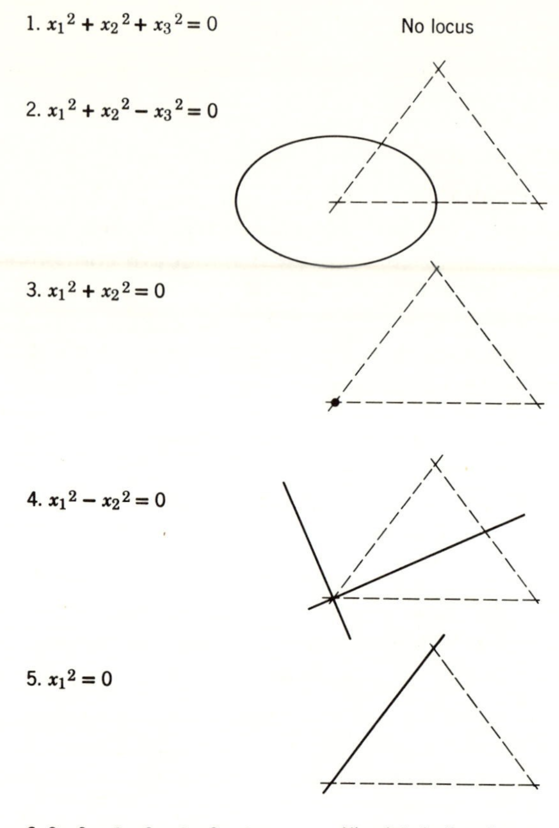


**Figure 9.16**

We thus have

**Theorem 9.45.** *The conic* $X^*AX = (0)$ *is* (a) *nondegenerate if* $A$ *is nonsingular,* (b) *a point or two lines if* $A$ *has rank two,* (c) *a line if* $A$ *has rank one, and* (d) *a plane of points if* $A$ *has rank zero.*

**Corollary 9.451.** *The conic* $X^*AX = (0)$ *is nondegenerate if an only if the determinant of* $A$ *is nonzero.*

It was the desire to identify nonsingularity of a matrix with nondegeneracy of the conic that prompted us to agree that an imaginary conic is nondegenerate.

The equations of the six cases of Figure 9.16 are said to be *canonical forms* for the equation of a conic. They are simple, and every conic can be written as one and only one of them for a suitable choice of coordinates. If we wish to prove a theorem on conics, it is sufficient to prove it for the canonical forms. Thus in proving any theorem about a nondegenerate, nonimaginary conic, we may assume the conic has equation

$$x_1^2 + x_2^2 - x_3^2 = 0.$$

## Exercises 9.4

1. a. Write each of the following in the form $\Sigma a_{ij}x_ix_j = 0$, where $a_{ij} = a_{ji}$, and

   *b*. Reduce each of the following to canonical form:

   (1) $x_1^2 - 4x_1x_3 + 3x_2^2 - x_3^2 = 0$,

   (2) $x_1^2 + 4x_1x_2 - 4x_1x_3 + 4x_2^2 - 8x_2x_3 + 4x_3^2 = 0$,

   (3) $x_1^2 - 4x_1x_3 + 3x_2^2 + 4x_3^2 = 0$,

   (4) $x_1x_2 + x_1x_3 + x_2x_3 = 0$,

   (5) $x_1^2 - x_2^2 - x_3^2 = 0$.

2. Find the equation of the conic through the points

   *a*. (1, 0, 0), (0, 1, 0), (0, 0, 1), (1, 1, 1), and (1, −1, 2),

   *b*. (1, 1, 0), (1, 0, 1), (0, 1, 1), (1, 2, 2), and (1, 2, 3),

   *c*. (1, 2, 3), (2, 3, 1), (3, 1, 2), (3, 5, 4), and (5, 4, 1).

3. Find the equation of the conic determined by pencils centered at (1, 1, 2) and (2, 2, 1) where [2, 0, −1] corresponds to [1, 0, −2], [1, −1, 0] corresponds to [1, 1, −4], and [1, 1, −1] corresponds to [1, 2, −6].

4. Prove that a conic for which $a_{lj} = 0, j = 1, 2, 3$, is degenerate.

5. Prove that a conic for which $a_{ii} = 0$, $i = 1, 2, 3$, $a_{ij} \neq 0$, $i \neq j$, is nondegenerate.

6. State the duals of all results of this section.

7. Prove directly the dual of Theorem 9.41.

8. Prove analytically Theorem 9.15.

## 9.5 POLARITIES

We have seen that a conic has an equation of second degree of the form $\Sigma a_{ij}x_ix_j = 0$. The expression $\Sigma a_{ij}x_ix_j$ is called a *quadratic form*, and the results of the preceding section are actually some of the elementary results of the algebraic theory of quadratic forms. Associated with such a quadratic form is a *bilinear form* $\Sigma a_{ij}x_iy_j$, $a_{ij} = a_{ji}$ in the two sets of variables, $x_1, x_2, x_3$ and $y_1, y_2, y_3$. The corresponding equation

$$\Sigma a_{ij}x_iy_j = 0$$

or in matrix form

$$X^*AY = (0)$$

can be thought of as a condition between two points, the condition being determined by the conic. Thus the points $(1, 1, 2)$ and $(1, 1, 1)$ satisfy such a condition with respect to the conic

$$x_1^2 + x_2^2 - x_3^2 = 0$$

since they satisfy the corresponding bilinear condition

$$x_1y_1 + x_2y_2 - x_3y_3 = 0.$$

In this section we investigate the relation defined by such bilinear conditions.

**Definition 9.51.** *Two points are* conjugate *with respect to the conic with equation* $\Sigma a_{ij}x_ix_j = 0$ *if their coordinates satisfy the condition* $\Sigma a_{ij}x_iy_j = 0$.

As an immediate consequence of this definition we have

**Theorem 9.51.** *A point is self-conjugate with respect to a conic if and only if it lies on the conic.*

Let us now fix a point $(x_1, x_2, x_3)$ and seek the locus of points conjugate to it. If the point were $(2, 3, 7)$ and the conic were

$$x_1^2 + x_1x_3 + x_3x_1 - 2x_2^2 + 4x_3^2 = 0,$$

a point $(y_1, y_2, y_3)$ would have to satisfy

$$2y_1 + 2y_3 + 7y_1 - 6y_2 + 28y_3 = 0$$

or

$$9y_1 - 6y_2 + 30y_3 = 0.$$

Thus the locus of points conjugate to $(2, 3, 7)$ would be a line. In general the locus of points conjugate to $(x_1, x_2, x_3)$ with respect to the conic with equation $\Sigma a_{ij}x_ix_j = 0$ would be a line, since the resulting bilinear form $\Sigma a_{ij}x_iy_j$ is linear in the $y$'s. It is possible that all of the coefficients of the $y$'s are zero. In this case the locus of conjugate points is the entire plane. This would be the case for the point $(0, 0, 7)$ with respect to the conic

$$x_1x_2 + x_2x_1 = 0$$

since the corresponding bilinear condition leads to

$$0y_1 + 0y_2 + 0y_3 = 0.$$

We summarize our results in the following theorem:

**Theorem 9.52.** *The locus of all points conjugate to a given point with respect to a conic is a line or the set of all points in the plane.*

In view of this result we make the following definition:

**Definition 9.52.** *If the locus of all points conjugate to a fixed point with respect to a conic is a line, the line is the* polar line *of the fixed point with respect to the conic, and the fixed point is the* pole *of the line with respect to the conic. The relation between lines and points so determined is a* polarity

A polarity need not be defined everywhere, since we have seen that the set of conjugate points of a given point may be the entire plane, in which case the point has no polar line. In the example above this occurred when the conic was degenerate. This this was no coincidence is shown in

**Theorem 9.53.** *The polarity defined by a nondegenerate conic is a one-to-one correspondence between all of the points and all of the lines of the plane of the conic.*[5]

PROOF. Let the conic have equation $\Sigma a_{ij}x_ix_j = 0$. Given a point $(x_1, x_2, x_3)$, the line coordinates, $[u_1, u_2, u_3]$ of its polar line, if it exists, must be given by[6]

$$\begin{aligned} u_1 &= a_{11}x_1 + a_{21}x_2 + a_{31}x_3 \\ u_2 &= a_{12}x_1 + a_{22}x_2 + a_{32}x_3 \\ u_3 &= a_{13}x_1 + a_{23}x_2 + a_{33}x_3. \end{aligned} \tag{9.51}$$

Since the conic is nondegenerate, we know by Corollary 9.451 that

$$\begin{vmatrix} a_{11} & a_{21} & a_{31} \\ a_{12} & a_{22} & a_{32} \\ a_{13} & a_{23} & a_{33} \end{vmatrix} \neq 0.$$

Now for a given point not all $x_i$ are zero, hence not all the $u_j$ can be zero, for if they were, equations (9.51) would have the unique solution $x_1 = x_2 = x_3 = 0$. We see each point has a unique polar line. Conversely, given any line, equations (9.51) become a system of equations for the coordinates of its pole. In view of the nonvanishing of the determinant of coefficients, this system will have a unique solution, and clearly not all of the $x_i$ will be zero.

Let us now investigate the relation between a point and its polar line. Let us suppose first that the point is on the conic, which is nondegenerate, and denote the coordinates of the point by the column matrix $X$. The condition that a point with coordinates given by the column matrix $Y$ lie on the polar line is $X^*AY = (0)$. We know by Theorem 9.51 that

[5]The result is valid even if the conic is imaginary.

[6]These equations establish a one-to-one correspondence between the points and the lines of a plane with coordinates linearly related. Thus a polarity is a projective transformation in the sense of Definition 8.82.

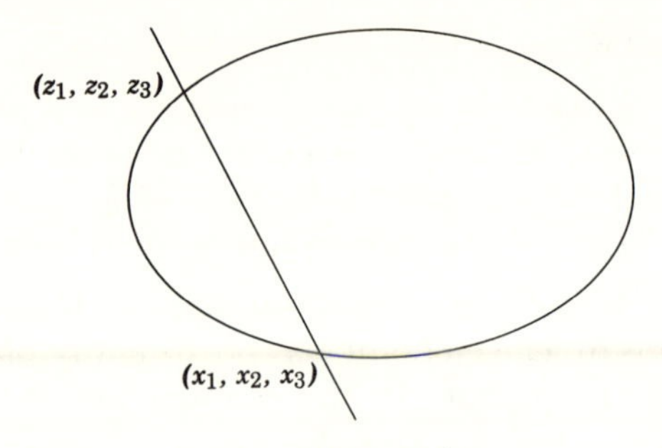


**Figure 9.17**

the given point is on its own polar line; that is, $X^*AX = (0)$. Suppose now that we have a second point on the conic and on the polar line. If $Z$ is the column matrix of coordinates of this point, then $Z^*AZ = (0)$ and $X^*AZ = (0)$. Now any point on the polar line has a column matrix of coordinates of the form $c_1X + c_2Z$, and we know

$$X^*A(c_1X + c_2Z) = c_1X^*AX + c_2X^*AZ = (0).$$

This polar line is also the polar line of the point with coordinates $Z$, since this line contains two points conjugate to the point with coordinates $Z$ (see Exercise 9.51). Thus $Z^*A(c_1X + c_2Z) = (0)$. It follows that

$$(c_1X + c_2Z)^*A(c_1X + c_2Z) = (0).$$

This says every point on the polar line would be on the conic. This is impossible for a nondegenerate conic, and we conclude the polar line is the tangent line to the conic. We have proved

**Theorem 9.54.** *The polar line to a point on a nondegenerate conic is the tangent line to the conic at that point.*

The development of the duals of the results of Section 9.4 was left as an exercise for the reader. It is obvious that a line conic must have equation of the form

$$\Sigma b_{ij}u_iu_j = 0,$$

or in matrix form

$$U^*BU = (0),$$

where $B$ is symmetric. With Theorem 9.54 proved we can now become more specific and find the equation of the particular conic $X^*AX = (0)$ regarded as a line conic. If $X$ is the matrix of coefficients of a point on the conic, then the tangent line to the conic at this point has line coordinates given by equations (9.51), which can be written in the matrix forms

$$U = AX$$

or

$$U^* = X^*A^* = X^*A.$$

From the former we infer

$$X = A^{-1}U,$$

whence

$$U^*A^{-1}U = X^*AX = (0).$$

Since $AA^{-1} = I$, we have $(A^{-1})^*A^* = (A^{-1})^*A = I$. However, $A^{-1}A = I$, and it thus follows that if $A$ is nonsingular, $A^{-1} = (A^{-1})^*$, showing that $A^{-1}$ is symmetric. We have proved

**Theorem 9.55.** *The equation of the nondegenerate point conic $X^*AX = (0)$ regarded as a line conic is $U^*A^{-1}U = (0)$, where $A^{-1}$ is symmetric.*

For the sake of completeness we state here the duals of Definition 9.51 and Theorems 9.51 and 9.52, interpreted in the light of Theorem 9.55. We note that Theorems 9.53 and 9.54 and Definition 9.52 are self-dual.

**Definition 9.53.** *Two lines $U$ and $V$ are* conjugate *with respect to the conic with equation $X^*AX = (0)$ if their coordinates satisfy the condition $U^*A^{-1}V = (0)$.*

**Theorem 9.56.** *A line is self-conjugate with respect to a conic if and only if it is a tangent to the conic.*

**Theorem 9.57.** *The locus of all lines conjugate to a given line with respect to a conic is a point or the set of all lines in the plane.*

We conclude by considering the pole-polar relation for points not on a conic.

**Theorem 9.58.** *If through a point not on a conic two distinct tangents to the conic exist, the polar line of the point is the line joining the points of tangency of the tangents.*

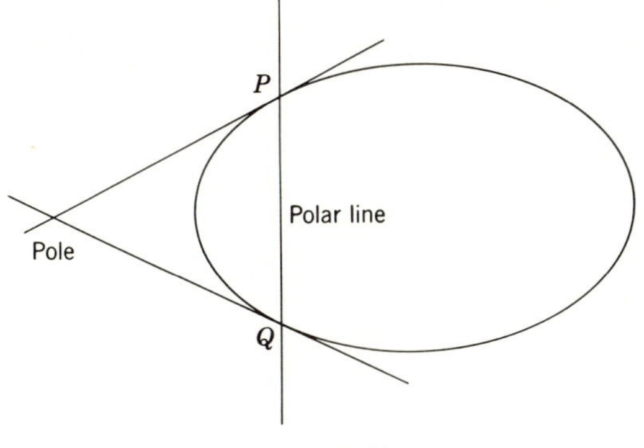


**Figure 9.18**

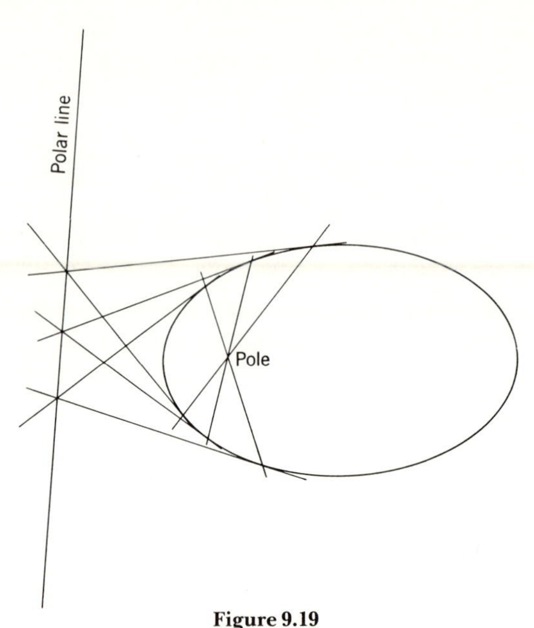


**Figure 9.19**

PROOF. In view of Theorems 9.16 and 9.39 there are no more than two tangents from such a point. The existence of tangents implies the conic is nondegenerate; hence every point has a polar line. Let the points of tangency be $P$ and $Q$. The given point is conjugate to each of these by Theorem 9.54. Conjugacy is symmetric (see Exercise 9.51); hence $P$ and $Q$ are conjugate to the given point and lie on its polar line.

**Theorem 9.59.** *The polar line of a point with respect to a nondegenerate conic is the locus of points of intersection of tangents drawn from the intersections with the conic of lines through the given point.*

PROOF. In view of Theorem 9.55 any point on this locus is conjugate to the given point. Thus the line is the desired polar line.

## Exercises 9.5

1. Prove that if $A$ is conjugate to $B$, then $B$ is conjugate to $A$ with respect to a conic.

2. Given the conic $x_1^2 + 2x_1x_2 + x_2^2 + 4x_2x_3 + 4x_3^2 = 0$,
   *a.* Find the equation of the tangent at $(1, -1, 0)$,
   *b.* Find the polar line of $(1, -2, 3)$,
   *c.* Find the pole of $[2, 4, -1]$,
   *d.* Find the tangents passing through $(0, 0, 1)$.

3. Given the conic $x_1^2 - 2x_1x_2 + 4x_2^2 - 4x_3^2 = 0$,
   *a.* Find the equation of the tangent through $(2, 2, -1)$,
   *b.* Find the polar line of $(2, 3, -1)$,
   *c.* Find the pole of $[3, -2, 1]$,
   *d.* Find the tangents passing through $(0, 0, 1)$.

4. Prove Theorem 9.51.

5. Prove Theorem 9.54 by writing out explicitly the relations given in matrix form in the proof of that theorem.

6. Prove that if a conic is a pair of lines, every point except the intersection of the lines has a polar line.

7. Prove that if a conic is a line, every point not on the conic has a polar line, which is the conic.

8. Prove that if a conic is a point, every point not on the conic has a polar line which passes through the conic.

9. Prove that the polar lines of all points on a line are concurrent.

10. The triangle $ABC$ is *self-polar* with respect to a conic if $A$ is the pole of $BC$, $B$ is the pole of $AC$, and $C$ is the pole of $AB$. Prove that the diagonal triangle of a complete quadrangle inscribed in a conic is self-polar with respect to the conic.

11. Prove that if a line through a point $P$ meets a conic at $A$ and $B$, it meets the polar line of $P$ at a point which is the harmonic conjugate of $P$ with respect to $A$ and $B$.

## REFERENCES


Adler, Claire Fisher, *Modern Geometry, An Integrated First Course*, 2nd ed., McGraw-Hill Book Company, New York, 1967, Chapter 7.

Artzy, Rafael, *Linear Geometry*, Addison-Wesley Publishing Company, Reading, Mass., 1965, Chapter 3.

Busemann, Herbert, and Paul J. Kelly, *Projective Geometry and Projective Metrics*, Academic Press, New York, 1953, Chapter 2.

Coxeter, H. S. M., *Non-Euclidean Geometry*, 5th ed., University of Toronto Press, Toronto, 1965, Chapters 3, 4.

Coxeter, H. S. M., *The Real Projective Plane*, McGraw-Hill Book Company, New York, 1949, Chapters 6, 7.

Gans, David, *Transformations and Geometries*, Appleton-Century-Crofts, New York, 1969, Chapters 5, 6, 8 to 10.

Graustein, William C., *Introduction to Higher Geometry*, The Macmillan Company, New York, 1945, Chapters 12, 14.

Gruenberg, K. W., and A. J. Weir, *Linear Geometry*, D. Van Nostrand Company, Princeton, 1967, Chapter 5.

Levy, Harry, *Projective and Related Geometries*, The Macmillan Company, New York, 1961, Chapter 4.

O'Hara, C. W., and D. R. Ward, *An Introduction to Projective Geometry*, Oxford University Press, New York, 1937, Chapters 5, 6.

Patterson, Boyd Crumrine, *Projective Geometry*, John Wiley and Sons, New York, 1937, Chapters 7 to 9.

Rainich, G. Y., and S. M. Dowdy, *Geometry for Teachers*, John Wiley and Sons, New York, 1968, Chapter 3.

Rosenbaum, Robert A., *Introduction to Projective Geometry and Modern Algebra*, Addison-Wesley Publishing Company, Reading, Mass., 1963, Chapters 8, 10.

Sanger, R. G., *Synthetic Projective Geometry*, McGraw-Hill Book Company, 1939, Chapter 5.

Seidenberg, A., *Lectures in Projective Geometry*, D. Van Nostrand Company, Princeton, 1962, Chapters 6, 10 to 12.

Springer, C. E., *Geometry and Analysis of Projective Spaces*, W. H. Freeman and Company, San Francisco, 1964, Chapter 8.

Struik, Dirk J., *Lectures on Analytic and Projective Geometry*, Addison-Wesley Publishing Company, Reading, Mass., 1953, Chapter 5.

Veblen, Oswald, and John Wesley Young, *Projective Geometry*, vol. I, Ginn and Company, Boston, 1910, Chapter 5.

CHAPTER TEN

# AXIOMATIC PROJECTIVE GEOMETRY

In the previous chapters real projective geometry was treated as an extension of Euclidean geometry, the synthetic development being started by the addition of new points to the Euclidean plane and the analytic development being started by the change to natural homogeneous coordinates. Projective geometry can be developed without any appeal to Euclidean geometry by basing it on its own set of axioms. When this is done, it can then be shown that Euclidean and many other geometries can be derived from projective geometry by suitable specialization. The goal of the present chapter is to redevelop axiomatically the foundations of real projective geometry and to examine some generalizations possible if only a part of the axioms are assumed. We confine ourselves exclusively to plane geometry. The extensions to higher dimensions are relatively simple, and the plane case contains interesting features not found in any other case.

## 10.1 AXIOMS FOR THE PROJECTIVE PLANE

We take as undefined terms *point* and *line* and as an undefined relation *incidence*. We assume the following:

1. *There exists at least one line.*

2. *There are at least three points on every line.*
3. *Not all points are on the same line.*
4. *Two distinct points lie on one and only one line.*
5. *Two distinct lines meet in one and only one point.*

These axioms would not enable us to prove all the results of Chapter 5. In particular we could not prove the plane case of the theorem of Desargues, since we would not be able to make the necessary constructions in space. It is pertinent to ask whether this theorem could be proved using only the five axioms above. To show that this is not the case, we exhibit a plane consistent with the five axioms in which the theorem of Desargues does not hold.

Let us take a real Euclidean plane and divide it by a vertical line. Points will be Euclidean points. Lines will be all Euclidean lines of nonpositive slope and all broken lines of positive slope, the slope of the ray to the left of the given vertical line being twice the slope of the ray to the right (see Figure 10.1). To the plane we add one ideal point for each family of "parallel" lines as in Chapter 4. The resulting plane is clearly consistent with the first three axioms; that it is consistent with the last two is left as an exercise for the reader. This plane does not allow a proof of the theorem of Desargues, as is shown in Figure 10.2. In the figure the entire configuration is to the left of the basic vertical line except the point $B''$. The points $A''$ and $C''$ determine a horizontal line, but $B''$ does not lie on this line.

To ensure the validity of the theorem of Desargues, we must add an axiom to our system. We shall assume

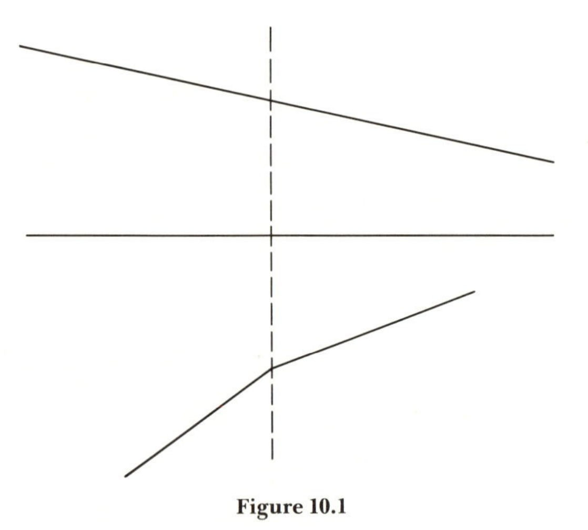

**Figure 10.1**

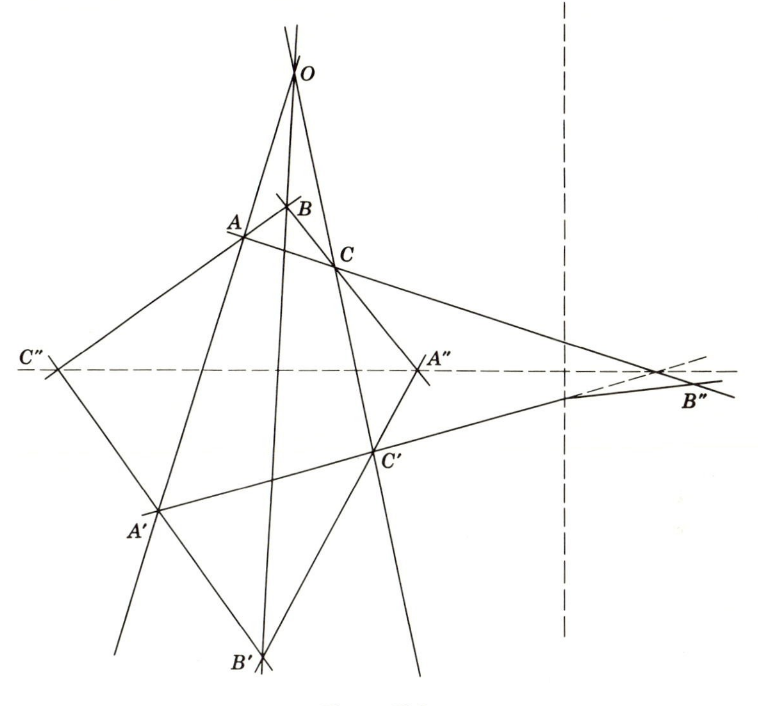


**Figure 10.2**

6. *If two triangles are perspective from a point, they are perspective from a line.*

This axiom is peculiar to plane projective geometry, for in space we could have proved the theorem of Desargues as in Chapter 5 and would not have needed the axiom. Projective planes are said to be *Desarguean* or *non-Desarguean* depending on whether Axiom 6 is or is not fulfilled. We confine ourselves to Desarguean projective planes in this chapter.

The axioms are still insufficient for the development of the material of Chapter 5. The theorem of Pappus cannot be proved using only the axioms above, and we take it as a final axiom.

7. *If $A$, $B$, and $C$, and $A'$, $B'$, and $C'$ are triples of distinct points on two distinct lines, the points of intersection of $AB'$ and $A'B$, of $AC'$ and $A'C$, and of $BC'$ and $B'C$ are collinear.*

This set of axioms is now sufficient for the development of all the material of Chapter 5. The only major change that would be required would be in the proof of the Fundamental Theorem. This theorem can be proved using Axiom 7. Indeed, Axiom 6 is now redundant, for it can be proved using Axiom 7. The interested reader should consult the references at the end of the chapter.

The set of axioms stated above does not exhibit the principle of duality, which we would like to have present. To be sure, Axioms 4 and 5 are duals of each other, but the duals of Axioms 1, 2, 3, 6, and 7 are not present in the system. We assert that these duals are readily proved using the seven axioms and that the principle of duality is present in the axiomatic introduction to the projective plane (see Exercises 10.3 to 10.7).

## Exercises 10.1

1. Prove that if Axiom 5 has "one and only one" replaced by "at least one," the axiom as stated can be proved as a theorem.

2. Prove that if $A$, $B$, and $C$ are noncollinear, $A$, $B$, and $D$ are collinear, and $B$, $C$, and $E$ are collinear, there exists a point $F$ such that $A$, $C$, and $F$ are collinear and $D, E$, and $F$ are collinear (Pasch's axiom).

3. Prove there exists at least one point.

4. Prove there are at least three lines on every point.

5. Prove that not all lines lie on the same point.

6. Prove the converse of Axiom 6 using only Axioms 1 to 6.

7. Prove the dual of Axiom 7.

8. Prove that there exist at least seven lines.

9. Prove the non-Desarguean plane of Figure 10.1 is consistent with Axioms 1 to 5.

## 10.2 INTRODUCTION OF COORDINATES ON A LINE

In the preceding section we showed a set of axioms sufficient to begin the development of synthetic projective geometry without any appeal to Euclidean geometry. In this section we consider the problem of analytic projective geometry. One of the remarkable facts about projective geometry is that the seemingly synthetic ideas of the previous section contain within themselves the concept of "number." We develop this concept and show how it leads to coordinates, hence to analytic projective geometry.

As a first step we demonstrate how the "number" concepts of addition and multiplication can be applied to points on a line. We consider first addition. Let us fix distinct points $O$ and $I$ on a line $l$ and consider arbitrary distinct lines $l'$ and $l''$, each distinct from $l$, through $I$. To define the sum $A+B$ of two points $A$ and $B$ on $l$, we choose an arbitrary point $P$ on $l''$, distinct from $I$, and let $Q$ be the intersection of $OP$ and $l'$. If $AQ$ meets $l''$ at $R$ and $BP$ meets $l'$ at $S$, we define $A+B$ to be the intersection of $RS$ and $l$. The resulting configuration is shown in Figure 10.3.

The point $A+B$ is determined by the points $P$, $Q$, $R$, and $S$, which we have chosen to determine in the order named. On occasion it will be convenient to associate a sum point with the four points determining it. We use the notation

$$P, Q, R, S \to A+B$$

to exhibit this association.

To show the sum is well defined, we must prove that the point $A+B$ is well determined (that is, the lines $RS$ and $l$ are distinct) and is independent of the choice of arbitrary elements $l'$, $l''$, and $P$ of the above construction. This we do in the following two theorems.

**Theorem 10.21.** *If $A \neq I$ and $B \neq I$, $A+B$ is well determined.*

PROOF. The point $A+B$ is determined as the intersection of the lines $l$ and $RS$. To prove it is well determined, we must show that the points $R$ and $S$ are distinct and that the lines $RS$ and $l$ are distinct. Since the points $O$, $I$, and $P$ are distinct and noncollinear, and since the lines $l$, $l'$, and $l''$ are concurrent at $I$, the point $Q$ is well determined (that is, $OP$ and $l'$ are distinct lines) and is not $I$. Now if $A \neq I$, the lines $AQ$, $l'$, and $l''$ are distinct, and $R$ is well determined and is not $I$. Finally, if $B \neq I$, $BP$ and $l'$ are distinct, from which $S$ is well determined and is not $I$. If neither $R$ nor $S$ is $I$, they are then distinct and determine a line distinct from $l$. It follows that $A+B$ is well determined.

**Corollary 10.211.** *If $A \neq I$ and $B \neq I$, then $A+B \neq I$.*

In view of the result of this theorem we restrict the addition of points to the set of points on $l$ other than $I$.

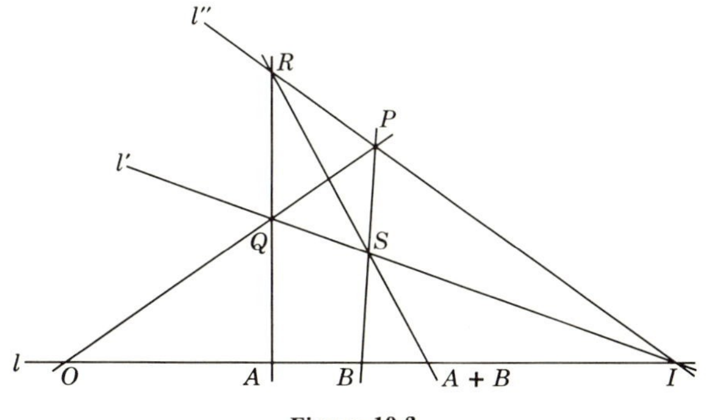


**Figure 10.3**

**Theorem 10.22.** *The point $A+B$ is uniquely determined.*

PROOF. We duplicate the construction for $A+B$ as shown in Figure 10.4. We must prove that $RS$ and $R'S'$ meet on $l$. The proof makes fundamental use of the theorem of Desargues and is similar to the proof of Theorem 5.21. We leave the details to the reader.

We turn now to a verification of the facts that addition of points defined here has the properties ordinarily associated with addition of "numbers." The following two theorems follow readily from the definition of addition (Figure 10.3) and their proofs are left to the reader.

**Theorem 10.23.** $A+O=O+A=A$ *for arbitrary* $A$.

**Corollary 10.231.** *For arbitrary* $A$ *there is a unique point* $-A$ *such that* $A+(-A)=O$.

**Theorem 10.24.** *For arbitrary* $A$ *and* $C$ *there is a unique point* $X$ *such that* $A+X=C$.

**Theorem 10.25 (Commutativity).** $A+B=B+A$ *for arbitrary* $A$ *and* $B$.

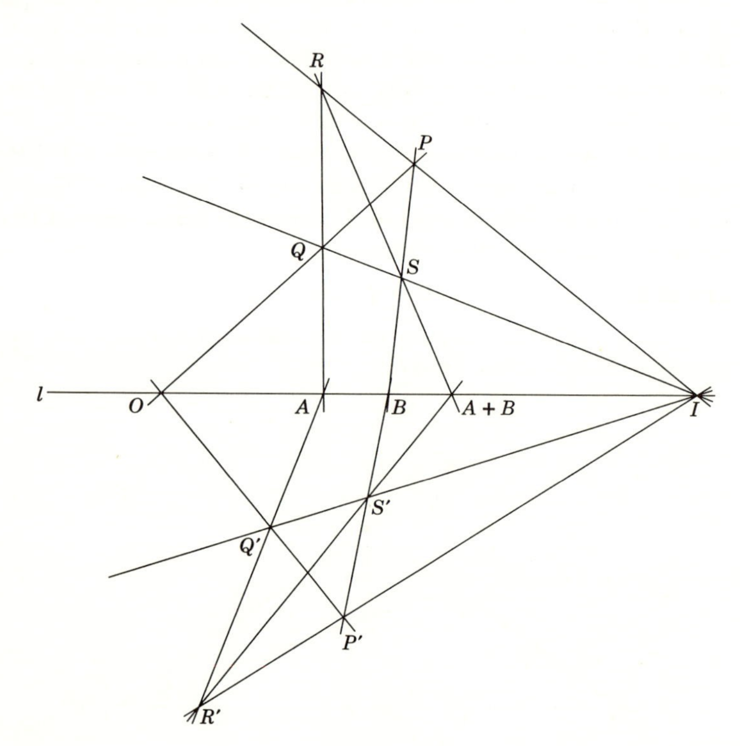


**Figure 10.4**

PROOF. Referring to Figure 10.3, we note that

$$P, Q, R, S \to A+B$$

and

$$Q, P, S, R \to B+A.$$

Thus $A+B$ and $B+A$ are both the intersection of $l$ with $RS$ (or $SR$).

**Theorem 10.26 (Associativity).** *$(A+B)+C = A+(B+C)$ for arbitrary $A$, $B$, and $C$.*

PROOF. The crux of the matter occurs after $A$ and $B$ have been added. To add this sum to $C$, we must choose an arbitrary point on $l''$ (or $l'$, or some other line through $I$). To obtain a simple proof, we take this point to be the intersection $T$ of $OS$ with $l''$, as shown in Figure 10.5. We have

$$P, Q, R, S \to A+B$$

and

$$T, S, R, V \to (A+B)+C.$$

We also have

$$T, S, P, V \to B+C$$

and

$$P, Q, R, V \to A+(B+C).$$

Thus $(A+B)+C$ and $A+(B+C)$ are both the intersection of $l$ and $RV$.

It should be noted that the axioms guarantee no more than three points on any line. In some of the figures more than three points are indicated on a line. The reader will find it instructive to verify in detail

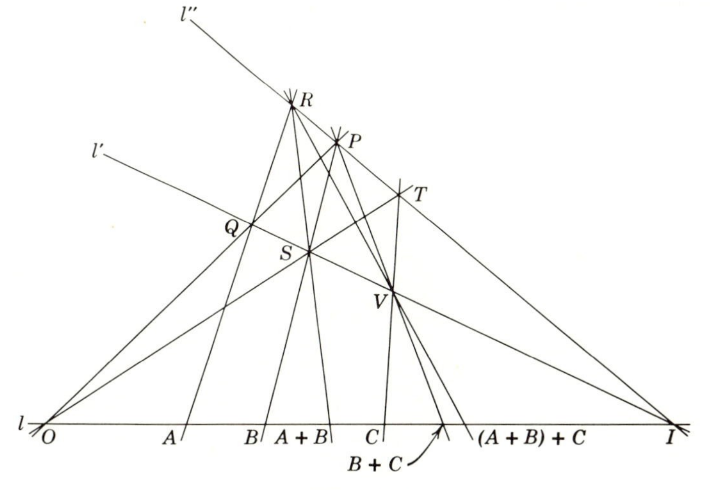


**Figure 10.5**

that the axioms have not been violated. Some of the points do have certain distinctness conditions attached, but many are determined as the intersection of lines and need not be distinct from some of the previously determined points.

We turn now to a definition of multiplication of points on a line. We fix a point $U$ on $l$, distinct from $O$ and $I$, and take arbitrary lines $m$ and $n$ through $O$ and $I$, respectively, each distinct from $l$. We choose a point $K$ on $n$, distinct from $I$ and not on $m$. Let $KU$ meet $m$ at $L$ and let $AL$ meet $n$ at $M$. If $BK$ meets $m$ at $N$, we define the intersection of $MN$ and $l$ to be the product point $A \cdot B$. The construction is shown in Figure 10.6. We emphasize the determination of $A \cdot B$ by the points $K$, $L$, $M$, and $N$ with the notation

$$K, L, M, N \rightarrow A \cdot B.$$

As in the addition case we must prove that the product is well and uniquely determined and obeys the familiar rules for multiplication. These results are contained in the following theorems, the proofs of which are left to the reader.

**Theorem 10.27.** *If $A \neq I$ and $B \neq I$, the point $A \cdot B$ is well determined.*

We continue the restriction of considering operations as applied only to the points other than $I$ on $l$.

**Theorem 10.28.** *The point $A \cdot B$ is uniquely determined.*

**Theorem 10.29.** *$A \cdot O = O \cdot A = O$ and $A \cdot U = U \cdot A = A$ for arbitrary $A$.*

**Theorem 10.210.** *For arbitrary points $A$ and $C$, $A \neq O$, there are unique points $X$ and $Y$ such that $A \cdot X = C$ and $Y \cdot A = C$.*

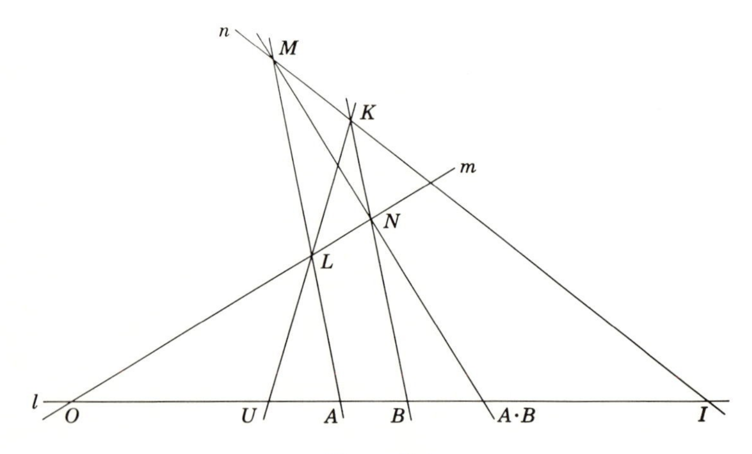


**Figure 10.6**

**Corollary 10.2101.** *For arbitrary $A \neq O$ there is a unique point $A^{-1}$ such that $A \cdot A^{-1} = A^{-1} \cdot A = U$.*

**Theorem 10.211 (Associativity).** $(A \cdot B) \cdot C = A \cdot (B \cdot C)$ *for arbitrary $A$, $B$, and $C$.*

As was the case with Theorem 10.26 it will be advisable to make a judicious choice of the arbitrary elements in the proof of this theorem. If $K$, $L$, $M$, $N \to A \cdot B$, when this product point is multiplied by $C$ it will be advisable to let the arbitrary point on $n$ be the intersection $P$ of $n$ with $UN$, as shown in Figure 10.7. The details of the proof are left to the reader.

It would be desirable to have some relation between the two operations. The "number" relation joining the operations is the familiar "parenthesis removal" law, versions of which we next prove.

**Theorem 10.212 (Distributivity).** $A \cdot (B+C) = A \cdot B + A \cdot C$ *and* $(B+C) \cdot A = B \cdot A + C \cdot A$ *for arbitrary $A$, $B$, and $C$.*

PROOF. We prove the first of the statements; a similar proof will justify the second. The configuration is shown in Figure 10.8. We have

$$K, L, M, N \to A \cdot B,$$
$$K, L, M, N' \to A \cdot C,$$

and

$$P, N, M, S \to A \cdot B + A \cdot C.$$

We also have

$$P, N, K, S' \to B + C$$

and

$$K, L, M, T \to A \cdot (B+C).$$

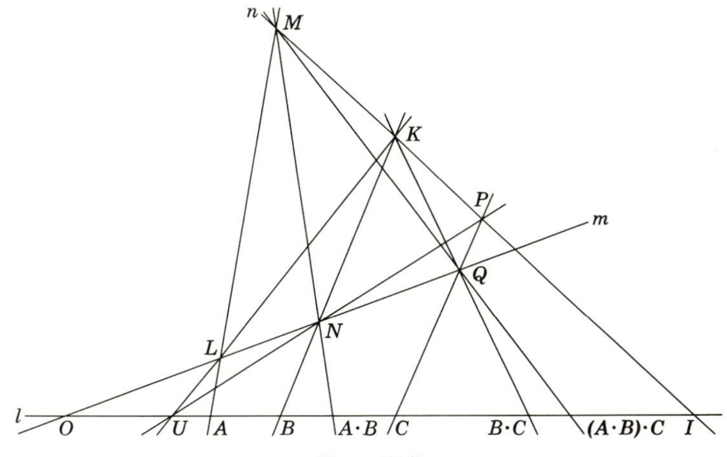


**Figure 10.7**

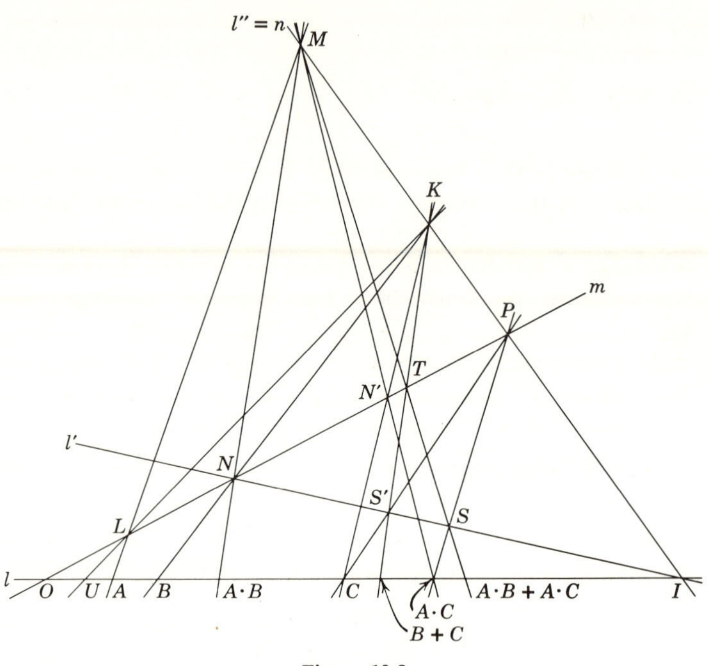


**Figure 10.8**

To prove the theorem, we must prove that $MS$ and $MT$ coincide, or equivalently that $MT$, $MS$, and $OP$ are concurrent. We claim that the triangles $MKN'$ and $SS'P$ are Desarguean, since $MK$ meets $SS'$ at $I$, $MN'$ meets $SP$ at $A \cdot B$, and $KN'$ meets $S'P$ at $C$, these points being collinear. The triangles thus must be perspective from the point $T$, and $MS$ does pass through $T$.

Let us summarize the properties that the operations have been shown to possess on the set of points other than $l$:

1. Closure: $a+b$ and $ab$ are uniquely defined and in the same set as $a$ and $b$ for arbitrary $a$ and $b$ of the set;

2. Associativity: $a+(b+c)=(a+b)+c$ and $a(bc)=(ab)c$ for arbitrary $a$, $b$, and $c$;

3. Commutativity: $a+b=b+a$ for arbitrary $a$ and $b$;

4. Identity: there exist elements 0 and 1 such that $a+0=0+a=a$ and $a1=1a=a$ for arbitrary $a$;

5. Inverses: for arbitrary $a$ there is an element $-a$ such that $a+(-a)=0$ and, if $a \neq 0$, an element $a^{-1}$ such that $aa^{-1}=a^{-1}a=1$;

6. Distributivity: $a(b+c)=ab+ac$ and $(b+c)a=ba+ca$ for arbitrary $a$, $b$, and $c$.

We define the system having these properties and summarize our analysis in the following theorem, noting that verification of the above properties required use of only Axioms 1 to 6.

**Definition 10.21.** *A set of elements together with two operations satisfying* 1 to 6 *above is a* division ring.

**Theorem 10.213.** *If one point is excluded from a line in a projective plane satisfying Axioms* 1 to 6, *addition and multiplication can be so defined that the points form a division ring under these operations.*

There is one gap in the properties of a division ring and in the sequence of theorems proved about the operations. We have not shown that multiplication is commutative; that is, we have not proved that $A \cdot B = B \cdot A$. The perceptive reader will note that this property was used in the analytic proof of the theorem of Pappus. This commutative law cannot be proved without the theorem of Pappus, Axiom 7, and the following theorem is the first in this chapter in which the axiom is used.

**Theorem 10.214.** $A \cdot B = B \cdot A$ *for arbitrary* $A$ *and* $B$.

PROOF. Referring to Figure 10.9 we see that

$$K, L, M, N \rightarrow A \cdot B$$

and

$$K, L, P, Q \rightarrow B \cdot A.$$

We must show that $P$, $Q$, and $A \cdot B$ are collinear. This follows from Axiom 7. The two sets of points are $A, L$, and $M$, and $N, K$, and $B$: $AK$

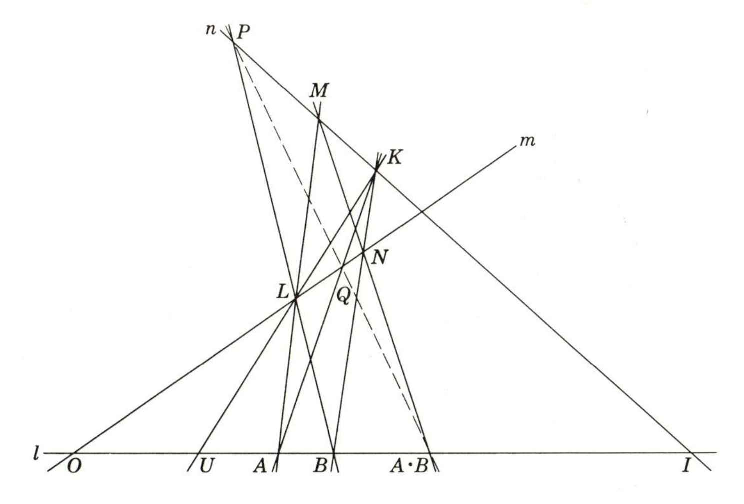


**Figure 10.9**

and $LN$ meet at $Q$, $LB$ and $MK$ meet at $P$, and $AB$ and $MN$ meet at $A \cdot B$. The desired collinearity is thus established.

If we assume Axiom 7, we conclude that we have a special type of division ring in which the commutative law of multiplication holds and that the points, excluding $I$, will form this special type of division ring. We define this special division ring and summarize the results in the following definition and theorem.

**Definition 10.22.** *A division ring in which the commutative law of multiplication is valid is a* field.

**Theorem 10.215.** *If one point is excluded from a line in a projective plane satisfying Axioms* 1 *to* 7, *addition and multiplication can be so defined that the points form a field under these operations.*

The set of all real numbers, the set of all complex numbers, and the set of all rational numbers are fields under ordinary addition and multiplication. Our use of the adjective "real" in real projective geometry in the previous chapters implied that the field of Theorem 10.215 was being taken as the field of real numbers. There are other subsets of the complex number field which are fields; some are discussed in the exercises at the end of this section. A field need not have an infinite number of elements; we consider finite fields further in a later section. In the rest of this chapter we assume Axiom 7 and work with fields rather than the more general division ring.

## Exercises 10.2

1. Prove Theorem 10.23.
2. Prove Theorem 10.24.
3. Prove Theorem 10.27.
4. Prove Theorem 10.28.
5. Prove Theorem 10.29.
6. Prove Theorem 10.210.
7. Prove Theorem 10.211.
8. Prove that a division ring is a group under the operation of addition.
9. Prove that the nonzero elements of a division ring form a group under the operation of multiplication.
10. Determine which of the following are fields:
    *a.* The set of positive integers;
    *b.* The set of all integers;
    *c.* The set of all rational numbers;

*d.* The set of all integers and reciprocals of integers;
*e.* The set of all numbers of the form $a+b\sqrt{2}$ where $a$ and $b$ are rational;
*f.* The set of all numbers of the form $a+b\sqrt[3]{2}$ where $a$ and $b$ are rational;
*g.* The set of all numbers of the form $a+b\sqrt{2}$ where $a$ and $b$ are integers;
*h.* The set of all numbers of the form $a+ib\sqrt{2}$ where $a$ and $b$ are real and $i=\sqrt{-1}$.

11. If $T$ is the involution on $OI$ with fixed point $I$ for which $T(A)=B$, prove that $T(O)=A+B$.

12. Prove that $H(O, I; A, -A)$.

13. Prove that $H(A, I; O, A+A)$.

14. If $T$ is the involution on $OI$ for which $T(O)=I$ and $T(A)=B$, prove that $T(U)=A\cdot B$.

## 10.3 COORDINATES IN THE PLANE

At this point we have shown the existence of "number" ideas in the synthetic axioms for a projective plane, but we have not as yet carried the idea to the point of introducing coordinates in the projective plane. We turn our attention to this introduction. Our coordinates will now come from an arbitrary field rather than the special field of real numbers.

For convenience we first introduce nonhomogeneous coordinates in all of the plane except one line, then make the transition to homogeneous coordinates for all of the plane.

We might be tempted to introduce coordinates in the plane by taking two distinct lines and considering the field of points on each. Presumably each point in the plane, except perhaps the points on the line $II'$ of Figure 10.10, would have coordinates of the form $(P, P')$ where $P$ is a point on $OI$ and $P'$ is a point on $OI'$. This would lead to an awkward

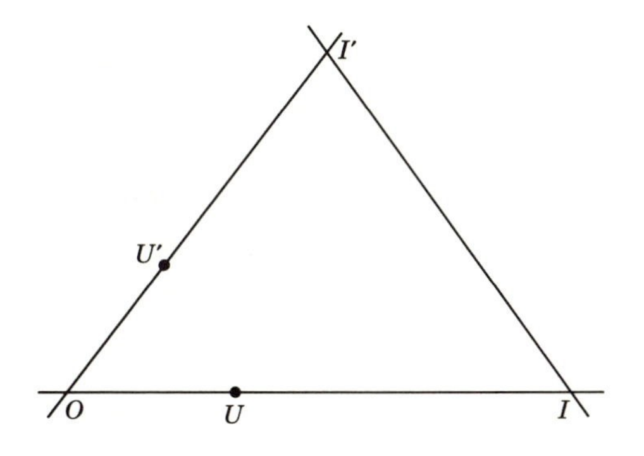


**Figure 10.10**

situation, for the coordinates of a point would then be elements of distinct fields. We would prefer to have just one field involved. In addition, we wish to introduce coordinates in the plane in such a way that we can prove that any line has a first-degree equation. The procedure discussed below does avoid these difficulties.

We suppose that a line of the plane has been chosen and the points $O$, $U$, and $I$ fixed on that line. We introduce addition and multiplication on this line as in Section 10.2 and obtain a field. We use the symbols of properties 1 to 6 of page 184 for the elements of the field; that is, we associate with each point on the line as a label or coordinate an element of a field, this field being a copy of the field of points itself.[1] In particular the point $O$ is associated with 0 and the point $U$ is associated with 1. We next choose lines through $O$ and $I$ meeting at $I'$ and choose an arbitrary point $U'$, distinct from $O$ and $I'$, on $OI'$. We now introduce nonhomogeneous coordinates for all points in the plane except those on the line $II'$.

If a point is on the line $OI$ and is associated with the field element $x$, we take $(x, 0)$ as its coordinates. If the point is on the line $OI'$, join it to the intersection $U^*$ of $II'$ and $UU'$. If the line so determined meets $OI$ at a point with coordinates $(y, 0)$, the coordinates of the point on line $OI'$ will be taken to be $(0, y)$. Finally, if the point is on neither line, consider the lines joining it to $I$ and $I'$. If these lines meet the previously

[1]The algebraist says these fields are isomorphic.

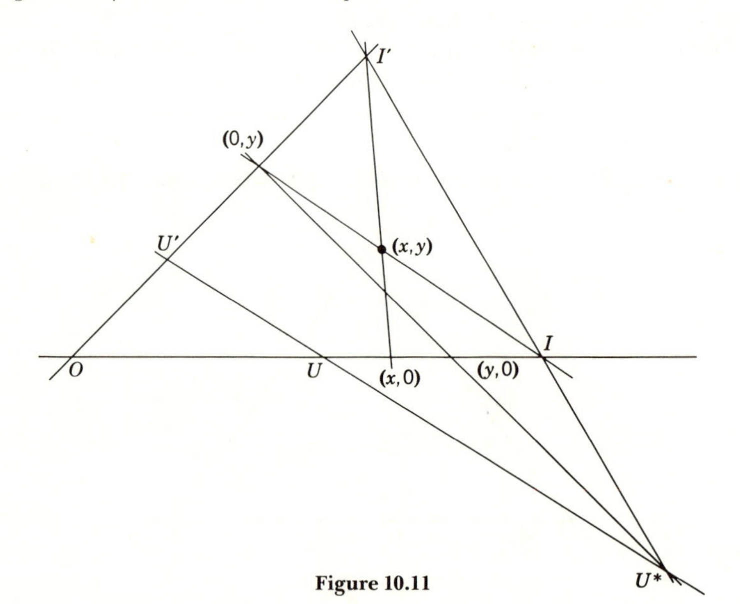


**Figure 10.11**

coordinatized lines at the points with coordinates $(x, 0)$ and $(0, y)$, the given point will be given coordinates $(x, y)$.

We can now extend these nonhomogeneous coordinates to homogeneous coordinates as we did in Chapter 6. Homogeneous coordinates of any point not on $II'$ will be $(x_1, x_2, x_3)$ if its nonhomogeneous coordinates are $(x_1/x_3, x_2/x_3)$. This implies, of course, that $x_3 \neq 0$ for any such point. To obtain homogeneous coordinates on the line $II'$, we consider the lines joining points of that line to $O$. If a line so determined for a point on $II'$ meets $UI'$ at the point with nonhomogeneous coordinates $(1, y)$, we take as homogeneous coordinates of the point on $II'$ $(x_1, x_2, 0)$ where $x_2/x_1 = y$. The point $I'$ will be given coordinates $(0, x_2, 0)$, $x_2 \neq 0$. As before a point will have many different sets of homogeneous coordinates, all proportional to each other. The constant of proportion will be a nonzero element of the field being used to create the coordinates.

The triangle $OII'$ is exactly the reference triangle of Chapter 8. The line $II'$ is analogous to the ideal line in the case of real natural homogeneous coordinates. The unit point as before would be taken to be the point with coordinates $(1, 1, 1)$, determined as the intersection of $IU'$ and $I'U$.

Now that coordinates have been introduced in the projective plane, we can undertake an analytic study of that plane. If we are to proceed in a manner like that of Chapter 6, we must first consider the equation of a line. We shall prove it has equation of the form $u_1x_1 + u_2x_2 + u_3x_3 = 0$; that is, there are field elements $u_1$, $u_2$, and $u_3$ such that the above equation is satisfied by the coordinates of those and only those points lying on a line.

**Theorem 10.31.** *A line has homogeneous first-degree equation, and conversely.*

PROOF. We note first that if the line is $II'$, its equation is $x_3 = 0$. If the line passes through $I'$ and not $I$, its equation is clearly $x = a$ or $x_1 - ax_3 = 0$.

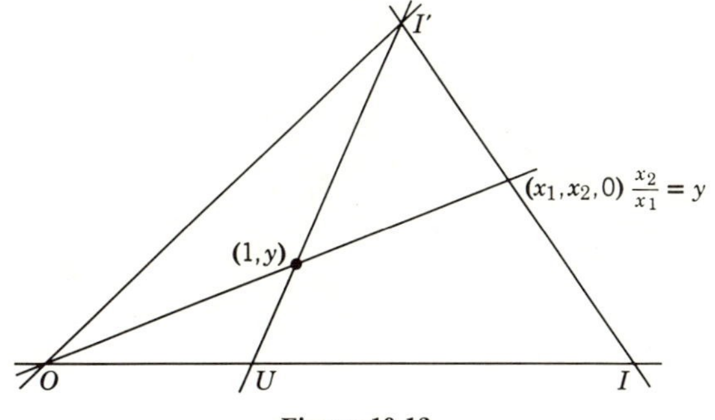


**Figure 10.12**

If it passes through $I$ and not $I'$, its equation is clearly $y = b$ or $x_2 - bx_3 = 0$. In the following we accordingly assume that the line contains neither $I$ nor $I'$.

We suppose the line meets $OI$, $OI'$, and $II'$ at $A$, $B'$, and $C^*$, respectively. Let $P$ be any point on the line except $C^*$; we suppose $P$ has coordinates $(x, y)$. Let $IP$ meet $OI'$ at $Y'$ with coordinates $(0, y)$, and let $I'P$ meet $OI$ at $X$ with coordinates $(x, 0)$. Let $IP$ meet $I'A$ at $Z$, and let $XZ$ meet $II'$ at $D^*$. We first assert that $D^*$ has constant coordinates; that is, it is independent of the choice of $P$ on the line. If $\bar{P}$ is a second point on the line with associated points $\bar{Y}'$, $\bar{X}$, and $\bar{Z}$, as shown in Figure 10.13, the triangles $Z\bar{Z}I$ and $X\bar{X}I'$ are perspective from the line of $A$, $P$, and $\bar{P}$. It follows that the lines $ZX$, $\bar{Z}\bar{X}$, and $II'$ are concurrent; that is, that $\bar{Z}\bar{X}$ passes through $D^*$.

Since the field operations are defined on the line $OI$, we "transfer" the points $Y'$ and $D^*$ to $OI$; that is, we find the points on $OI$ with coordinates associated with those of these points. Corresponding to $Y'$ we find the point with coordinates $(y, 0)$ which is the intersection of $OI$ and $U^*Y'$. To find the $D$ associated with $D^*$, we let $OD^*$ meet $UI'$ at $D''$, let $ID''$ meet $OI'$ at $D'$, and let $U^*D'$ meet $OI$ at $D$; we may assume $D$ has coordinates $(d, 0)$ where $d$ is independent of $x$ and $y$ in view of the preceding argument. The configuration is shown in Figure 10.14.

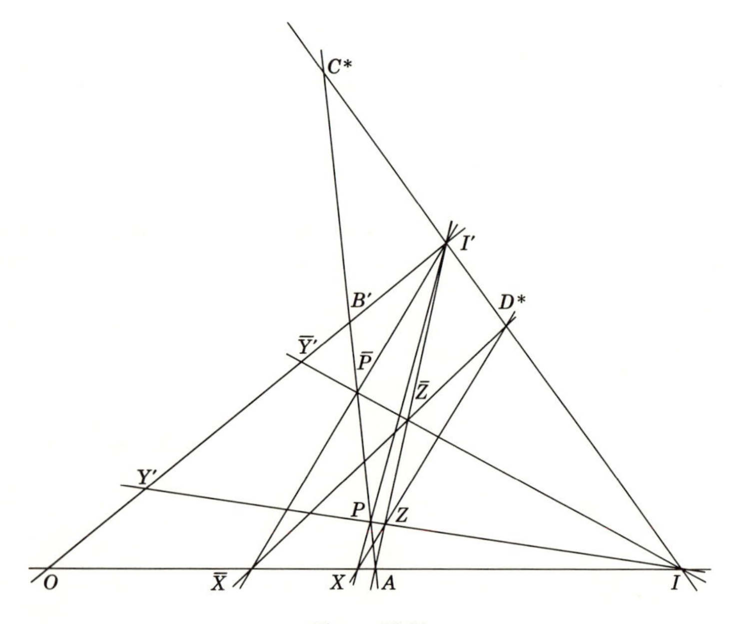


**Figure 10.13**

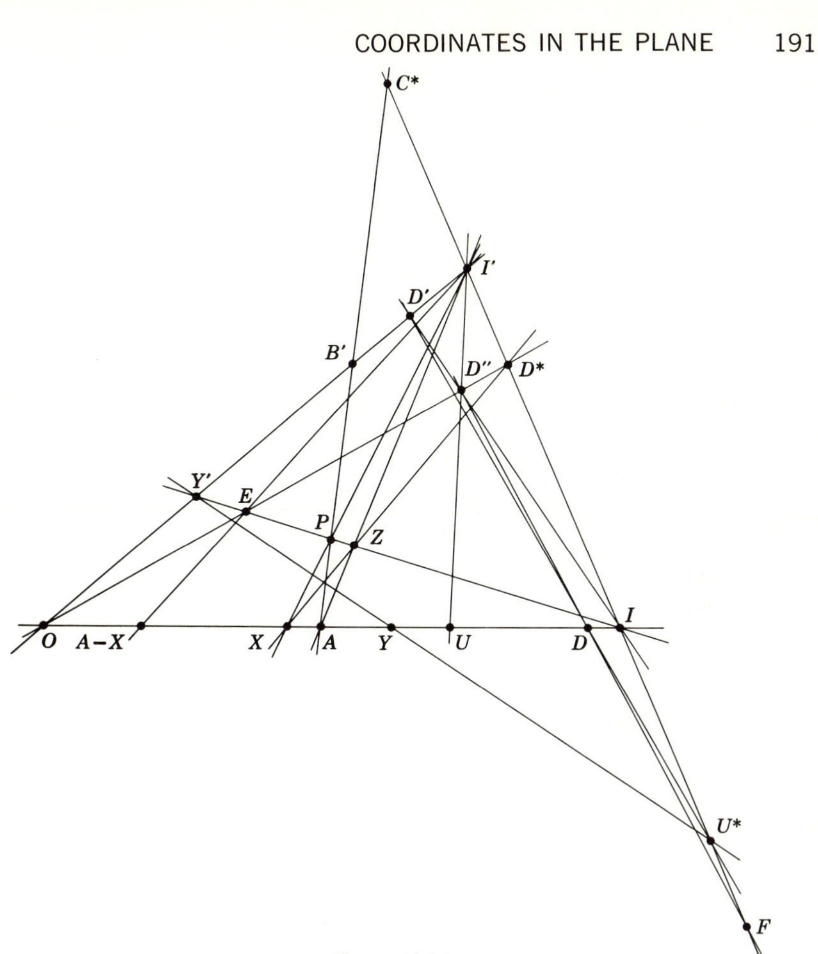


**Figure 10.14**

Let $OD^*$ meet $IP$ at $E$, and let $DD''$ meet $II'$ at $F$. We claim next that $E$, $F$, and $Y$ are collinear since these are the points of intersection of corresponding sides of the triangles $D''DO$ and $IU^*Y'$, which are perspective from the point $D'$. We assert that $I'E$ meets $OI$ at the point $A-X$ since

$$D^*, E, I', Z \to (A-X)+X = A.$$

We next see that

$$I', D'', F, E \to D \cdot (A-X) = Y.$$

This shows that the coordinates of $P$ satisfy the equation

$$y = d(a-x),$$

or in homogeneous form

$$x_2 = d(ax_3 - x_1).$$

Next, if $C^*$ has coordinates $(x_1, x_2, 0)$, we must show $x_2 = -dx_1$ or $x_2/x_1 = -d$. We transfer $C^*$ to $OI$ by letting $C''$ be the intersection of $OC^*$ with $I'U$, letting $C'$ be the intersection of $IC''$ with $OI'$, and letting $C$ be the intersection of $U^*C'$ with $OI$ (see Figure 10.15). The point $C$ then has coordinates $(x_2/x_1, 0)$. We see that $H(I, I'; C^*, D^*)$, the quadrangle determining this harmonic sequence being the one with vertices at $A$, $P$, $X$, and $Z$. If $T_1$ is the central perspectivity from $II'$ to $UI'$ with center at $O$, we have $T_1(I, I', D^*, C^*) = U, I', D'', C''$. If $T_2$ is the central perspectivity from $UI'$ to $OI'$ with center $I$, we find $T_2(U, I', D'', C'') = O, I', D', C'$. Finally, if $T_3$ is the central perspectivity from $OI'$ to $OI$ with center $U^*$, then $T_3(O, I', D', C') = O, I, D, C$. Thus $[T_3T_2T_1](I, I', D^*, C^*) = O, I, D, C$ and $H(O, I; D, C)$, from which it follows that $D + C = O$, $C = -D$, or $x_2/x_1 = -d$ (see Exercise 10.12).

We have shown that any point $P$ on the line satisfies a linear homogeneous equation. Suppose now that $Q$ is a point not on the line. We assert that its coordinates cannot satisfy

$$x_2 = d(ax_3 - x_1). \tag{10.31}$$

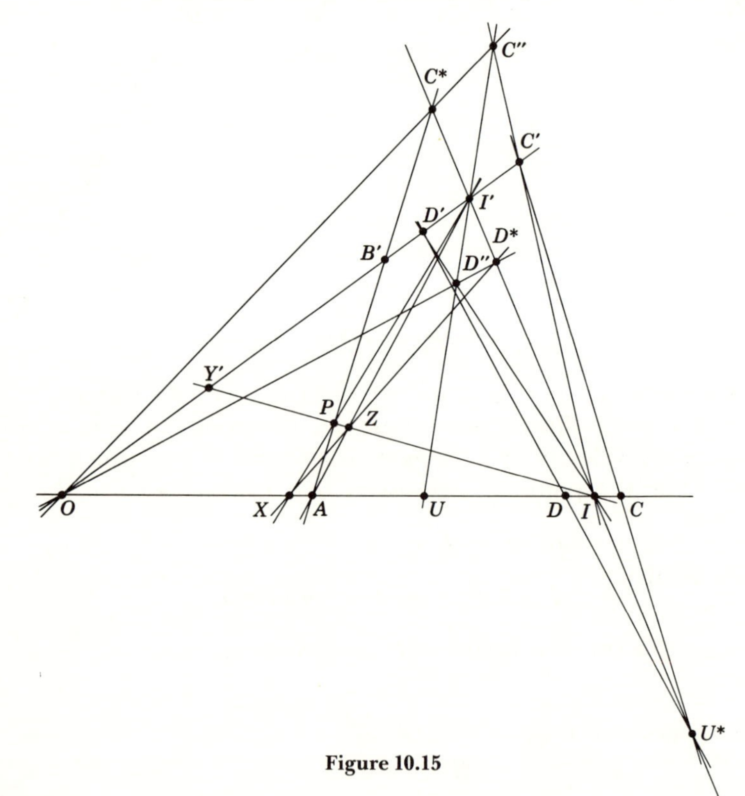


**Figure 10.15**

The line determined by $Q$ and $A$ determines points $\bar{B}'$, $\bar{C}^*$, and $\bar{D}^*$, distinct from $B'$, $C^*$, and $D^*$, respectively. By the previous argument the coordinates of $Q$ satisfy

$$x_2 = \bar{d}(ax_3 - x_1) \tag{10.32}$$

where $d \neq \bar{d}$. It is readily verified that $A$ is the only point whose coordinates simultaneously satisfy equations (10.31) and (10.32). Hence the coordinates of $Q$ do not satisfy equation (10.31).

Conversely, consider the equation

$$u_1x_1 + u_2x_2 + u_3x_3 = 0.$$

If $u_1 = 0$, this is clearly the equation of a line through $I$; if $u_2 = 0$, this is clearly the equation of a line through $I'$. If $u_1u_2 \neq 0$, we have

$$\begin{aligned} x_2 &= -\frac{u_1}{u_2}x_1 - \frac{u_3}{u_2}x_3 \\ &= \frac{u_1}{u_2}\left(-\frac{u_3}{u_1}x_3 - x_1\right), \end{aligned}$$

which we know is the line determined by the points $A$ with coordinates $(-u_3/u_1, 0, 1)$ and $D^*$ with coordinates $(1, u_1/u_2, 0)$. This completes the proof of the theorem.

At this point we have developed the preliminaries of analytic projective geometry; in particular we have carried out the development of the basic ideas of Chapter 8 as applied to the projective plane. Most of the analysis of previous chapters now follows for projective planes with coordinates in an arbitrary field. In the development of Chapter 7 we now let scalars be elements of the field, and we let elements of the matrices be from this field, but all results of the chapter are still valid. The entire theory of dependent and independent points remains valid, as does the development of the change of coordinates and projective transformations. The basic results of Chapters 6, 8, and 9 in analytic projective geometry all remain valid. Some specific results may become invalid because of some property of a particular field. In the following two sections we consider two fields other than the real numbers and show where some specific results do fail or require alteration.

## 10.4 THE SEVEN-POINT PLANE

In the preceding section analytic projective geometry was extended to the point where coordinates could be taken from an arbitrary field. The most familiar fields are the rational numbers, the real numbers, and the complex numbers. All of these fields contain an infinite number of elements; this need not be so for a field. If a field contains a finite number of elements, the resulting projective plane will contain a finite number of points and lines. In this section we consider some finite fields and see that the smallest of these leads to a projective plane containing only seven points.

The reader has probably sometimes encountered some version of the puzzle, "When do 10 and 4 add to 2?" The solution is that this happens when we consider time on a clock, for there 4 hours after 10 o'clock will indeed be 2 o'clock. In general, when we add on a clock, we form the sum and, if it is greater than 11, we find the remainder when the sum is divided by 12 (we agree here that 12 o'clock will be called 0 o'clock). Thus $7+9=4$, $11+8=7$, $4+8=0$, and $4+6=10$. Although we do not do so when computing times, we could also define multiplication in a similar manner. If we wish to find 5 times 7, we find the product 35 and determine its remainder after division by 12. We thus say $5\times 7=11$. In a similar manner we find $7\times 9=3$, $11\times 8=4$, $4\times 6=0$, and $2\times 5=10$. The resulting number system is different from the familiar one and contains a finite number of elements, namely, 0, 1, 2, 3,4, 5, 6, 7, 8, 9, 10, and 11. This number system is said to be the system of *integers modulo 12*.

We can build up the integers modulo any integer $n>1$ in a similar manner. The multiplication and addition tables for the integers modulo 5 are shown in Table 10.41. We see that $2+4=1$, $1+2=3$, $3\times 3=4$, and $2\times 3=1$.

***Table 10.41 Integers Modulo 5***

ADDITION

| | 0 | 1 | 2 | 3 | 4 |
|---|---|---|---|---|---|
| 0 | 0 | 1 | 2 | 3 | 4 |
| 1 | 1 | 2 | 3 | 4 | 0 |
| 2 | 2 | 3 | 4 | 0 | 1 |
| 3 | 3 | 4 | 0 | 1 | 2 |
| 4 | 4 | 0 | 1 | 2 | 3 |

MULTIPLICATION

| | 0 | 1 | 2 | 3 | 4 |
|---|---|---|---|---|---|
| 0 | 0 | 0 | 0 | 0 | 0 |
| 1 | 0 | 1 | 2 | 3 | 4 |
| 2 | 0 | 2 | 4 | 1 | 3 |
| 3 | 0 | 3 | 1 | 4 | 2 |
| 4 | 0 | 4 | 3 | 2 | 1 |

Our concern here is whether the set of integers modulo $n$ form a field. It is clear from the way in which the operations of addition and multiplication have been defined and from the properties of the positive integers that the commutative law of multiplication and properties 1, 2, 3, 4, and 6 of page 184 are satisfied. We are thus concerned with property 5. If we seek an additive inverse $-a$ of an element $a$, we are seeking a solution of $a+x=0$. This equation has the solution $-a=n-a$; that is, the desired inverse is the element we denote by the symbol for the positive integer $n-a$.

The problem of finding a multiplicative inverse $a^{-1}$ for $a \neq 0$ is not so simply settled. We are now seeking a solution of $ax = 1$. For the integers modulo 5 this can be done by inspection of the multiplication table of Table 10.41. Here 1 appears in each row and each column. We see that 1, 2, 3, and 4 have inverses 1, 3, 2, and 4, respectively. The situation is not so simple for the integers modulo 12. By trial and error we find $4x = 1$ has no solution while $5x = 1$ has solution 5. We conclude the integers modulo 5 form a field while the integers modulo 12 do not.

We claim the integers modulo $n$ cannot form a field if $n$ is a composite number. If $n$ is composite, we can write the positive *integer* $n$ as the product of positive *integers* $a$ and $b$, where $a < n$ and $b < n$. In the set of *integers modulo* $n$ we then have $ab = 0$ where neither $a$ nor $b$ is zero. This is impossible in a field, for if $ab = 0$, $a \neq 0$, we can multiply by $a^{-1}$ to obtain $b = a^{-1}0 = 0$.[2]

On the other hand we claim that the integers modulo $p$ where $p$ is a prime form a field. Proofs of this involve some elementary number theory. We give a proof here which assumes only that if a prime $p$ divides the product $ab$, it must divide either $a$ or $b$. Let us choose a positive integer $a < p$ where $a \neq 0$ and consider the set of integers $0a$, $1a, 2a, \cdots, (p-1)a$ modulo $p$. We assert these are distinct, for if $na = ma$ where $n < p$, $m < p$, and where we may assume $m < n$, then $(n-m)a = 0$. This would require that the integer $p$ divide either the integer $a$ or the integer $n-m$. These are both less than $p$, and thus this is impossible. We conclude that this set is the entire set of integers modulo $p$, so that $ax = 1$ or $x = a^{-1}$ for some $x$, where $0 < x < p$. We have proved

**Theorem 10.41.** *The integers modulo n are a field if and only if n is a prime.*

The simplest such field is the system of integers modulo 2. This field contains only the elements 0 and 1 where $0+0=0$, $1+1=0$, $1+0=0+1=1$, $0\times 0=0$, $1\times 1=1$, and $1\times 0=0\times 1=0$. If we consider the projective plane with coordinates taken from this field, each point will have coordinates containing only the symbols 0 and 1. If we exclude (0, 0, 0), there are only seven triples possible, and hence only seven points in this projective plane. This plane has the advantage that the points have unique coordinates since the only possible nonzero proportionality factor is 1.

The seven-point plane is shown in Figure 10.16. By duality this plane contains only seven lines, also shown in the figure. One of these lines

[2]It is easily proved that in a field $a0 = 0$.

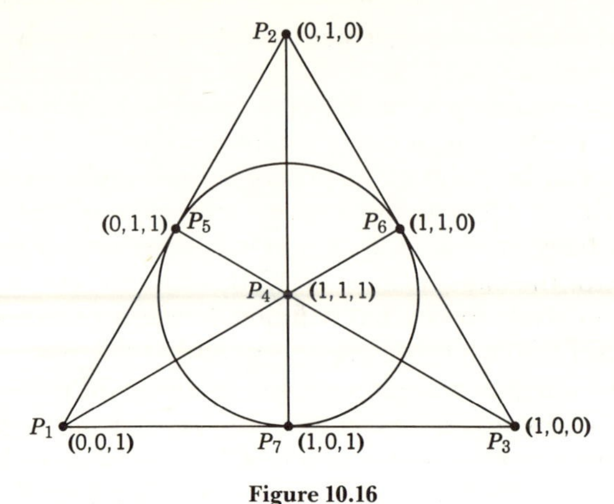


**Figure 10.16**

cannot be indicated as a "straight line." It should be pointed out that this figure is a very poor model for the plane, for the plane contains only the seven points and lines with nothing "between" the points on the lines. The line might better be identified with the set of points it contains; thus the line determined by $P_1$ and $P_2$ in the figure might better be denoted by $[P_1, P_2, P_5]$.

The analytic projective geometry in Chapters 6, 8, and 9 was developed for the real field. Most of the analysis is still valid in an arbitrary field. When considering a field other than the reals, we must take care that no peculiar property of the field will affect the analysis. In the case of the integers modulo 2 we must be sure that an even nonzero integer was not an inherent part of the analysis of these chapters, for now this integer must be replaced by zero and this change may affect the conclusion drawn. There are two occasions where the results break down in the seven-point plane.

In Theorem 8.22 we proved that the diagonal points of a complete quadrangle are not collinear. The diagonal points were proved independent in view of the fact that a certain determinant was $2 \neq 0$. This 2 was not coincidental, and in the present case the determinant will be zero. Thus in the seven-point plane the diagonal points of a complete quadrangle are collinear. This is also readily verified in Figure 10.16; for example, the diagonal points of the quadrangle with vertices $P_2$, $P_4$, $P_5$, and $P_6$ are the collinear points $P_1$, $P_7$, and $P_3$.

A second difficulty in the seven-point plane arises in the derivation of the equation of a conic. A breakdown occurs on page 161 where division by 2 is used to make the equation of the conic symmetric. In the seven-point plane we cannot assert in Theorem 9.41 that the matrix $A$ is symmetric. The process of square completion also involves division by 2 and is not valid here. This invalidates Theorem 9.42. As a result of these breakdowns the analysis of Section 9.4 does not hold in the seven-point plane.

Planes analogous to the seven-point plane can be described using the integers modulo a prime $p$, $p > 2$. All such planes have a finite number of points. These, however, are not the only finite planes. We note that any field must contain $1+1$, $1+1+1$, etc. Let us denote by $n1$, where $n$ is a positive integer, the result in a field of adding 1 to itself $n$ times. It is readily verified that if $n$ is the least integer for which $n1 = 0$, then $n$ must be a prime $p$. When this is the case, the field contains the integers modulo $p$ as a subset (and subfield) and is said to have *characteristic* $p$. When this is not the case, the field contains the rational numbers as a subset and is said to have *characteristic zero*. The reals and the complex numbers are fields of characteristic zero. It can be shown that given any prime $p$ and any positive integer $n$ there is essentially one field of characteristic $p$ with $p^n$ elements and that these are all of the finite fields. There are also infinite fields with finite characteristic $p$. It is apparent that the limitations of the seven-point plane discussed above apply in any field of characteristic 2.

## Exercises 10.4

1. Draw the figure analogous to Figure 10.16 for the projective plane with coordinates from the field of integers to modulo 3.

2. Show that if elements $a$ and $a+1$ are added to the field of integers modulo 2 where $a^2+a+1=0$, we obtain the field with $2^2$ elements.

3. Show that the finite projective plane with coordinates from the field with $p^n$ elements contains $p^{2n}+p^n+1$ points.

4. How many points are on a nondegenerate conic in the seven-point plane?

## 10.5 COMPLEX PROJECTIVE GEOMETRY

Let us now consider the projective plane where points have coordinates of the form $(z_1, z_2, z_3)$, the $z_i$ being complex numbers. This complex projective plane should not be confused with the complex plane of analysis where the point with real coordinates $(x, y)$ represents the complex number $z = x + yi$. This complex plane is actually the set of ordinary points on a complex projective line with homogeneous coordinates $(z_1, z_2)$ where $z_1/z_2 = z$. Indeed, it is customary in complex variable theory to add just one "point at infinity" to this complex plane rather than a whole ideal line of such points. This single point so added is just the ideal point of the complex projective line. Much of complex variable theory corresponds to projective geometry on a complex projective line. In this theory, for example, we study cross ratio and linear (projective) transformations just as we have studied them in previous chapters.

The advantage of using complex numbers for coordinates lies in the fact that any algebraic equation of degree $n$ with complex coefficients has exactly $n$ solutions. This leads to simplification of some of the previous results for the case of the complex projective plane.

An example of this simplification occurs in Theorem 8.71. The quadratic equation considered in the proof of that theorem will always have a solution, and the distinction between hyperbolic and elliptic transformations becomes meaningless. We can replace Theorem 8.71 by

**Theorem 10.51.** *A projective transformation of a complex projective line onto itself has at least one fixed point.*

The same principle leads to the following theorem, the proof of which is left to the reader.

**Theorem 10.52.** *A line not tangent to a conic meets the conic in exactly two points.*

The entire theory of conics is simpler in the complex case. Any complex number has a square root which is a complex number. Thus on page 164 we may set $b_i = \sqrt{a_{ii}}$ if $a_{ii} \neq 0$ and $b_i = 1$ if $a_{ii} = 0$. Theorem 9.42 then becomes

**Theorem 10.53.** *For a suitable choice of coordinates a point conic has equation*

$$c_1 x_1^2 + c_2 x_2^2 + c_3 x_3^2 = 0,$$

*where each $c_i$ is* 1 *or* 0.

In matrix form this means Theorem 9.43 can be replaced by

**Theorem 10.54.** *Given a symmetric matrix $A$, there exists a nonsingular matrix $B$ such that $B^*AB$ is a matrix with zeros off the main diagonal and only ones and zeros on the main diagonal.*

Now there can be no imaginary conics and we can dispense with Definition 9.41. For example, the conic

$$x_1^2 + x_2^2 + x_3^2 = 0$$

does contain points, one such being $(3, 4, 5i)$. The conic

$$x_1^2 + x_2^2 = 0$$

is now satisfied by many points and is actually the pair of lines $x_1 + ix_2 = 0$ and $x_1 - ix_2 = 0$. Thus in Figure 9.16 we can now dispense with cases 1 and 3 as canonical forms and write Theorem 9.45 in the form of

**Theorem 10.55.** *The conic* $X^*AX = (0)$ *is* (a) *nondegenerate if* $A$ *is non-singular,* (b) *two lines if* $A$ *has rank two,* (c) *a line if* $A$ *has rank one, and* (d) *a plane of points if* $A$ *has rank zero.*

Associated with each complex number $x + yi$ is its *complex conjugate* $\bar{z} = x - yi$. This suggests that with each point $P$ we can associate a *complex conjugate point* $\bar{P}$ whose coordinates are the complex conjugates of the coordinates of $P$. We denote the column matrices of coordinates of these points by $X$ and $\bar{X}$. We can now consider a mapping $T$ where $T(P') = P$ given by $X = A\bar{X}'$. Such a mapping is not a projectivity and is sometimes called an *antiprojectivity*. Antiprojectivities have many of the properties of projectivities. An antiprojectivity of a line onto itself preserves the harmonic relation. An antiprojectivity of a plane onto a plane carries one-dimensional primitive forms onto one-dimensional primitive forms The complex conjugate concept also permits a generalization of the study of conics and polarities by using matrix relations of the form $\bar{X}^*AX = (0)$ and $\bar{Y}^*AX = (0)$; in these cases instead of requiring that $A$ be symmetric we require that $a_{ji} = \bar{a}_{ij}$.

Some of the points in the complex projective plane have coordinates whose ratios are real. Such a point would be one whose coordinates are $(2, 3, -5)$, $(2i, 3i, -5i)$, etc. Such points are said to be *real points*. Real lines are similarly defined, and we can even define real conics by requiring that the coefficients of their equations have real ratios. In a certain sense the real projective plane can be thought of as a subset of the complex projective plane. In the following chapters it will be convenient on occasion to so regard the real projective plane.

## Exercises 10.5

1. Prove Theorem 10.52.
2. Find the line determined by each of the following pairs of points:
   *a.* $(1, 8, -1)$ and $(1+i, 1-i, 0)$;
   *b.* $(2i, 3, -5i)$ and $(2, 3i, -5)$;
   *c.* $(2, 3, -5)$ and $(3, 2i, 5)$;
   *d.* $(2+i, 3-2i, 4-3i)$ and $(3+i, 4+2i, 3-2i)$;
   *e.* $(1+i, 1-i, 1)$ and $(i, 1, 1+i)$.
3. Prove that if a line contains two real points, it is real.

4. Reduce each of the conics of Exercise 9.41 to canonical form in the complex projective plane.

5. Prove that if $P$ is on a real line, so is $\overline{P}$.

6. Prove that if $P$ is on $l$, then $\bar{P}$ is on $\bar{l}$.

7. Take the equation of a circle in the Euclidean plane and rewrite it in homogeneous form, regarding it as a conic in the complex projective plane. Show that the points $(1, \pm i, 0)$ (the circular points at infinity) lie on this conic.

## 10.6 THE REAL PROJECTIVE PLANE

We have seen that Axioms 1 to 7 lead to a projective plane with coordinates lying in a field. In the previous two sections we have considered briefly projective planes with coordinates from fields other than the reals. In previous chapters we have considered the projective plane with real coordinates. We would like to conclude an axiomatic introduction to projective geometry by creating the real projective plane. We could do this, of course, by brute force, taking a final axiom which states that the coordinate field is to be the field of real numbers. It would be more consistent with the previous seven synthetic axioms if we could add additional similar axioms which do indeed create the real projective plane. In this section we outline briefly the axioms required to achieve this end.

In Section 5.4 we introduced the concept of a net of rationality, the set of all points obtained by taking three distinct collinear points and repeatedly completing to a harmonic sequence in all possible ways. Our first axiom makes an assumption about nets of rationality which precludes at least all finite coordinate fields. It is a stronger assumption than we need, but is frequently used as the next axiom for the real projective plane.

8. *(Fano) There exists an infinite net of rationality.*

The separation concept introduced in Section 8.6 can now be examined axiomatically. We take an undefined relation of *separation* applied to sets of four distinct collinear points $A$, $B$, $C$, and $D$. If the points fulfill the relation in the order stated, we say $A$ and $B$ separate $C$ and $D$, which we write as $AB \parallel CD$. If the points do not fulfill the relation, we write $AB \nparallel CD$. We introduce axiom

9. *For distinct collinear points $A, B, C, D$, and $E$,*
   a. *$AB \parallel CD$ implies $AB \parallel DC$;*
   b. *$AB \parallel CD$ implies $AC \nparallel BD$;*

c. *$AB \parallel CD$ or $AC \parallel BD$ or $AD \parallel BC$;*
d. *$AB \parallel CD$ and $AC \parallel BE$ imply $AB \parallel DE$;*
e. *If $T$ is a projectivity and if $AB \parallel CD$, then $T(A)T(B) \parallel T(C)T(D)$.*

Axiom 9d is illustrated in Figure 10.17. The reader will do well to note the pattern in which the points involved appear in this statement. We use this part of the axiom frequently in what follows and call it the *principle of conjunctivity*.

We note some elementary consequences of this axiom.

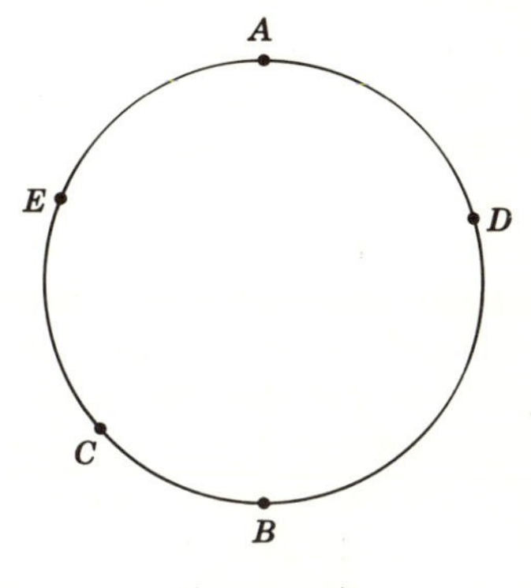

Figure 10.17

**Theorem 10.61.** *Only one of the separation statements of Axiom 9c can be valid.*

The result follows readily from parts a and b of the axiom, and the details of the proof are left to the reader.

**Theorem 10.62.** *If $AB \parallel CD$, then $CD \parallel AB$.*

PROOF. Let $T$ be the involution for which $T(A,B) = C,D$. The result now follows from part e of the axiom.

**Theorem 10.63.** *If $AB \parallel CD$ and $AB \parallel CE$, then $AB \not\parallel DE$.*

PROOF. We first note that we cannot have $AC \parallel DE$. If this were so, by Axiom 9a and Theorem 10.62 we would have $CA \parallel DE$ and $CD \parallel AB$. By conjunctivity we would then have $CA \parallel BE$ and hence $AB \not\parallel CE$, contrary to hypothesis. If $AC \not\parallel DE$, we must have $AD \parallel CE$ or $AE \parallel CD$. If the former holds, by Axiom 9a and Theorem 10.62 we have $DA \parallel CE$ and $DC \parallel AB$. By conjunctivity these imply $DA \parallel BE$, from which $AB \not\parallel DE$ by Theorem 10.61. If the latter holds, in a similar manner we have $EA \parallel CD$ and $EC \parallel AB$ implying $EA \parallel BD$, from which $AB \not\parallel DE$.

**Theorem 10.64.** *$H(A,B;C,D)$ implies $AB \parallel CD$.*

PROOF. We know $H(A,B;C,D)$ implies $H(A,B;D,C)$ and that there is an involution $T$ for which $T(A,B,C,D) = A,B,D,C$. If $AC \parallel BD$ or $AD \parallel BC$, application of Axiom 9e leads to a contradiction with Theorem 10.61. We conclude that $AB \parallel CD$.

It should be noted that we made use of Fano's axiom here in assuming $H(A,B;C,D)$ implies $A$, $B$, $C$, and $D$ are distinct.

Let us now suppose that the points $O$, $U$, and $I$ have been chosen on a line and that the points on this line form a field under the operations of addition and multiplication defined in Section 10.2. Let $\mathscr{P}$ be the set of points $P$ on this line for which $OI \| P(-U)$. We prove the following three theorems about $\mathscr{P}$.

**Theorem 10.65.** *If $A$ is neither $O$ nor $I$, then one and only one of the points $A$ and $-A$ is in $\mathscr{P}$.*

PROOF. Figure 10.18 shows the construction for the point $-A$. We see that $H(O, I; A, -A)$. Thus $OI \| A(-A)$, and in particular $OI \| U(-U)$. Moreover, if $T$ is the involution on the line with fixed points $O$ and $I$, we know by Theorem 8.65 that $T(A) = -A$. Suppose that $-A$ is not in $\mathscr{P}$, so that $OI \nparallel (-A)(-U)$. If $OA \| I(-U)$, by conjunctivity we would have $OI \| (-A)(-U)$ contrary to assumption. If $O(-U) \| IA$, then $OI \| AU$ by conjunctivity. If we apply $T$ and Axiom 9e, this again implies $OI \| (-A)(-U)$. We conclude that $OI \| A(-U)$ and $A$ is in $\mathscr{P}$. We have proved that $A$ or $-A$ is in $\mathscr{P}$.

We must still show only one of the points $A$ and $-A$ is in $\mathscr{P}$. If $A$ is in $\mathscr{P}$, then $OI \| A(-U)$ and $OI \| U(-U)$. By Theorem 10.63 we have $OI \nparallel AU$. If we apply the involution $T$, we conclude by Axiom 9e that $OI \nparallel (-A)(-U)$ and $-A$ is not in $\mathscr{P}$. A similar argument shows that if $-A$ is in $\mathscr{P}$, then $A$ is not in $\mathscr{P}$.

**Theorem 10.66.** *If $A$ is in $\mathscr{P}$ and $B$ is in $\mathscr{P}$, then $A+B$ is in $\mathscr{P}$.*

PROOF. We have $OI \| A(-U)$ and $OI \| B(-U)$. Hence by Theorem 10.63 $OI \nparallel AB$. Without loss of generality we may assume $OB \| AI$. Let $T$ be the involution for which $T(O, B) = (A+B), A$. If we apply Theorem 8.77 to the quadrangle with vertices $P$, $Q$, $R$, and $S$ shown in Figure 10.19, we conclude that $T(I) = I$. We have

$$T(O, B, A, I) = (A+B), A, B, I.$$

Hence $(A+B)A \| BI$ since $OB \| AI$. By conjunctivity these relations imply

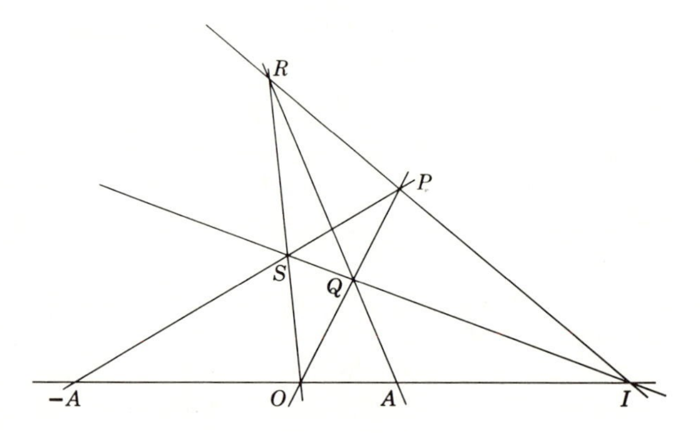


Figure 10.18

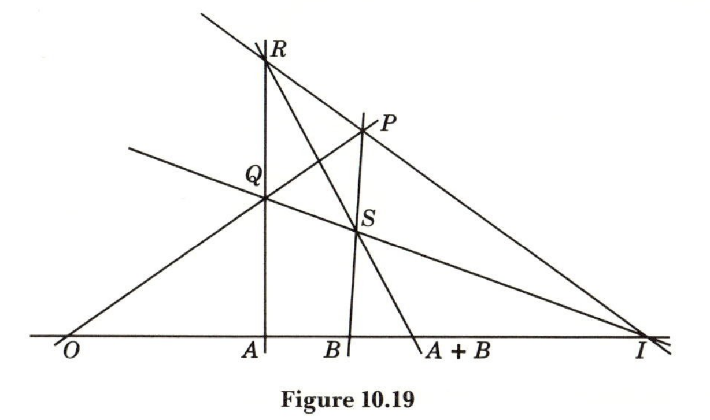


Figure 10.19

$(A+B)O \parallel AI$. This, together with $(-U)A \parallel OI$, implies $(A+B)(-U) \parallel OI$ or $OI \parallel (A+B)(-U)$, whence $A+B$ is in $\mathscr{P}$.

**Theorem 10.67.** *If $A$ is in $\mathscr{P}$ and $B$ is in $\mathscr{P}$, then $A \cdot B$ is in $\mathscr{P}$.*

PROOF. If $T$ is the involution for which $T(A,O)=B,I$, we see $T(U) = A \cdot B$ by applying Theorem 8.77 to the quadrangle with vertices $K, L, M$, and $N$ shown in Figure 10.20. Since $OI \parallel A(-U)$ we know $OI \parallel AU$ (see the proof of Theorem 10.65), whence $OA \parallel IU$ or $OU \parallel AI$. If we apply $T$ to the former, we conclude $IB \parallel O(A \cdot B)$. This, together with $IO \parallel B(-U)$, implies $IO \parallel (A \cdot B)(-U)$ or $A \cdot B$ is in $\mathscr{P}$. Like reasoning shows that if $OU \parallel AI$, we must still have $IO \parallel (A \cdot B)(-U)$ and $A \cdot B$ must be in $\mathscr{P}$.

The three previous theorems derived from Axioms 8 and 9 have placed a considerable restriction of the field of points on the line $OI$. We introduce

**Definition 10.61.** *A field is* ordered *if it contains a subset $\mathscr{P}$ of elements such that*

a. *if $a \neq 0$, one and only one of the elements $a$ and $-a$ is in $\mathscr{P}$;*

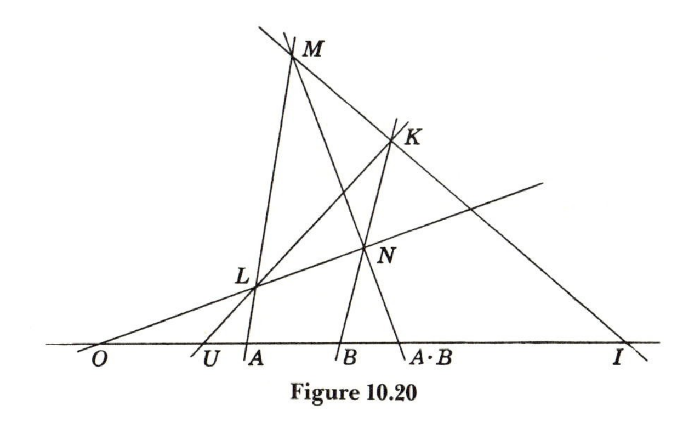


Figure 10.20

b. *if a and b are in $\mathscr{P}$, then $a+b$ and $ab$ are in $\mathscr{P}$.*[3]

We thus have

**Theorem 10.68.** *A projective plane fulfilling Axioms* 1 to 9 *has coordinates from an ordered field.*

Let us consider the nature of an ordered field. First we note that $-1$ cannot be in $\mathscr{P}$; if it were, 1 would not be in $\mathscr{P}$ while $(-1)\ (-1) = 1$ would have to be in $\mathscr{P}$. We conclude that 1 is in $\mathscr{P}$. Now we must have $1+1+\cdots+1$ in $\mathscr{P}$, whence $1+1+\cdots+1 \neq 0$ for all such finite sums of 1*s*. This says an ordered field has characteristic zero (page 197). Such finite sums of 1*s* are essentially the positive integers and can be identified with them. The least field containing the positive integers is the field of rational numbers. In short, an ordered field must contain the rational numbers, and we have

**Theorem 10.69.** *A projective plane fulfilling Axioms* 1 to 9 *has coordinates from a field containing the rational numbers.*

To ensure that this field is the field of real numbers, we again add the axiom of Archimedean order and the axiom of completeness The axiom of Archimedean order used in Hilbert's axioms employs the concept of order. The reader should have no difficulty in perceiving that the following statement of the axiom in terms of separation carries the same force. We denote $A+A+\cdots+A$ ($n$ times) by $nA$.

10. *If $OI \parallel A(-U)$ and $OI \parallel B(-U)$, there is a positive integer n such that $O(nA) \parallel BI$.*

We again assert without proof that an Archimedean ordered field is a subfield of the reals. Assuming this, we have

**Theorem 10.610.** *A projective plane fulfilling Axioms* 1 to 10 *has coordinates from a subfield of the field of real numbers.*

Finally, we take

11. *No additional points or lines can be added to the system without violating one of the preceding axioms.*

[3]In an ordered field we can introduce the inequality relation simply be defining $a > b$ if $a-b$ is in $\mathscr{P}$. Thus $a$ in $\mathscr{P}$ is equivalent to stating $a > 0$, and $\mathscr{P}$ is the set of "positive" elements of the field.

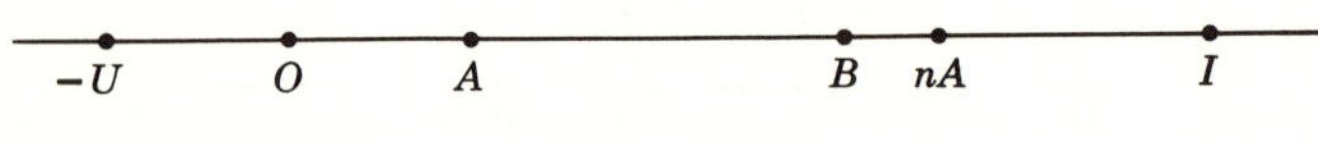


**Figure 10.21**

We then have

**Theorem 10.611.** *A projective plane fulfilling Axioms* 1 to 11 *has coordinates from the field of real numbers.*

While Axioms 1 to 11 thus create the real projective plane, it should not be inferred that these axioms constitute a minimal set for this plane. We have already noted that Axioms 1 to 5 and 7 imply Axiom 6, which is thus not needed as an axiom. It should be apparent that a far weaker assumption than Axiom 8 would have been sufficient to enable the separation concept to lead to Theorem 10.69.

Finally, we note we can now show the identity of separation as introduced here with the concept of Section 8.6. If we compute $R(I, O, -U, P)$, we find that it is $-x$, where $x$ is the coordinate of $P$. We thus see that this cross ratio is negative if and only if $x > 0$ and $IO \parallel (-U)P$. Since both cross ratio and separation are preserved under a projectivity, we have proved

**Theorem 10.612.** $R(A, B, C, D) < 0$ *if and only if* $AB \parallel CD$.

## Exercises 10.6

1. Prove there is a one-to-one correspondence between the points on a circle and the points of a real projective line.

2. Prove Theorem 10.61.

3. Prove that the inequality definition of the footnote of page 204 leads to a relation which is transitive and fulfills the trichotomy (see page 23).

## REFERENCES

Adler, Claire Fisher, *Modern Geometry, An Integrated First Course,* 2nd ed., McGraw-Hill Book Company, New York, 1967, Chapters 4, 5.

Artzy, Rafael, *Linear Geometry,* Addison-Wesley Publishing Company, Reading, Mass., 1965, Chapter 4.

Blumenthal, Leonard M., *A Modern View of Geometry,* W. H. Freeman and Company, San Francisco, 1961, Chapters 4–6.

Coxeter, H. S. M., *Non-Euclidean Geometry,* 5th ed., University of Toronto Press, Toronto, 1965, Chapters 2, 4.

Coxeter, H. S. M., *The Real Projective Plane,* McGraw-Hill Book Company, New York, 1949, Chapter 11.

Dorwart, Harold L., *The Geometry of Incidence,* Prentice-Hall, Englewood Cliffs, N. J., 1966, Chapter 4.

Eves, Howard, *A Survey of Geometry,* vol. I, Allyn and Bacon, Boston, 1963, Chapter 8.

Gruenberg, K. W., and A. J. Weir, *Linear Geometry,* D. Van Nostrand Company, Princeton, 1967, Chapter 2.

Hartshorne, Robin, *Foundations of Projective Geometry.* W. A. Benjamin, New York, 1967, Chapters 6, 7.

Hodge, W. V. D., and D. Pedoe, *Methods of Algebraic Geometry,* vol. 1, Cambridge University Press, Cambridge, 1947, Chapter 6.

O'Hara, C. W., and D. R. Ward, *An Introduction to Projective Geometry,* Oxford University Press, New York, 1937, Chapter 7.

Robinson, Gilbert deB., *The Foundations of Geometry,* The University of Toronto Press, Toronto, 1952, Chapters 6 to 9.

Rosenbaum, Robert A., *Introduction to Projective Geometry and Modern Algebra,* Addison-Wesley Publishing Company, Reading, Mass., 1963, Chapter 3.

Seidenberg, A., *Lectures in Projective Geometry,* D. Van Nostrand Company, Princeton, 1962, Chapters 2 to 5.

Veblen, Oswald, and John Wesley Young, *Projective Geometry,* vol. I, Ginn and Company, Boston, 1910, Chapters 6, 7.

Young, John Wesley, *Projective Geometry,* The Open Court Publishing Company, Chicago, 1930, Chapter 8.

CHAPTER ELEVEN

# AFFINE AND EUCLIDEAN GEOMETRY

In the preceding chapter real projective geometry was developed without any prior assumption of the existence and results of Euclidean geometry This chapter is devoted to a consideration of the relation of the abstract plane projective geometry of that chapter to Euclidean geometry and some related geometries.

## 11.1 GROUPS AND SUBGROUPS OF TRANSFORMATIONS

The reader has been invited to consider the concept of *group* in several previous exercises. We turn now to a formal investigation of this concept. Given a set of elements, we say a *binary operation* "*" is defined on the set if any ordered pair $(a, b)$ of elements of the set determines a unique element $c$ of the set, which we denote by $c = a*b$; thus a binary operation is a function whose domain is the set of ordered pairs of elements of the set and whose range is all or part of the set. Addition or multiplication in a division ring or a field is an example of a binary operation. We are concerned here with cases where the binary operation fulfills the following conditions:

1. (*Closure*) *$a*b$ is defined and in the set for any $a$ and $b$;*

2. (*Associativity*) $a*(b*c) = (a*b)*c$ *for all* $a, b,$ *and* $c$;

3. (*Identity*) *there is an element* $e$ *such that* $a*e = e*a = a$ *for all* $a$;

4. (*Inverse*) *given any* $a$, *there is an element* $a^{-1}$ *such that* $a*a^{-1} = a^{-1}*a = e$.

We make the following definition:

**Definition 11.11.** *A set of elements on which a binary operation fulfilling Axioms* 1 *to* 4 *is defined is* a group *under this operation. If the operation is commutative* ($a*b = b*a$), *the group is* commutative *or* abelian.

Many of the sets we have been considering are groups under an appropriate operation. We consider here those pertinent for our investigation and leave some of the others for verification by the reader in the exercises.

**Theorem 11.11.** *The set of all* $n \times n$ *nonsingular matrices with elements from a field is a group under the operation of matrix multiplication.*

The proof is left as an exercise for the reader.

Many sets of projective transformations are groups. We consider the projective transformations of a plane of points onto itself given by the matrix relations $kX = AX'$ where $X$ and $X'$ are column matrices with three elements and $A$ is a nonsingular $3 \times 3$ matrix. Such a transformation also induces a mapping of lines onto lines and pencils of lines onto pencils of lines. It is convenient to think of the plane simultaneously as a plane of points, a plane of lines, and a plane of pencils of lines; we say the plane is a *planar field.* The above transformation of points of the projective planes induces mappings of lines onto lines and pencils onto pencils. We thus carry one-dimensional primitive forms of this planar field onto like one-dimensional primitive forms. Such a transformation is said to be a *projective collineation.* It is clear that a projective transformation of a line onto a line is the restriction of a collineation to the domain of collinear points. Now as a simple consequence of Theorem 11.11 we have

**Theorem 11.12.** *The set of all projective collineations of a projective plane is a group under function multiplication.*

The proof is left to the reader.

Let $R$ be a relation; that is, $R$ is a set of ordered $n$-tuples $(P_1, P_2, \cdots, P_n)$ of points of the projective plane. The relation is said to be *invariant* under a transformation $T$ if whenever $(P_1, P_2, \cdots, P_n)$ is in the relation, then $(T(P_1), T(P_2), \cdots, T(P_n))$ is in the relation. The relation is said to be invariant under a group of transformations if it is invariant under every transformation of the group.

Suppose the relation $H$ is the harmonic relation; thus $H$ is the set of all $(A, B, C, D)$ for which $H(A, B; C, D)$. Then $H$ is invariant under the group of all collineations since $H(A, B; C, D)$ implies $H(T(A), T(B); T(C), T(D))$. If $R_x$ is the set of all $(A, B, C, D)$ for which $x = R(A, B, C, D)$, this relation is also invariant under the group of collineations. If $L$ is the set of all $(A, B, C)$ for which $A$, $B$, and $C$ are collinear, this relation is also invariant under the group of collineations. Indeed, all fundamental relations of projective geometry which we have developed can be shown to be invariant under the group of collineations.

Projective geometry can be characterized as the study of relations invariant under the group of collineations. Felix Klein (1849 to 1925) in his famous *Erlangen Program* pointed out that any geometry can be characterized as the study of relations invariant under some group of transformations. Thus Euclidean geometry has associated with it some group of transformations. If we are concerned with a comparison of projective and Euclidean geometry, it would be pertinent to compare the group of collineations of the projective plane with the group of transformations associated with Euclidean geometry. We see later that the latter can be taken to be a subset of the former.

The idea of a subset of a group which is itself a group suggests the following definition:

**Definition 11.12.** *A subset of a group is a* subgroup *of the given group if it is a group under the same binary operation.*

If we have a subset of a group, what must be done to verify that it is a group? Clearly the associative property holds, but the other three properties must be verified. We claim that it is sufficient to verify closure within the subset and the presence of inverses in the subset, for if these are verified and if $a$ is in the subset, $a^{-1}$ is in the subset as is $a * a^{-1} = e$. We thus have

**Theorem 11.13.** *A subset $S$ of a group $G$ is a subgroup of $G$ if for every $a$ and $b$ in $S$, $a * b$ and $a^{-1}$ are in $S$.*

Using this theorem, we can readily find many subgroups of the group $G$ of collineations of the projective plane. Thus the set $H$ of transformations leaving a given line in the plane fixed is a subgroup, for if the transformations $T$ and $S$ leave the line fixed, certainly the transformations $ST$ and $T^{-1}$ also leave the line fixed. In a like manner, the set $J$ of transformations leaving this line pointwise fixed (mapping each point of the line onto itself) is a subgroup; indeed $J$ is a subgroup of $G$ and also a subgroup of $H$.

We claim that the geometry associated with any subgroup of transformations of this group $G$ is a specialization of projective geometry. The elements of the subgroup are all elements of the original group; thus ideally all relations invariant in projective geometry will remain invariant in the geometry associated with the subgroup: lines will transform into lines, cross ratio will be preserved, harmonic sequences will be mapped onto harmonic sequences, and so on. On the other hand, there will be presumably additional invariant relations, associated with the subgroup, which are not invariant in the larger group. We conclude that the theorems of the geometry associated with the subgroup are those of projective geometry plus others peculiar to the specialized geometry.

The argument above assumes that the subgroup of $G$ is considered to be a group of transformations whose domains are the points of the projective plane. If we choose to restrict the domain, as we shall, to a part of the projective plane, some modification of this ideal situation will be required.

In the following sections we consider geometries associated with certain subgroups of the group of collineations of the real projective plane.

## Exercises 11.1

1. Prove that each of the following is a group:
   *a.* A division ring, under addition;
   *b.* The nonzero elements of a division ring, under multiplication;
   *c.* A vector space, under addition;
   *d.* The set of all $m \times n$ matrices, under matrix addition;
   *e.* The integers, under addition;
   *f.* The set of all hyperbolic transformations of a line onto itself with the same fixed points.

2. Prove that each of the following is a subgroup of the group of projective collineations:
   *a.* The transformations leaving a given conic fixed;
   *b.* The transformations leaving a given line fixed and having two given fixed points on this line.

3. Prove that a projectivity of a line onto a line is the restriction of a collineation to a domain of collinear points.

## 11.2 AFFINE GEOMETRY

We consider the geometry associated with the subgroup of collineations leaving a given line fixed. We choose coordinates so that this given line has equation $x_3 = 0$. We formally introduce the distinction between ideal and ordinary point in

**Definition 11.21.** *The points on* $x_3 = 0$ *are* ideal *while all other points in the real projective plane are* ordinary. *The line* $x_3 = 0$ *is* ideal; *all other lines are* ordinary.

The subgroup of transformations we are considering will carry ordinary points into ordinary points and ordinary lines into ordinary lines. We restrict our consideration to the ordinary points and ordinary lines, the ideal points and line being deleted; we also restrict the transformations to their effect on ordinary points. In short, we restrict the domain of a transformation $T$ of the subgroup to the ordinary points of the plane. The subgroup is so defined that the range of such a $T$ will be the ordinary points of the plane. Since this restriction leaves us with a one-to-one transformation with smaller range and domain, the relations invariant under a projective transformation will remain invariant here, although the set of elements in a relation may be reduced. Thus if $H$ is the set of $(A, B, C, D)$ for which $H(A, B; C, D)$, we here restrict the relation to the set $H'$ for which $A$, $B$, $C$, and $D$ are all ordinary. Now if $T$ is an element of the subgroup, $T(A)$, $T(B)$, $T(C)$, and $T(D)$ are all ordinary and $(T(A), T(B), T(C), T(D))$ is in $H'$. Other subsets of $H$ will lead to new invariant relations. The set $H''$ of $(A, B, C)$, all points ordinary, for which $H(A, B; C, D)$ with $D$ ideal is a relation invariant under the subgroup of transformations.

**Definition 11.22.** *A real projective plane from which a line has been deleted is a* real affine plane. *The projective transformations leaving the line fixed, restricted to the real affine plane, are* real affine transformations. *The study of the affine plane and real affine transformations is* real plane affine geometry.

Where there is no confusion, we shall omit the word "real." The points and lines of the affine plane are just the points and lines of the Euclidean plane. While we have deleted a line and some points of the projective plane to obtain the affine plane, we can regard the latter as imbedded in the former and can use results of projective geometry to prove theorems of affine geometry. In particular it should be noted that an affine transformation $T'$ is a projective transformation $T$ restricted to the domain of ordinary points; for an ordinary point $P$, $T(P) = T'(P)$ while if $I$ is ideal, we are sure that $T(I)$ is ideal.

All points in the affine plane have coordinates of the form $(x_1, x_2, x_3)$ where $x_3 \neq 0$. Each such point has a well-defined pair of real numbers $(x, y)$, $x = x_1/x_3$, $y = x_2/x_3$, associated with it, these numbers being called the *nonhomogeneous coordinates* of the point. It is convenient to use these coordinates in studying affine transformations. We easily show that affine transformations have the form given in

**Theorem 11.21.** *An affine transformation has the form*

$$x = b_{11}x' + b_{12}y' + b_{13}$$
$$y = b_{21}x' + b_{22}y' + b_{23}$$

*where*

$$\begin{vmatrix} b_{11} & b_{12} \\ b_{21} & b_{22} \end{vmatrix} \neq 0.$$

The proof is left as an exercise for the reader.

Affine transformations are more general than the rigid motions of Euclidean geometry and contain them as a subgroup, as we see later. Affine transformations can lead to translations or rotations, but also to stretches and shears in the plane; that is, the figure we call a square can be mapped onto a rectangle or even a parallelogram. Some special types of affine transformations are considered in the exercises at the end of the section.

Let us see which of Hilbert's axioms are valid in affine geometry. It is clear that with the deletion of the ideal line the axioms of incidence are valid. Parallelism can be introduced by

**Definition 11.23.** *Lines are* parallel *if they have no common point.*

With this definition made we can establish the axiom of parallels by proving

**Theorem 11.22.** *Given a line and a point not on the line, there is one and only one line containing the given point and parallel to the given line.*

PROOF. The unique parallel containing the given point is the unique line determined by the point and the ideal point of the given line.

It is apparent now that Theorem 2.41 is a theorem of affine geometry as well as of Euclidean geometry. That parallelism is an affine property is shown in the next theorem, the proof of which is left to the reader.

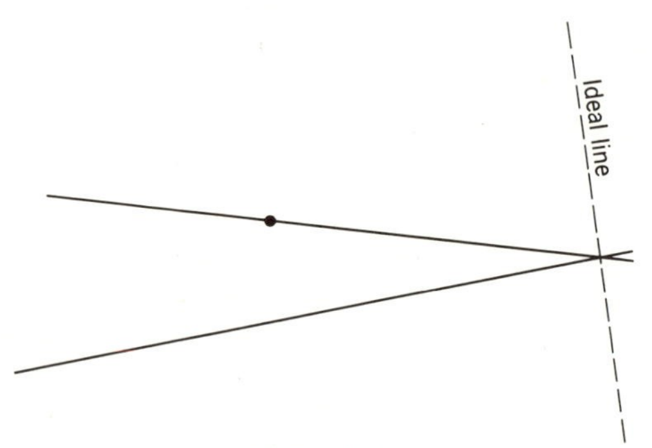


**Figure 11.1**

We denote by $T(a)$ the line containing the transforms of all points of the line $a$.

**Theorem 11.23.** *If $T$ is an affine transformation and if $m$ is parallel to $n$, then $T(m)$ is parallel to $T(n)$.*

We assert the order concept of Hilbert's axioms is affine. We introduce the betweenness relation in

**Definition 11.24.** *If $A$, $B$, and $C$ are distinct points on a line with ideal point $I$, then $B$ is* between *$A$ and $C$, or $ABC$, if $AC\|BI$.*

That betweenness is invariant under an affine transformation follows at once from Axiom 9e of separation and that fact that an affine transformation is the restriction of a projective transformation carrying ideal points into ideal points. Thus if $AC\|BI$ and if $T$ is affine, then $T(A)T(C)$ $\|T(B)T(I)$. We have proved

**Theorem 11.24.** *If $ABC$ and if $T$ is affine, then $T(A)T(B)T(C)$.*

We must show that the order concept introduced in Definition 11.24 is just the order concept of Hilbert's axioms. This is shown in

**Theorem 11.25.** *Affine geometry satisfies Hilbert's axioms of order under the definition of betweenness of Definition 11.24.*

PROOF. Hilbert's Axioms 5, 6, and 8 follow at once from the axiom of separation and Theorem 10.61. Verification is left to the reader.

To verify Hilbert's Axiom 7, apply Theorem 10.64. The points $B$ and $D$ can be taken as the points for which $H(A,C;B,I)$ and $H(A,D;C,I)$, $I$ being the ideal point of the line of $A$ and $C$.

We consider next Hilbert's Axiom 9. Let $A$, $B$, $C$, and $D$ be distinct collinear points. We assume they are so labeled that $ACD$; that is, $AD\|CI$ or $DA\|CI$. We assert we may also assume $AC\|BD$. If this is not the case, we must have either $AB\|CD$ or $AD\|BC$; we show that in either of these cases appropriate relabelling of points leads to the desired relations. If $AB\|CD$ or $DC\|AB$, by separation Axiom 9d $DA\|BI$ and the desired relations hold if we interchange $B$ and $C$. If $AD\|CB$, then since $AD\|CI$, $AD\not\|BI$ by Theorem 10.63. If $AB\|DI$, we achieve the desired relations by replacing $B$ by $D$, $C$ by $B$, and $D$ by $C$. If $AI\|BD$, or $DB\|AI$, this with $DA\|BC$ implies $DB\|CI$ by conjunctivity. The desired relations are then obtained by interchanging $A$ and $B$.

We now have $AD\|CI$ and $AC\|DB$. By conjunctivity we have $AD\|BI$, from which $ABD$, and $AC\|BI$, from which $ABC$. If we apply the same principle to $DA\|CI$ and $DB\|AC$, we have $DB\|CI$, from which $BCD$. We have thus verified Hilbert's Axiom 9.

Finally we must prove Pasch's axiom. Suppose a line meets the sides of the triangle with vertices $A$, $B$, and $C$ at the points $D$, $E$, and $F$, as shown in Figure 11.2. We suppose that $ADB$ or $AB\|DI$. We assume $E$ (which may be ideal) is not between $B$ and $C$. We must prove $AFC$. We assume $BE\|CI''$; the proof for the case where $BI''\|EC$ is similar. Let $FI$ meet $BC$ at $G$. If $T$ is the central perspectivity from $AB$ to $CB$ with center $F$, we have

$$T(A, B, D, I) = C, B, E, G,$$

from which $CB\|EG$ or $BC\|EG$. If we apply the principle of conjunctivity to this relation and to $BE\|CI''$, we see $BC\|GI''$. Now if $S$ is the central perspectivity from $BC$ to $AC$ with center $I$, we have

$$S(B, C, I'', G) = A, C, F, I'.$$

Thus $AC\|FI'$ and $AFC$. This completes the proof of Theorem 11.25.

With Hilbert's axioms of order now established, it follows that segment and angle may be defined and that all the results of Section 2.2 are properly a part of affine geometry as well as Euclidean geometry. With segment and angle defined we could attempt to define length of a segment and measure of an angle by the usual Euclidean formulas. It is readily verified (Exercise 11.26) that these measures are not preserved under an affine transformation. Congruence is not a concept of affine geometry. This concept is the distinctive difference between affine and Euclidean geometry. In the former we preserve neither "size" nor "shape" under an affine transformation.

While segment length is not defined in affine geometry, we can use the result of Exercise 5.211 to define the midpoint of a segment. In the following definition midpoint is defined, but there is no claim being made that the midpoint divides the segment into two congruent segments.

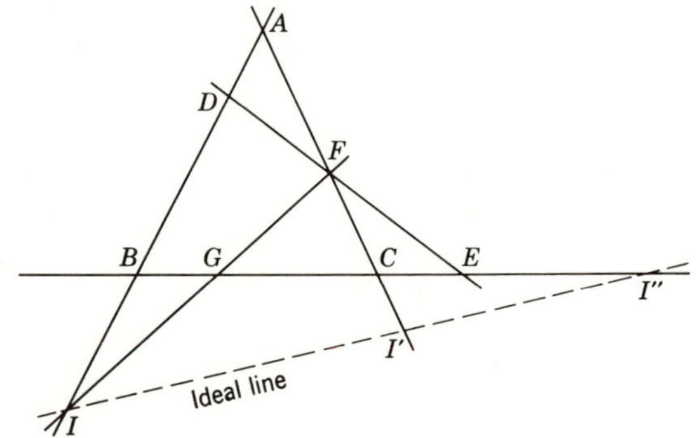


**Figure 11.2**

**Definition 11.25.** *$C$ is the* midpoint *of the segment $AB$ if $H(A, B; C, I)$ where $I$ is the ideal point of the line of $A$ and $B$.*

The reader should find it interesting to use this definition to prove the following theorems, which are actually affine as well as Euclidean.

**Theorem 11.26.** *The line joining the midpoints of two sides of a triangle is parallel to the remaining side of the triangle.*

**Theorem 11.27.** *The diagonals of a parallelogram bisect each other.*

Since segment is an affine concept, we can assume in affine geometry that the sides of a triangle are segments; that is, Definition 5.11 need not be used. A parallelogram is defined to be a simple quadrilateral (Definition 9.21) with opposite sides parallel.

We noted in Section 9.1 that the classification of conics into ellipse, parabola, and hyperbola seemed to involve the incidence relations between the conic and the ideal line. Thus the classification should be affine. We introduce

**Definition 11.26.** *A nondegenerate conic is an* ellipse, parabola, *or* hyperbola *according as it meets the ideal line in no, one, or two points.*

Using this definition, we can simply prove the discriminant condition of

**Theorem 11.28.** *The nondegenerate conic with equation*

$$Ax^2 + 2Bxy + Cy^2 + 2Dx + 2Ey + F = 0 \tag{11.21}$$

*is an ellipse, parabola, or hyperbola according as $B^2 - AC$ is negative, zero, or positive.*

PROOF. We first rewrite the equation of the conic in homogeneous form and consider the nature of its intersection with $x_3 = 0$. The details of the proof are left to the reader.

Since affine transformations are more restrictive than the projective ones, the reductions to canonical form are not the same as those of Section 9.4 for conics in the projective plane.

If the nondegenerate conic of equation (11.21) is a parabola so that $B^2 - AC = 0$, then $A \neq 0$ or $C \neq 0$, for otherwise $B = 0$ and

$$\begin{vmatrix} A & B & D \\ B & C & E \\ D & E & F \end{vmatrix} = \begin{vmatrix} 0 & 0 & D \\ 0 & 0 & E \\ D & E & F \end{vmatrix} = 0,$$

violating the assumption of nondegeneracy. We may assume without loss of generality that $A \neq 0$ and that $A > 0$.[1] Then the affine change of coordinates

$$\begin{aligned} x' &= x + \left(\frac{B}{A}\right) y \\ y' &= \qquad\quad y \end{aligned} \tag{11.22}$$

where

$$\begin{vmatrix} 1 & B/A \\ 0 & 1 \end{vmatrix} = 1 \neq 0$$

reduces the conic to

$$Ax'^2 + 2D'x' + 2E'y' + F' = 0, \tag{11.23}$$

where $D'$, $E'$, and $F'$ are found by solving equations (11.22) for $x$ and $y$ and substituting in equation (11.21). Nondegeneracy requires that

$$\begin{vmatrix} A & 0 & D' \\ 0 & 0 & E' \\ D' & E' & F' \end{vmatrix} = E'^2 A \neq 0$$

or $E' \neq 0$. The affine change of coordinates

$$\begin{aligned} x'' &= Ax' \qquad\qquad\qquad + D' \\ y'' &= \qquad 2E'y' + (D'^2 + F') \end{aligned}$$

where

$$\begin{vmatrix} A & 0 \\ 0 & 2E' \end{vmatrix} \neq 0$$

reduces equation (11.23) to

$$x''^2 + y'' = 0.$$

This is the canonical form for a parabola. If we have an ellipse or hyperbola with $B^2 - AC \neq 0$, we may use the methods of Section 9.4 to reduce equation (11.21) to the form

$$x'^2 \pm y'^2 = \pm 1. \tag{11.24}$$

Thus in view of Theorem 11.28 we see that ellipses have canonical forms

$$x^2 + y^2 = +1$$

and

$$x^2 + y^2 = -1$$

for the real and imaginary cases, respectively. The remaining two possibilities in equation (11.24) may be written, using interchange of variables if necessary, in the single canonical form

$$x^2 - y^2 = 1$$

[1]If $A = 0$, then $C \neq 0$, and the affine change of coordinates $x' = y$, $y' = x$ in effect interchanges the variables.

for a hyperbola. Thus we have

**Theorem 11.29.** *Under an affine change of coordinates a nondegenerate conic can be reduced to the canonical forms*

$$x^2 + y^2 = \pm 1$$

*for an ellipse,*

$$x^2 + y = 0$$

*for a parabola, and*

$$x^2 - y^2 = 1$$

*for a hyperbola.*

## Exercises 11.2

1. Prove Theorem 11.21.
2. Prove Theorem 11.23.
3. Complete the proof of Theorem 11.25.
4. Prove Theorem 11.27.
5. Prove Theorem 11.28.
6. Prove that the distance function $\sqrt{(x_2 - x_1)^2 + (y_2 - y_1)^2}$ is not invariant under an affine transformation.
7. Prove that the set of affine transformations leaving the ideal pointwise fixed is a subgroup of the group of affine transformations (these transformations are called *homothetic*).
8. Prove that the set of affine transformations for which $b_{12} = b_{21} = 0$, $b_{11} = b_{22} = 1$ is a subgroup. Interpret geometrically.
9. Prove that the set of affine transformations for which $b_{12} = b_{21} = 0$ is a subgroup. Interpret geometrically.
10. Prove that the set of affine transformations for which $b_{21} = 0$, $b_{22} = 1$, $b_{12} = 0$ is a subgroup. Interpret geometrically.
11. Reduce to canonical form:
    *a.* $4x^2 - 12xy + 9y^2 - 36 = 0$;
    *b.* $4x^2 + 4xy + y^2 - 8x + 18y - 36 = 0$;
    *c.* $4x^2 + 2xy + y^2 - 8x + 18y - 72 = 0$;
    *d.* $4x^2 + 2xy - y^2 - 8x + 18y - 72 = 0$;
    *e.* $4x^2 - 2xy + y^2 - 6x - 12y + 72 = 0$;
    *f.* $4x^2 - 2xy - 2y^2 - 5x + 16y - 72 = 0$.

## 11.3 SIMILARITY GEOMETRY

Let us now fix an elliptic involution on the ideal line and call it the *absolute involution*. Some affine transformations, as extended to the projective plane, will have the property that they carry each pair of ideal points corresponding under this involution into a pair (possibly a different pair) of points also corresponding under this involution. Such an affine transformation is said to leave the absolute involution invariant. We have the following definitions and theorem:

**Definition 11.31.** *An affine transformation leaving the absolute involution invariant is a* similarity transformation.

**Theorem 11.31.** *The set of similarity transformations is a subgroup of the group of affine transformations.*

**Definition 11.32.** *The study of similarity transformations and the relations invariant under these transformations is* similarity geometry.

If the real projective plane were imbedded in the complex projective plane, the absolute involution would have two fixed points. Let us choose coordinates so that these fixed points are $(1, i, 0)$ and $(1, -i, 0)$. It is clear then that if we extend the similarity transformation to the complex projective plane, we must either leave these two points fixed or interchange these points. Application of this fact leads to the following theorem, the proof of which is left to the reader.

**Theorem 11.32.** *A similarity transformation has equations of the form*

$$\begin{aligned} x &= b_{11}x' + b_{12}y' + b_{13} \\ y &= eb_{12}x' - eb_{11}y' + b_{23} \end{aligned} \qquad (11.31)$$

*where* $b_{11}{}^2 + b_{12}{}^2 \neq 0$ *and* $e^2 = 1$.

Since Hilbert's axioms of incidence and order are satisfied in affine geometry, they will also be satisfied in similarity geometry. At this point we may also introduce angle measure and congruence and show that the axioms for angle congruence are satisfied in similarity geometry. We could introduce the appropriate angle measure formula and proceed by brute force. In the following chapters we shall find it convenient to introduce measures as logarithms of cross ratios, and we follow the same attack here.

Let us consider two intersecting lines $a$ and $b$ with equations $u_1x + u_2y + u_3 = 0$ and $u_1'x + u_2'y + u_3' = 0$. Their extensions to the projective plane will contain the ideal points $(u_2, -u_1, 0)$ and $(u_2', -u_1', 0)$, respectively, and these will be distinct if the lines intersect. We consider also the imaginary lines $j$ and $k$ joining the point of intersection to the fixed points $(1, i, 0)$ and $(1, -i, 0)$ of the absolute involution. Then $R(a, b, j, k)$ will be the same as the cross ratio of the four stated points on the ideal line. We apply Theorem 8.35 and find that

$$R(a,b,j,k) = \frac{(u_1+u_2i)(u_1'-u_2'i)}{(u_1-u_2i)(u_1'+u_2'i)}$$
$$= \frac{(u_1u_1'+u_2u_2')+(u_2u_1'-u_1u_2')i}{(u_1u_1'+u_2u_2')-(u_2u_1'-u_1u_2')i}.$$

This is not a real number if $a$ and $b$ are distinct. We next ask what happens if we take the natural logarithm of such a number. The logarithm function is indeed defined for a complex number, and we shall here use its properties in the complex domain, referring the reader to any text on complex analysis for the details and derivation. Any complex number $x+yi$ has associated with it a modulus $r = \sqrt{x^2+y^2}$ and arguments $\theta$ where $\tan\theta = y/x$. The $r$ and $\theta$ are just polar coordinates of the point $x+yi$ in the complex plane. We note that $\theta$ is determined only to within multiples of $2\pi$. We assert that

$$\log_e(x+yi) = \log_e r + i\theta.$$

The logarithm of a complex number is thus not uniquely determined unless we choose a principal value by putting an appropriate restriction on $\theta$. One customarily requires that $-\pi < \theta \leq \pi$. If we make this restriction, we also have

$$\log_e(x-yi) = \log_e r - i\theta$$

unless $x < 0$ and $y = 0$. All familiar logarithm properties still hold (if we take suitable care with principal values), whence

$$\log_e \frac{x+yi}{x-yi} = \log_e(x+yi) - \log_e(x-yi)$$
$$= 2\theta i$$

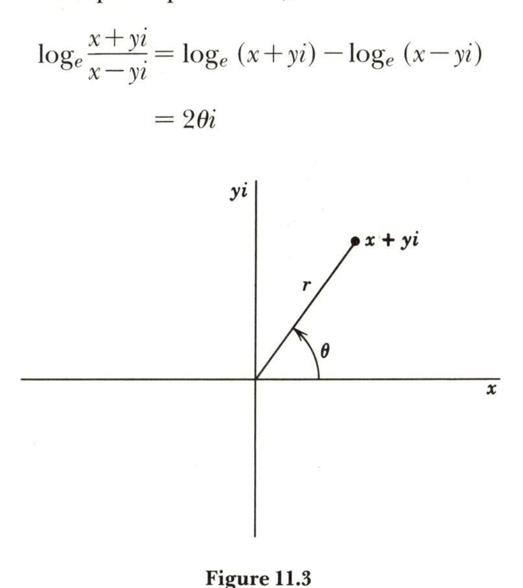


**Figure 11.3**

if $y \neq 0$. In the case at hand we thus have

$$\log_e R(a, b, j, k) = 2\theta i$$

where

$$\tan \theta = \frac{u_2 u_1' - u_i u_2'}{u_1 u_1' + u_2 u_2'}$$

or

$$\cos \theta = \frac{u_1 u_1' + u_2 u_2'}{\sqrt{(u_1 u_1' + u_2 u_2')^2 + (u_2 u_1' - u_1 u_2')^2}}$$

$$= \frac{u_1 u_1' + u_2 u_2'}{\sqrt{(u_1{}^2 + u_2{}^2)(u'^2_1 + u'^2_2)}}.$$

In view of this result we make

**Definition 11.33.** *The* measure *m(a, b) of two intersecting lines a and b is* $|(i/2) \log_e R(a, b, j, k)|$, *where j and k are the lines concurrent with a and b through the fixed points of the absolute involution.*

We have at once

**Theorem 11.33.**

$$\cos m(a, b) = \frac{|u_1 u_1' + u_2 u_2'|}{\sqrt{(u_1{}^2 + u_2{}^2)(u_1'^2 + u_2'^2)}}.$$

Definition 11.33 gives us a measure for a pair of intersecting lines, not for an angle, which is determined by a pair of rays lying on two intersecting lines. We must use the definition to define the measure of an angle. We first introduce

**Definition 11.34.** *Two intersecting lines are* perpendicular *if their measure is* $\pi/2$.

It is readily verified that the absolute involution has defining equation $x = -1/x'$. From this we find at once

**Theorem 11.34.** *Lines are* perpendicular *if and only if the ideal points of their extensions to the projective plane correspond under the absolute involution.*

We can now use the plane separation property of Theorem 2.22 to classify angles.

**Definition 11.35.** *If h and k are opposite rays on a line, (h, k) is a* straight angle. *If h and k lie on perpendicular lines, (h, k) is a* right angle. *If k lies in the same half plane as does h with respect to the line perpendicular to the line of h through the vertex of the angle, (h, k) is an* acute angle. *If it lies in the opposite half plane, (h, k) is an* obtuse angle.

The definition of an acute angle does not use symmetrically the rays, and we must accordingly prove

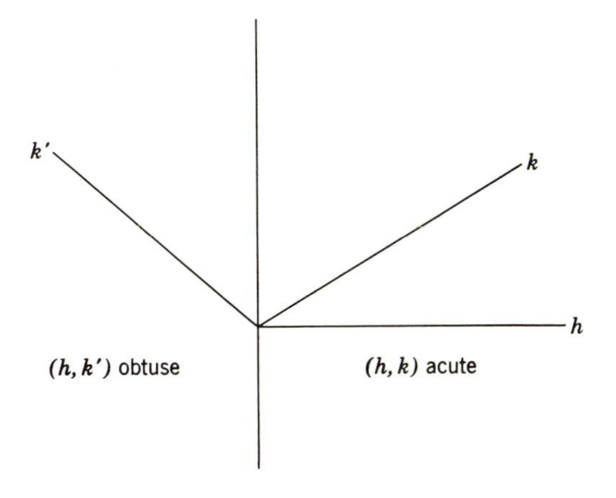


Figure 11.4

**Theorem 11.35.** *An acute angle is well defined.*

PROOF. Let the rays $h$ and $k$ lie on the same side of the line $h'$ through their vertex and perpendicular to $h$. Let $k'$ be the line through the vertex perpendicular to $k$. We choose points $H$ and $K$ on $h$ and $k$, respectively. We may suppose that the line meets $h'$ at $H'$ (If $HK$ is parallel to $h'$, choose a different $K$.) Since all points on each of the rays $h$ and $k$ lie in one of the half planes determined by $k'$, it will be sufficient to show the segment $H$-$K$ does not contain a point of $k'$. If $HK$ is parallel to $k'$, we are done. We assume that $HK$ meets $k'$ at $K'$. We let $I, J, M$, and $N$ be the ideal points on the extensions of $h, k, h'$, and $k'$, respectively. We know that $I$ and $M$, and $J$ and $N$, correspond under the absolute involution; thus if $I$ has coordinates $(x_1, x_2)$ on the ideal line, $M$ has coordinates $(x_2, -x_1)$, while if $J$ has coordinates $(y_1, y_2)$, then $N$ has coordinates $(y_2, -y_1)$. We find that $R(I, M, J, N) = R(H, H', K, K') < 0$. By hypothesis $H'$ is not between $H$ and $K$. We assert that this fact and the fact that $H$ and $H'$ separate $K$ and $K'$ lead in every possible case to the conclusion that $K'$ cannot be between $H$ and $K$. The details are left to the reader.

A similar argument shows that obtuse angles are well defined. We now finally can define angle measure.

**Definition 11.36.** *If the rays $h$ and $k$ lie on the intersecting lines $a$ and $b$ and form an angle $(h, k)$ then the* measure *$m(h, k)$ of the angle is:*

(1) $m(a, b)$ if $(h, k)$ is acute or right,
(2) $\pi - m(a, b)$ if $(h, k)$ is obtuse,
(3) $\pi$ if $(h, k)$ is straight.

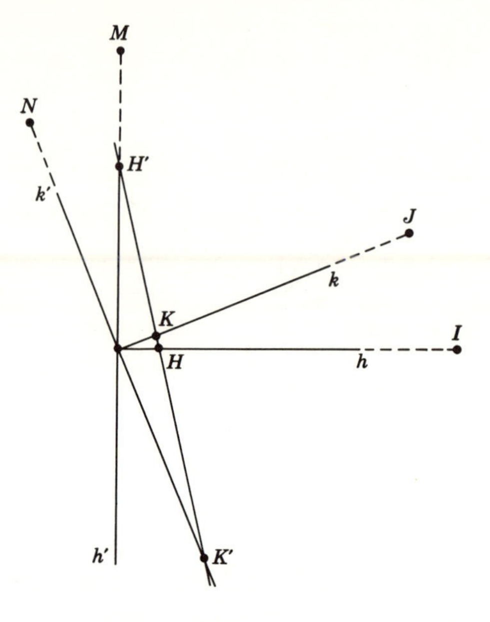


**Figure 11.5**

Since our development has been based on cross ratio and order consideration, both are preserved under a similarity transformation, and, on use of the absolute involution, we have

**Theorem 11.36.** *If $T$ is a similarity transformation, then $m(h, k) = m[T(h), T(k)]$.*

We finally introduce congruence in

**Definition 11.37.** *The angles $(h, k)$ and $(h', k')$ are* congruent $[(h, k) \cong (h', k')]$ *if $m(h, k) = m(h', k')$.*

We find

**Theorem 11.37.** *Similarity geometry satisfies Hilbert's axioms of congruence for angles.*

PROOF. Axioms 17, 18, and 19 follow at once from real number properties. In Axiom 16, if $(h, k)$ is a right angle, the conclusion follows readily from Theorem 11.34. If the given angle is not a right angle, there will lie two rays on each side of the line of $h'$ for which the line measure equals the line measure associated with the angle $(h, k)$. Just one of the rays on each side will be correct in terms of matching the acuteness or obtuseness of $(h, k)$, hence creating a congruent angle. The details are left to the reader.

Segment length is not preserved under a similarity transformation (see Exercise 11.38), and similarity geometry is not yet Euclidean geometry, but is intermediate between the latter and affine geometry. We can define similarity of triangles and polygons in similarity geometry by requiring that corresponding angles have the same measure. Thus we have similar triangles in similarity geometry, but we cannot consider the lengths of their sides or the ratios of lengths. Loosely speaking, in similarity geometry we have arrived at the position where we preserve "shape" of a configuration but not "size."

## Exercises 11.3

1. Prove Theorem 11.31.
2. Prove Theorem 11.32.
3. Complete the proof of Theorem 11.35.
4. Complete the proof of Theorem 11.37.
5. Prove that if corresponding sides of two triangles are parallel, the triangles are similar.
6. Prove that if a transversal intersects two parallel lines, the alternate interior angles are congruent, and conversely.
7. Prove that the sum of the measures of the angles of a triangle has the measure of a straight angle.
8. Prove that segment length is not preserved under an arbitrary similarity transformation (see Exercise 11.26).

## 11.4 EQUIAREAL GEOMETRY

If we apply the criterion of Theorem 7.41 to points represented in terms of their nonhomogeneous coordinates, we know that points with coordinates $(x_1, y_1)$, $(x_2, y_2)$, and $(x_3, y_3)$ are collinear if and only if

$$\begin{vmatrix} x_1 & y_1 & 1 \\ x_2 & y_2 & 1 \\ x_3 & y_3 & 1 \end{vmatrix} = 0.$$

If this determinant is not zero, the points are vertices of a triangle. In this case it is a result of elementary analytic geometry that the absolute value of this determinant is the area of this triangle. We anticipate this result in

**Definition 11.41.** *If the vertices of a triangle have coordinates $(x_1, y_1)$, $(x_2, y_2)$, and $(x_3, y_3)$, the* measure of the triangle *is*

$$\text{abs val} \begin{vmatrix} x_1 & y_1 & 1 \\ x_2 & y_2 & 1 \\ x_3 & y_3 & 1 \end{vmatrix}.$$

**Definition 11.42.** *An affine transformation leaving the measure of all triangles invariant is an* equiareal transformation.

**Theorem 11.41.** *The set of equiareal transformations is a subgroup of the group of affine transformations.*

**Definition 11.43.** *The study of equiareal transformations and the relation invariant under these transformations is* equiareal geometry.

If the affine transformation

$$\begin{aligned} x &= b_{11}x' + b_{12}y' + b_{13} \\ y &= b_{21}x' + b_{22}y' + b_{23} \end{aligned}$$

is to be equiareal, we must have

$$\text{abs val} \begin{vmatrix} x_1' & y_1' & 1 \\ x_2' & y_2' & 1 \\ x_3' & y_3' & 1 \end{vmatrix} = \text{abs val} \begin{vmatrix} x_1 & y_1 & 1 \\ x_2 & y_2 & 1 \\ x_3 & y_3 & 1 \end{vmatrix}$$

$$= \text{abs val} \begin{vmatrix} b_{11}x_1' + b_{12}y_1' + b_{13} & b_{21}x_1' + b_{22}y_1' + b_{23} & 1 \\ b_{11}x_2' + b_{12}y_2' + b_{13} & b_{21}x_2' + b_{22}y_2' + b_{23} & 1 \\ b_{11}x_3' + b_{12}y_3' + b_{13} & b_{21}x_3' + b_{22}y_3' + b_{23} & 1 \end{vmatrix}.$$

By elementary column operations the last determinant can be reduced to

$$\text{abs val} \begin{vmatrix} (b_{11} - b_{12}b_{21})x_1' & b_{22}y_1' & 1 \\ (b_{11} - b_{12}b_{21})x_2' & b_{22}y_2' & 1 \\ (b_{11} - b_{12}b_{21})x_3' & b_{22}y_3' & 1 \end{vmatrix} = |b_{11}b_{22} - b_{12}b_{21}| \text{ abs val} \begin{vmatrix} x_1' & y_1' & 1 \\ x_2' & y_2' & 1 \\ x_3' & y_3' & 1 \end{vmatrix}.$$

Thus we have

**Theorem 11.42.** *An equiareal transformation has equations of the form*

$$\begin{aligned} x &= b_{11}x' + b_{12}y' + b_{13} \\ y &= b_{21}x' + b_{22}y' + b_{23} \end{aligned}$$

*where* $|b_{11}b_{22} - b_{12}b_{21}| = 1$.

We still have not put a severe enough restriction on an affine transformation to enable us to create Euclidean geometry. Again equiareal geometry is intermediate between affine and Euclidean geometry. It is a geometry preserving "size" but not "shape."

## Exercises 11.4

1. Prove Theorem 11.41.

2. Prove that segment length is not invariant under an arbitrary equiareal transformation.

## 11.5 EUCLIDEAN GEOMETRY

To complete the cycle and return to Euclidean geometry, we need only combine the features of similarity and equiareal geometry. Thus we introduce

**Definition 11.51.** *An affine transformation which is both a similarity transformation and an equiareal transformation is a* Euclidean transformation.

**Theorem 11.51.** *The set of Euclidean transformations is a subgroup of the group of affine (and similarity and equiareal) transformations.*

**Definition 11.52.** *The study of Euclidean transformations and the relations invariant under these transformations is* Euclidean geometry.

From Theorems 11.32 and 11.42 we have at once

**Theorem 11.52.** *A Euclidean transformation has equations of the form*

$$\begin{aligned} x &= b_{11}x' + b_{12}y' + b_{13} \\ y &= eb_{12}x' - eb_{11}y' + b_{23} \end{aligned} \tag{11.51}$$

*where* $b_{11}{}^2 + b_{12}{}^2 = 1$ *and* $e^2 = 1$.

Since $b_{11}{}^2 + b_{12}{}^2 = 1$, we may set $b_{11} = \cos\theta$ and $b_{12} = \sin\theta$. The above equations then represent a combination of a translation of axes (determined by $b_{13}$ and $b_{23}$) and a rotation of axes if $e = -1$. If $e = +1$, we must also introduce a reflection of the plane about the $x$-axis.

We now define measure of a segment in

**Definition 11.53.** *If* $P_1$ *and* $P_2$ *have coordinates* $(x_1, y_1)$ *and* $(x_2, y_2)$, *respectively, the* measure of the segment $P_1$-$P_2$ *is*

$$m(P_1\text{-}P_2) = \sqrt{(x_2 - x_1)^2 + (y_2 - y_1)^2}.$$

At last we can readily prove

**Theorem 11.53.** *Measure of a segment is invariant under a Euclidean transformation.*

The proof is left to the reader.

We introduce congruence in

**Definition 11.54.** *The segments $A$-$B$ and $A'$-$B'$ are* congruent [$A$-$B \cong A'$-$B'$] *if* $m(A\text{-}B) = m(A'\text{-}B')$.

We find

**Theorem 11.54.** *Euclidean geometry satisfies Hilbert's axioms of congruence for segments.*

PROOF. Axioms 12, 13, and 14 follow from real number properties. To prove Axiom 11, we choose coordinates so that the line $A'B'$ is the $x$-axis and $A'$ is the origin.[2] The desired point $B'$ is one of the two points $B'$ whose coordinates $(x, 0)$ satisfy $|x| = m(A\text{-}B)$. In Axiom 15 the hypotheses will be fulfilled only if we have the order relations $ABC$ and $A'B'C'$. We apply Corollary 11.551 below to find a Euclidean transformation $T$ for which $T(A, B) = A', B'$. In view of Theorems 11.24 and 11.53, we must have $T(C) = C'$. The desired result then follows from Theorem 11.53.

We continue our considerations with

**Theorem 11.55.** *If $A$ is a point on line $a$ and $A'$ is a point on line $a'$, there is a Euclidean transformation $T$ for which $T(A, a) = A', a'$.*

PROOF. We let $a'$ be the $x$-axis and $A'$ the origin. We suppose that $a$ has equation $u_1x + u_2y + u_3 = 0$ where $u_1{}^2 + u_2{}^2 = 1$ and that $A$ has coordinates $(a_1, a_2)$. If we set

$$b_{11} = -eu_2,$$
$$b_{12} = eu_1,$$
$$b_{13} = eu_2a_1 - eu_1a_2,$$

and

$$b_{23} = u_3,$$

we can verify at once that the conditions of equations (11.51) are met and that the transformation has the desired effect.

We note in the proof above that there are two choices of $u_1$ and $u_2$ if $u_1{}^2 + u_2{}^2 = 1$ and two choices of $e$. There are thus four such transformations. Since the transformation preserves order, the four choices are determined by determining which rays on $a$ with terminal point $A$ are to correspond with which rays on $a'$ with terminal point $A'$ and which half planes determined by $a$ are to correspond to which half planes determined by $a'$. We thus have

**Corollary 11.551.** *If $A$-$B \cong A'$-$B'$, there are two Euclidean transformations for which $T(A, B) = A', B'$.*

[2]We choose the reference triangle in the projective plane so that its vertices are $A'$ and the ideal points associated with the coordinate axes (these points correspond under the absolute involution).

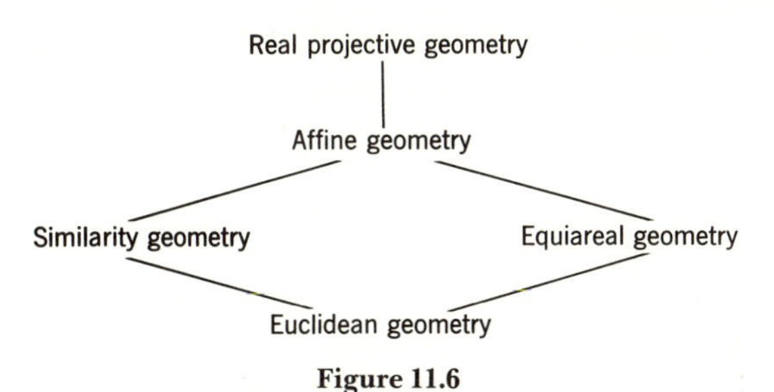


**Figure 11.6**

**Corollary 11.552.** *If $A\text{-}B \cong A'\text{-}B'$, there is a unique Euclidean transformation for which $T(A, B) = A', B'$ and which maps a given half plane determined by $AB$ onto a given half plane determined by $A'B'$.*

**Corollary 11.553.** *If $A\text{-}B \cong A'\text{-}B'$, there are four Euclidean transformations mapping the segment $A\text{-}B$ onto the segment $A'\text{-}B'$.*

We conclude with

**Theorem 11.56.** *Euclidean geometry satisfies Hilbert's congruence axiom for triangles.*

PROOF. By Corollary 11.552 and Hilbert's Axiom 16 we can find a Euclidean transformation $T$ for which $T(A, B, C) = A', B', C'$. The transformation thus maps one triangle onto the other, and the desired angle congruence follows at once. Note that we have, using Theorem 11.55 and its corollary, proved this triangle congruence by "superposition."

We have finally completed our journey and have returned to Euclidean geometry! The scheme of geometries considered is shown in Figure 11.6. We have considered here only two of the many geometries intermediate between affine and Euclidean geometry. Each of these has a wealth of theorems and invariant properties of its own. The reader is urged to explore these geometries further via the references at the end of the chapter.

## Exercises 11.5

1. Prove Theorem 11.51.
2. Prove Theorem 11.53.

## REFERENCES

Adler, Claire Fisher, *Modern Geometry, an Integrated First Course*, 2nd ed., McGraw-Hill Book Company, New York, 1967, Chapter 10.

Artzy, Rafael, *Linear Geometry*, Addison-Wesley Publishing Company, Reading, Mass., 1965, Chapter 2.

Blumenthal, Leonard M., *A Modern View of Geometry*, W. H. Freeman and Company, San Francisco, 1961, Chapters 4, 5, 7, 8.

Busemann, Herbert, and Paul J. Kelly, *Projective Geometry and Projective Metrics*, Academic Press, New York, 1953, Chapters 3,4.

Coxeter, H. S. M., *Non-Euclidean Geometry*, 5th ed., University of Toronto Press, Toronto, 1965, Chapter 9.

Coxeter, H. S. M., *The Real Projective Plane*, McGraw-Hill Book Company, New York, 1949, Chapter 12.

Gans, David, *Transformations and Geometries*, Appleton-Century-Crofts, New York, 1969, Chapters 1 to 4.

Gruenberg, K. W., and A. J. Weir, *Linear Geometry*, D. Van Nostrand Company, Princeton, 1967, Chapters 2, 6.

Hartshorne, Robin, *Foundations of Projective Geometry*, W. A. Benjamin, New York, 1967, Chapter 1.

Levy, Harry, *Projective and Related Geometries*, The Macmillan Company, New York, 1961, Chapter 5.

Modenov, P. S., and A. S. Parkhomenko, translated by Michael B. P. Slater, *Geometric Transformations*, vol. I: *Euclidean and Affine Transformations*, Academic Press, New York, 1965, Chapters 1 to 4.

O'Hara, C. W., and D. R. Ward, *An Introduction to Projective Geometry*, Oxford University Press, New York, 1937, Chapters 9, 10.

Rainich, G. Y., and S. M. Dowdy, *Geometry for Teachers*, John Wiley and Sons, New York, 1968, Chapters 7, 8.

Robinson, Gilbert deB., *The Foundations of Geometry*, The University of Toronto Press, Toronto, 1952, Chapters 4, 5.

Rosenbaum, Robert A., *Introduction to Projective Geometry and Modern Algebra*, Addison-Wesley Publishing Company, Reading Mass., 1963, Chapter 3.

Tuller, Annita, *A Modern Introduction to Geometries*, D. Van Nostrand Company, Princeton, 1967, Chapters 4, 6.

Veblen, Oswald, and John Wesley Young, *Projective Geometry*, vol. II, Ginn and Company, Boston, 1910, Chapters 3 to 5, 8.

Young, John Wesley, *Projective Geometry*, The Open Court Publishing Company, Chicago, 1930, Chapter 9.

## CHAPTER TWELVE

# HYPERBOLIC GEOMETRY

In the preceding chapter we considered a series of descendants of real projective geometry culminating in Euclidean geometry, which thus could be considered as based on an understanding of projective geometry. In this and the next chapter we show how analogous procedures can lead to a development of the two classical non-Euclidean geometries.

## 12.1 THE HYPERBOLIC PLANE

We developed affine-Euclidean geometries by choosing a line in the projective plane for special consideration, the line becoming the ideal line of the resulting geometries. A line is a special type of conic; indeed, it is the fifth of the six canonical forms listed in Figure 9.16. Similar procedures with different conics lead to different geometries. Let us now choose a nondegenerate real conic, which we shall call the *absolute conic*, or the *absolute*. We assume that coordinates are so chosen that the conic has the equation

$$x_1^2 + x_2^2 = x_3^2.$$

This conic, or any real nondegenerate conic, divides the projective plane into two regions. To determine this, we first prove

**Theorem 12.11.** *If a point $P$ has coordinates $(x_1, x_2, x_3)$ such that $x_1^2 + x_2^2 < x_3^2$, then this property holds for all coordinates of $P$.*

The proof is left to the reader. It is obvious that a like conclusion follows if we replace $<$ by $>$. With this theorem established, we can then make the following definition:

**Definition 12.11.** *The set of points whose coordinates satisfy* $x_1^2+x_2^2 < x_3^2$ *constitutes the* interior *of the conic* $x_1^2+x_2^2=x_3^2$. *The set of points satisfying* $x_1^2+x_2^2 > x_3^2$ *constitutes the* exterior.

One can now prove

**Theorem 12.12.** *The set of collineations leaving the absolute fixed and carrying interior points into interior points is a subgroup of the collineation group.*

In a manner similar to that used in the beginning of our development of affine geometry we then make the following definitions.

**Definition 12.12.** *The points interior to the absolute are* ordinary; *those on the absolute are* ideal; *those exterior to it are* ultraideal.

**Definition 12.13.** *A real projective plane from which the absolute conic and its exterior have been deleted is a* hyperbolic plane. *The collineations leaving the absolute fixed and carrying interior points onto interior points, restricted to the hyperbolic plane, are* hyperbolic isometries. *The study of the hyperbolic plane and hyperbolic isometries is* hyperbolic geometry.

The distinction between ideal and ordinary line is not so simple as in the affine plane, and clarification of the issue requires a more detailed consideration of points interior to and exterior to the absolute. We first note that any point with coordinates of the form $(x_1, x_2, 0)$ is exterior to the conic; hence, since any projective line contains such a point, any projective line will contain ultraideal points. In particular, the line $x_3 = 0$ contains only ultraideal points. It is also readily verified that any line of the form

$$u_1x_1 + u_2x_2 = 0$$

contains ordinary points, two ideal points, and ultraideal points. Excluding these special cases, we may introduce nonhomogeneous coordinates and take the equation of the line in the form

$$ax + by = 1$$

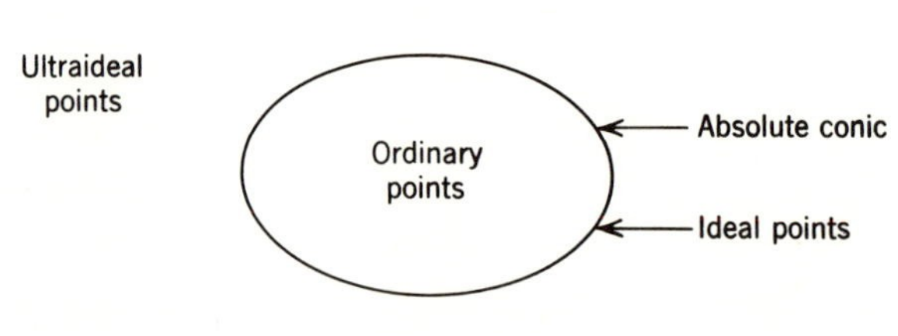


**Figure 12.1**

while the absolute will have equation

$$x^2+y^2 = 1.$$

If we attempt to find the intersection of the line and the absolute, we are led to a quadratic equation in $x$ or in $y$, the sign of whose discriminant is the sign of $a^2+b^2-1$. If $a^2+b^2 < 1$, we easily find by noting that

$$(a-x)^2+(b-y)^2 > 0$$

that any point on the line must satisfy $x^2+y^2 > 1$. Hence any line not meeting the conic contains only ultraideal points. If $a^2+b^2 = 1$, the line is tangent to the conic, and it contains all ultraideal points with the exception of one ideal point. If $a^2+b^2 > 1$, the line meets the conic twice. Let these points have homogeneous coordinates $(x_1, y_1, 1)$ and $(x_2, y_2, 1)$. Any point on the line then has coordinates

$$m_1(x_1, y_1, 1)+m_2(x_2, y_2, 1).$$

If we compare the sum of the squares of the first two coordinates with the square of the third coordinate, we must compare $m_1m_2(x_1x_2+y_1y_2)$ with $m_1m_2$. Since $x_1x_2+y_1y_2 \neq 0$ (if it were zero, the line would lie on the absolute), both inequality senses are possible for appropriate choices of $m_1$ and $m_2$. Thus such a line contains two ideal points, ordinary points, and ultraideal points. We have outlined a proof of

**Theorem 12.13.** *A line not meeting the absolute contains only ultraideal points. A line tangent to it contains one ideal point and ultraideal points. A line meeting it twice contains two ideal points, ordinary points, and ultraideal points.*

Lines in the hyperbolic plane thus come only from those projective lines that meet the absolute conic twice.

**Figure 12.2**

The reasoning above also indicates a redundancy in Definition 12.13. A collineation leaving the absolute fixed must map exterior points onto exterior points, for otherwise it would map a line not meeting the conic onto one meeting it. Likewise, if an interior point were mapped onto an exterior point, it would be possible to map a line meeting the conic onto a line not meeting it or tangent to it. Thus we have

**Theorem 12.14.** *A collineation, restricted to the hyperbolic plane, mapping the absolute conic onto itself is a hyperbolic isometry.*

The word "isometry" suggests equal measure, and we shall see in Section 12.4 why this word was chosen. We might suspect that at this point we have a geometry analogous to affine geometry and that further restrictions are needed to introduce metric ideas and develop an analogue to Euclidean geometry. This is actually not the case, and no further subgroup of this group of hyperbolic isometries will be considered in this chapter. The adjective "hyperbolic" in the phrase "hyperbolic isometry" indicates that we are considering isometries in the hyperbolic plane. Since this chapter is concerned exclusively with hyperbolic geometry, we omit the adjective and describe the transformations simply as isometries.

We might well ask how the group of isometries can be characterized analytically. No simple characterization like that of the affine transformations is possible. It is apparent, however, that the group is not empty. It certainly contains the identity collineation. It is also easily verified that it contains all collineations with matrices of the form

$$\begin{pmatrix} \cos\theta & \sin\theta & 0 \\ -\sin\theta & \cos\theta & 0 \\ 0 & 0 & 1 \end{pmatrix}. \tag{12.11}$$

It also contains many more. We shall prove a theorem asserting the existence of a useful type of isometry, but first we must prove

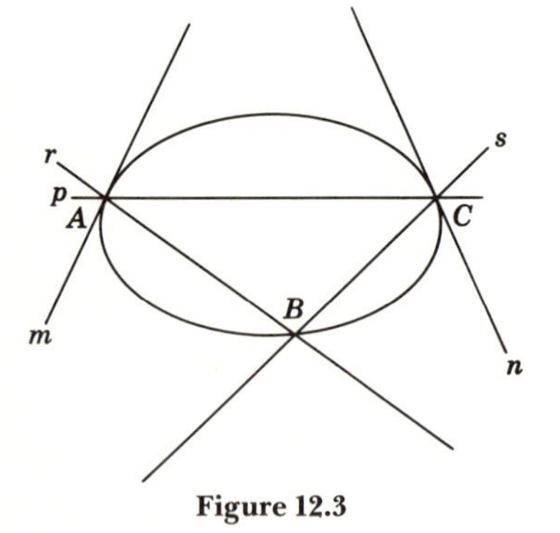

**Figure 12.3**

**Lemma 12.11.** *A conic is uniquely determined given three distinct noncollinear points on the conic and the tangents at two of these points, the third point being on neither tangent.*

PROOF. Let $A$, $B$, and $C$ be the given points; $m$ and $n$ the tangents at $A$ and $C$, respectively; and $p$, $r$, and $s$ the lines $AC$, $AB$, and $BC$, respectively. The noncollinearity of the points assures the distinctness of $p$, $r$, and $s$. By the Fundamental Theorem there is a unique projective transformation $T$ from the pencil at $A$ to the pencil at $C$ for which $T(m, p, r) = p, n, s$. The unique conic is the one determined by this projectivity.

We may now prove our theorem.

**Theorem 12.15.** *If $l$ and $l'$ are lines containing the points $P$ and $P'$, respectively, there exists an isometry $T$ such that $T(P, l) = P', l'$.*

PROOF. In Figure 12.4 we show two copies of the plane for simplicity; actually one part of the figure lies on top of the other. Let $l$ meet the absolute at $I$ and $J$, while $l'$ meets it at $I'$ and $J'$. Let $O$ be the intersection of the tangents to the absolute at $I$ and $J$, while $O'$ is to be the intersection of the tangents to the absolute at $I'$ and $J'$. Let the line $OP$ meet the absolute at $K$, while the line $O'P'$ meets the absolute at $K'$. The desired transformation is the one for which $T(I, J, O, K) = I', J', O', K'$. In view of the lemma it is an isometry. Since it is a collineation, it must map $IJ$ onto $I'J'$ and the intersection $P$ of $IJ$ and $OK$ onto the intersection $P'$ of $I'J'$ and $O'K'$.

We note that the isometry is not unique. We could obtain another by interchanging $I'$ and $J'$ or by choosing $K'$ as the other intersection of $O'P'$ with the absolute.

This theorem will be used frequently in the following sections of this chapter; for example, if we wish to prove a theorem about a point and a line on which it lies and if we can prove it in the special case where the line has equation $x_2 = 0$ and the point has coordinates $(0, 0, 1)$, the proof will be valid if it is concerned only with relations invariant under an isometry.

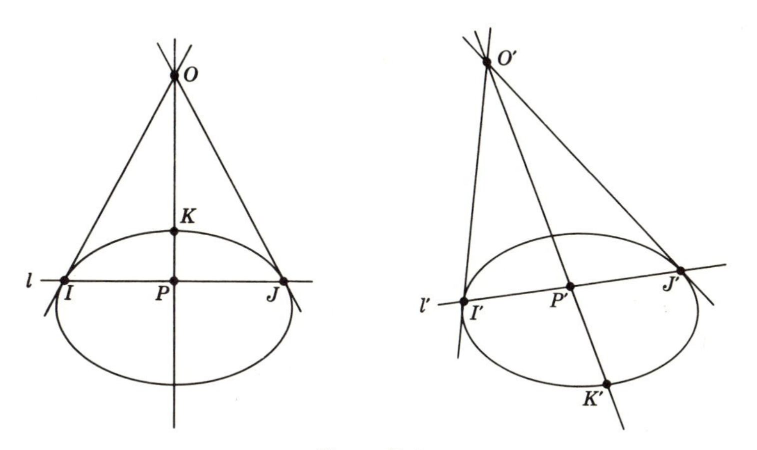


**Figure 12.4**

We might ask how similar hyperbolic geometry is to Euclidean. One way to answer this question is to determine the extent to which Hilbert's axioms can or cannot be proved in this new geometry. In the remaining sections of the chapter we consider the possibility of proving these axioms and investigate briefly the basic properties of triangles and certain simple quadrilaterals in the hyperbolic plane.

## Exercises 12.1

1. Prove Theorem 12.11.
2. Prove Theorem 12.12.
3. Complete the proof of Theorem 12.13.
4. Prove that matrix (12.11) is the matrix of an isometry.
5. Let

$$H = \begin{pmatrix} 1 & 0 & 0 \\ 0 & 1 & 0 \\ 0 & 0 & -1 \end{pmatrix}.$$

Prove that if a nonsingular matrix $A$ is such that $A^*HA = kA$, $A$ is the matrix of an isometry.

6. Prove that two distinct tangents to the absolute pass through each ultraideal point.

7. Prove that the polar line of an ordinary point contains only ultraideal points.

8. Prove that the polar line of an ultraideal point contains ordinary points.

## 12.2 INCIDENCE AND PARALLELISM

We find as an immediate consequence of the discussion of the preceding section

**Theorem 12.21.** *Hyperbolic geometry satisfies Hilbert's axioms of incidence.*

All of the theorems of Section 2.1 are thus valid in hyperbolic geometry.

Let us consider the question of parallelism, which is actually an incidence property. If we choose a line and a point not on the line, we see that in the sense of Definition 2.41 there are many parallels to the line passing through the given point, these being determined by the lines meeting the extension of the given line in ideal or ultraideal points. Exactly two of these meet the line in ideal points. We are primarily interested in these two lines and will change our terminology by adopting

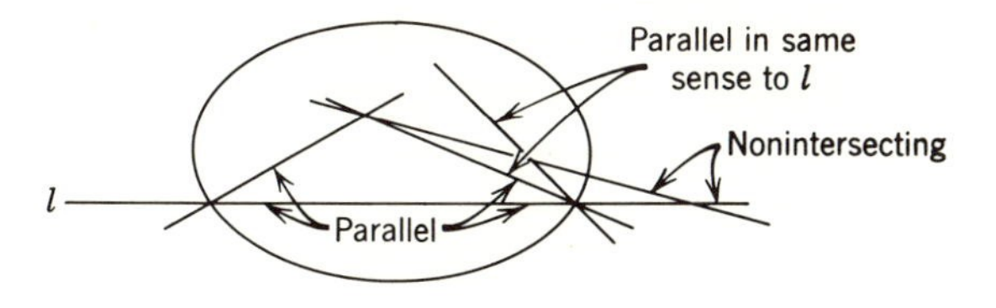


**Figure 12.5**

**Definition 12.21.** *Lines are* parallel *if the corresponding projective lines meet in an ideal point. Lines are* nonintersecting *if the corresponding projective lines meet in an ultraideal point. Distinct lines are said to be* parallel to a given line in the same sense *if they meet the corresponding projective line at the same ideal point.*

Instead of Axiom 21 of Hilbert's system we thus have

**Theorem 12.22.** *Given a line and a point not on the line, there are exactly two lines containing the given point and parallel to the given line.*

The following theorems also follow at once from the definition.

**Theorem 12.23.** *An isometry maps parallel lines onto parallel lines and nonintersecting lines onto nonintersecting lines.*

The properties of parallelism and nonintersection are thus invariants in hyperbolic geometry.

**Theorem 12.24.** *Distinct lines parallel to the same line in the same sense are parallel to each other.*

**Theorem 12.25.** *There exist lines parallel to each of a pair of intersecting lines.*

The fact that Theorem 12.22 has replaced Hilbert's Axiom 21 is the heart of the distinction between Euclidean and hyperbolic geometry. We are now considering a geometry of more than one parallel, and all differences that we shall find between the geometries can be traced to this characteristic, for we shall in succeeding sections be able to prove all of the rest of Hilbert's axioms.

## Exercises 12.2

1. Prove Theorem 12.21.
2. Prove Theorem 12.23.
3. Prove Theorem 12.24.
4. Prove Theorem 12.25.
5. Suppose that the absolute had been chosen as an imaginary conic. Deter-

mine to what extent the results of the previous two sections can be proved. If a theorem is not valid, see if you can prove a corresponding theorem.

6. Do the same if the absolute is a pair of intersecting lines.

7. Do the same if the absolute is a point.

8. Prove that through a point not on a given line there pass an infinite number of nonintersecting lines.

## 12.3 ORDER

Our definition of betweenness in the hyperbolic plane is identical to the one made for the affine plane, and with the exception of two details, all arguments will proceed as in that case.

**Definition 12.31.** *If $A$, $B$, and $C$ are distinct points on a line with ideal point $I$, then $B$ is* between *$A$ and $C$, or $ABC$, if $AC||BI$.*

The difficulty with this definition is that now every line contains two ideal points, and the definition will not be acceptable unless we show that the separation conclusion is independent of the ideal point chosen. This we do in our first theorem.

**Theorem 12.31.** *If $I$ and $J$ are the ideal points on the line containing $A$, $B$, and $C$, then $AC||BI$ if and only if $AC||BJ$.*

PROOF. Since separation is invariant under an isometry, we may map the given line onto the line $x_2 = 0$, hence assume that the points are on this line. The nonzero homogeneous coordinates of the points will then, by Theorem 8.35, be homogeneous coordinates on the line. Suppose that $A, B, C, I$, and $J$ have coordinates $(a, 1)$, $(b, 1)$, $(c, 1)$, $(1, 1)$, and $(-1, 1)$, respectively. We find that

$$R(A, C, B, I) = \frac{(a-b)(c-1)}{(c-b)(a-1)}$$

and

$$R(A, C, B, J) = \frac{(a-b)(c+1)}{(c-b)(a+1)}.$$

Since $a^2 < 1$ and $c^2 < 1$, the factors $(c-1)/(a-1)$ and $(c+1)/(a+1)$ are both positive. The two cross ratios thus have the same sign, and the assertion follows.

**Corollary 12.311.** *If $A$, $B$, and $C$ have projective coordinates $(a, 1)$, $(b, 1)$, and $(c, 1)$, respectively, derived in the natural manner from their coordinates in projective space on the line $x_2 = 0$ or $x_1 = 0$, then $ABC$ if and only if $a < b < c$ or $a > b > c$.*

**Corollary 12.312.** *If $P$ is an ultraideal point on the line of $A$, $B$, and $C$, then $ABC$ if and only if $AC \| BP$.*

If we let $P$ have coordinates $(p, 1)$ where $p^2 > 1$, the arguments applied to the cross ratios in the proof of the theorem will hold, and the equivalence of betweenness to the separation will follow from Corollary 12.311.

We have already noted that betweenness is invariant under an isometry, since separation is invariant under a collineation. Thus we have

**Theorem 12.32.** *If $ABC$ and $T$ is an isometry, then $T(A)T(B)T(C)$.*

Finally, we can prove all of Hilbert's axioms of order.

**Theorem 12.33.** *Hyperbolic geometry satisfies Hilbert's axioms of order.*

PROOF. In view of the identical nature of Definitions 11.24 and 12.31 the proof of all of the axioms except the last is the same as in Theorem 11.25. The proof of Pasch's axioms does not follow at once, since the points $I$, $I'$, and $I''$ of Figure 11.2 are not collinear, but points on the absolute conic. We note, however, that in view of Corollary 12.312 we can take these points to be the points where the extensions of the given lines meet the line $x_3 = 0$. The proof will then carry through as in Theorem 11.25.

As an alternate method of proof, we note that all of the order axioms except the last also follow at once from Corollary 12.311 and the order properties of the real number system. A direct computational proof leading to the last axiom is outlined in Exercise 12.34.

With all of Hilbert's order axioms established, we now know that all of the definitions and theorems of Section 2.2 hold in hyperbolic geometry. Thus we can define segment, ray, and angle, and we can be sure that the line and plane separation properties of Theorems 2.21 and 2.22 are valid in the hyperbolic plane.

In the hyperbolic plane each line is associated with two ideal points, and we even loosely say that each line contains two ideal points. With ray defined we may now associate a single ideal point with a ray, or say loosely that there is a single ideal point on a ray. In Figure 12.6 we feel that the ideal point $I$ is the one to be placed on the ray with terminal

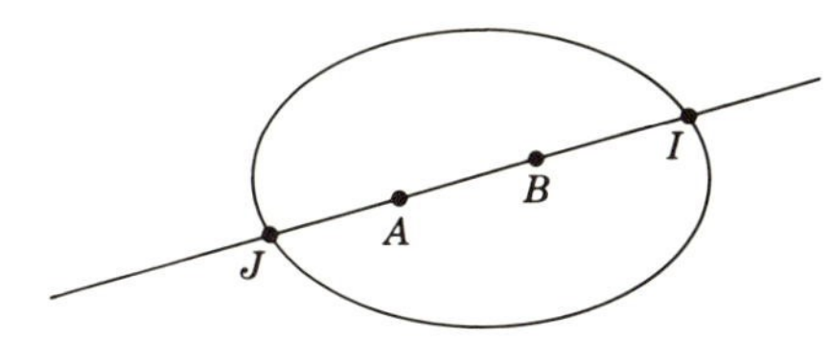


**Figure 12.6**

point $A$ and containing point $B$. We can see a criterion for this assignment by again applying Theorem 12.15 to place the points on the line $x_2 = 0$. Let $A$ have coordinates $(a, 0, 1)$ and $B$ have coordinates $(b, 0, 1)$ where $b > a$. We would like to place the ideal point $I$ with coordinates $(1, 0, 1)$ on the ray rather than the ideal point $J$ with coordinates $(-1, 0, 1)$. We find

$$R(A, I, B, J) = \frac{2(a-b)}{(1-b)(1+a)} < 0.$$

On the other hand,

$$R(A, J, B, I) = \frac{2(a-b)}{(b+1)(a-1)} > 0.$$

Thus $AI \| BJ$ but $AJ \not\| BI$. This suggests

**Definition 12.32.** *The* ideal point on a ray *with terminal point $O$ is the ideal point $I$ of the line of the ray for which $OI \| PJ$, where $P$ is any point on the ray and $J$ is the second ideal point on the line of the ray.*

We noted in the proof of Theorem 12.15 that we had two choices as to where to map the ideal points $I$ and $J$ of Figure 12.4. These choices correspond to choices of direction or to matching of rays on the lines involved. We see now that the theorem can be strengthened to

**Theorem 12.34.** *There exists an isometry mapping a given ray onto any given ray.*

The reader may well feel that there is something dishonest in our work to date, particularly the proof of Pasch's axiom. If we consider the Euclidean plane and then look at all of our previous work in this chapter, we find that the absolute conic is the unit circle with center at the origin and that lines in the hyperbolic plane are merely the line segments interior to the circle. In this interpretation the peculiar parallelism properties of the hyperbolic plane are immediately evident. The order properties of the hyperbolic plane are merely the order properties of the Euclidean (or affine) plane confined to the interior of the circle.

It is possible to interpret hyperbolic geometry in this Euclidean manner, making the plane the interior of the circle and taking lines to be line segments interior to the circle. The resulting model is known as the Cayley-Klein model of the hyperbolic plane. Other models can also be found in the Euclidean plane (see the references at the end of the chapter). On this basis the equivalent consistency of Euclidean and hyperbolic geometry can be proved; that is, if our axioms for Euclidean geometry are consistent, so are those for hyperbolic geometry, and conversely.

Whether we view the hyperbolic plane from a Euclidean or projective point of view, we find the differences accentuated when we consider the congruence relation. We are going to be able to prove all of Hilbert's axioms of congruence, but the associated segment and angle measures are not going to be Euclidean, and segments that would be congruent in the Euclidean plane are not going to be congruent in the hyperbolic plane.

## Exercises 12.3

1. Carry out the cross ratio computations for the cross ratios used in the proof of Theorem 12.31.

2. Prove that if $A$ and $B$ are ordinary and $I$ and $J$ are ideal, all points being collinear, then $AB \nparallel IJ$.

3. Prove that if $A$ and $B$ are ordinary and $P$ and $Q$ are ultraideal, all points being collinear, then $AB \nparallel PQ$.

4. It is a simple matter to prove that Theorem 2.22 implies Pasch's axiom. We indicate a direct proof of this theorem. By Theorems 12.15 and 12.32, it will be sufficient to prove the theorem for the line $x_2 = 0$. Let $I$ and $J$ be the ideal points $(0, 1, 1)$ and $(0, -1, 1)$, respectively.
   *a.* Prove that if $K$ is an ideal point distinct from $I$ and $J$ and not on $x_2 = 0$, one of the lines $IK$ and $JK$ meets $x_2 = 0$ at an ordinary point and the other meets it at an ultraideal point.
   *b.* Let $P$ and $Q$ have coordinates $(x_1, y_1, 1)$ and $(x_2, y_2, 1)$, respectively, where $y_1 y_2 \neq 0$. Let $PQ$ have ideal point $K$. Suppose that $IK$ meets $x_2 = 0$ at the ultraideal point $O$. If $T$ is the central perspectivity from $PQ$ to $x_1 = 0$ with center $O$, prove that the second coordinates of $T(P)$ and $T(Q)$ have the same sign as those of $P$ and $Q$ (if the third coordinates are 1).
   *c.* Prove Theorem 2.22.

## 12.4 MEASURE AND CONGRUENCE OF SEGMENTS

As in our development of Euclidean geometry in Chapter 11, we first introduce the concept of measure of a segment and then proceed to the concept of congruence by defining congruent segments to be those with equal measures. When we introduced line and angle measure in Chapter 11, we took it to be the logarithm of a cross ratio; we follow the same approach here both for segment and angle measure. In the hyperbolic plane we know that each line contains two ideal points. Let us suppose the ideal points on the line $AB$ are $J$ and $K$. We would like to investigate $R(A, B, J, K)$. Without loss of generality, in view of Theorem 12.15, we may assume that the line is $x_2 = 0$. Suppose that $A, B, J$, and $K$ have coordinates $(a, 0, 1)$, $(b, 0, 1)$, $(1, 0, 1)$, and $(-1, 0, 1)$, respectively. Then

$$R(A, B, J, K) = \frac{(a-1)(b+1)}{(a+1)(b-1)}.$$

If the point $A$ is ordinary, then $a^2 < 1$ and $(a-1)/(a+1) < 0$. If $B$ is also ordinary, the other factor is negative and the cross ratio is positive; if $B$ is ultraideal, then $b^2 > 1$, $(b+1)/(b-1) > 0$, and the cross ratio is negative [this is also the case if $B$ is the point with coordinates $(1, 0, 0)$]. We thus conclude, using Theorem 8.53, that for a fixed ordinary point $A$ the cross ratio assumes all positive values except 1 for suitable choices of ordinary point $B$ distinct from $A$ and the value 1 for $A = B$.

Let us next consider $k \log_e R(A, B, J, K)$. For a fixed $A$ this quantity will assume all real number values for an appropriate choice of $B$ and will have value zero when $A = B$. We note also that as $B$ approaches $J\,(b, 1)$ the quantity will become positively infinite, while as $B$ approaches $K(b, -1)$ it will become negatively infinite. We have thus put a number

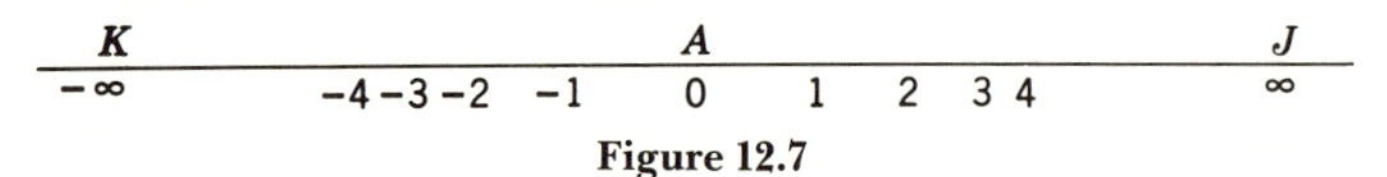


**Figure 12.7**

scale on the line which is unbounded as we approach either ideal point on the line. The quantity as it stands can thus be defined to be the directed distance from $A$ to $B$. If we wish to consider measure, we need only take the absolute value of the quantity. We can choose $k$ as we please as long as $k \neq 0$. We shall shortly see that $k = \frac{1}{2}$ is a convenient choice, and we introduce

**Definition 12.41.** *The* measure *in hyperbolic geometry* of the segment *A-B is the number*

$$m(A\text{-}B) = |\tfrac{1}{2} \log_e R(A, B, J, K)|,$$

*where J and K are the ideal points on the line AB.*

Since cross ratio is preserved under a collineation, we have at once

**Theorem 12.41.** *If $T$ is an isometry, then $m(A\text{-}B) = m[T(A)\text{-}T(B)]$.*

The word "isometry" means equal measure and has been chosen because it preserves measure of segments and also measure of angles as defined in the next section.

Any measure should conform to the conditions of the next theorem. All of these conditions are obvious from Definition 12.41 except the last. We note the theorem here and defer the proof of the last assertion until Theorem 12.91.

**Theorem 12.42.** (1) $m(A\text{-}B) \geqslant 0$,
(2) $m(A\text{-}B) = 0$ *if and only if* $A = B$,
(3) $m(A\text{-}B) = m(B\text{-}A)$,
(4) $m(A\text{-}C) \leqslant m(A\text{-}B) + m(B\text{-}C)$.

The last assertion is the "triangle inequality" and does not assume the collinearity of the three points.

We may now define congruence of segments by

**Definition 12.42.** *The segments $A\text{-}B$ and $A'\text{-}B'$ are* congruent ($A\text{-}B \cong A'\text{-}B'$) *if* $m(A\text{-}B) = m(A'\text{-}B')$.

We at once have

**Theorem 12.43.** *Hyperbolic geometry satisfies Hilbert's axioms of congruence for segments.*

PROOF. Axioms 12, 13, and 14 follow at once from the definition and equality properties of the real numbers. We note that by Theorem 8.53 for a given $A'$ and value for $m(A\text{-}B)$ there are two points $B'$ for which $m(A'\text{-}B') = m(A\text{-}B)$, one being the points $B'$ for which

$$m(A\text{-}B) = \tfrac{1}{2}\log_e R(A', B', J', K')$$

and the other being the point $B'$ for which

$$m(A\text{-}B) = -\tfrac{1}{2}\log_e R(A', B', J', K').$$

This proves Axiom 11. We now consider Axiom 15. If $A\text{-}B$ and $B\text{-}C$ are segments with no points in common, we have the order relation $ABC$. By Corollary 12.311 and the discussion preceding Definition 12.41 we conclude that $\log_e R(A, B, J, K)$ and $\log_e R(B, C, J, K)$ have the same sign. The same remarks apply to the points $A'$, $B'$, and $C'$. The conclusion then follows from the fact that $\log_e MN = \log_e M + \log_e N$. The details of the proof are left to the reader.

Before considering congruence for angles, let us consider in more detail the measure function of the hyperbolic plane. We first restrict ourselves to the line $x_2 = 0$. If we let $A$ be the point with coordinates $(0, 0, 1)$ and $B$ be the point with coordinates $(x, 0, 1)$, the number $x$ can be taken to be the coordinate of $B$ on this line, and we have a one-to-one correspondence between the points of the line and the numbers $x$ for which $-1 < x < 1$, as indicated in Figure 12.8.

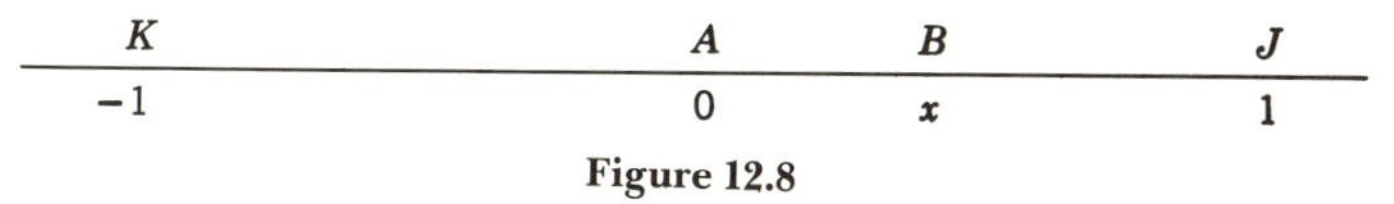


**Figure 12.8**

If we let

$$z = \tfrac{1}{2}\log_e R(A,B,J,K) = \tfrac{1}{2}\log_e \frac{1+x}{1-x}, \tag{12.41}$$

then we have a new coordinate scale on the line as in Figure 12.9 where there is a one-to-one correspondence between the points of the line and the set of all real numbers $z$. From this point of view our line is "infinite in extent," as we might like it to be. We note, moreover, that in terms of $z$-coordinates we have $m(A\text{-}B) = |z|$ and that if the points $B$ and $C$ have $z$-coordinates $z_1$ and $z_2$, then $m(B\text{-}C) = |z_2 - z_1|$.

Equation (12.41) gives us the $z$-coordinate in terms of the $x$-coordinate naturally associated with coordinates in the projective plane. If we solve this equation for $x$, we find

$$x = \frac{e^{2z}-1}{e^{2z}+1} = \frac{e^{z}-e^{-z}}{e^{z}+e^{-z}}$$

or

$$x = \tanh z. \tag{12.42}$$

Note that the choice of $\frac{1}{2}$ in the definition of measure leads to the simplicity of this relation. We shall call $x$-coordinates *projective coordinates* and $z$-coordinates *hyperbolic coordinates.*

A coordinate change similar to that above could also be carried out on the line $x_1 = 0$. The results could then be combined to lead to the conclusion that any ordinary point with projective coordinates $(x_1, x_2, 1)$ also has hyperbolic coordinates $(z_1, z_2)$. These hyperbolic coordinates make the hyperbolic plane seem more "like" the Euclidean plane. It should be pointed out, however, that lines have linear equations in the projective coordinates but do not have linear equations in the hyperbolic coordinates. Moreover, the simple distance formula of Euclidean geometry, based on the Theorem of Pythagoras, holds neither for the projective nor the hyperbolic coordinates.

We have simplified our analysis by using Theorem 12.15 and transforming points to one of the coordinate axes. It is a fairly simple matter to find $M(A\text{-}B)$ for points $A$ and $B$ in arbitrary position, in terms of projective coordinates. We suppose that the points have coordinates $(x_1, y_1, 1)$ and $(x_2, y_2, 1)$. The ideal points $J$ and $K$ on the same line will then be of the form $A + kB$ for suitable choices of $k$, and the cross ratio required will, by Exercise 8.53, be the ratio of these two $k$ values. The point $A + kB$ has coordinates

$$(x_1 + kx_2,\ y_1 + ky_2,\ 1 + k)$$

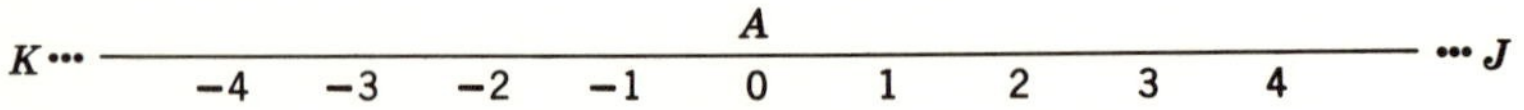

**Figure 12.9**

which satisfy the equation of the absolute; that is, we must have

$$(x_1+kx_2)^2+(y_1+ky_2)^2=(1+k)^2,$$

which can be written in the form

$$ak^2+bk+c=0$$

where

$$a=1-x_2^2-y_2^2,$$
$$b=2(1-x_1x_2-y_1y_2),$$

and

$$c=1-x_1^2-y_1^2.$$

If the roots of this quadratic are $r$ and $s$ with $r/s>1$, then

$$m(A\text{-}B)=\tfrac{1}{2}\log_e\frac{r}{s}.$$

From this we infer that

$$e^{2m(A\text{-}B)}=\frac{r}{s}$$

and

$$e^{-2m(A\text{-}B)}=\frac{s}{r}.$$

If we add these and also add $\frac{1}{2}$ to each side of the resulting equation, we find after simplification that

$$\frac{e^{2m(A\text{-}B)}+2+e^{-2m(A\text{-}B)}}{4}=\frac{1}{4}\frac{r^2+2rs+s^2}{rs}$$

or

$$\left(\frac{e^{m(A\text{-}B)}+e^{-m(A\text{-}B)}}{2}\right)^2=\left(\frac{r+s}{2\sqrt{rs}}\right)^2.$$

If we note that the left side is the square of a hyperbolic cosine, we may extract square roots to find

$$\cosh m(A\text{-}B)=\frac{(r+s)}{2\sqrt{rs}}.$$

Now $r+s=-b/a$ and $rs=c/a$, whence this becomes

$$\cosh m(A\text{-}B)=\frac{|b/a|}{2\sqrt{c/a}}$$

$$=\frac{|b/2|}{\sqrt{ac}},$$

or

$$\cosh m(A\text{-}B)=\frac{|1-x_1x_2-y_1y_2|}{\sqrt{(1-x_1^2-y_1^2)(1-x_2^2-y_2^2)}} \tag{12.43}$$

giving us the general distance formula, in projective coordinates, for the hyperbolic plane.

## Exercises 12.4

1. Complete the details of the proof of Hilbert's Axiom 15.

2. *a.* Let the points $A$ and $B$ have projective coordinates $(\frac{1}{4}, \frac{1}{4}, 1)$ and $(\frac{1}{2}, \frac{1}{2}, 1)$. Show that the Euclidean measure of $A$-$B$ is $\sqrt{2}/4 = 0.35$.

   *b.* Show that the hyperbolic measure of the same segment is $(\frac{1}{2}) \log_e [(15+2\sqrt{2})/(15-2\sqrt{2})] = 0.19$.

3. Find the hyperbolic measure of the segment determined by the points with projective coordinates $(\frac{1}{2}, \frac{1}{2}, 1)$ and $(\frac{2}{3}, \frac{2}{3}, 1)$.

4. Are the segments in Exercises 12.2 and 12.3 congruent in Euclidean geometry? In hyperbolic geometry?

5. What is the equation of the line $u_1x_1 + u_2x_2 + u_3 = 0$, $x_1{}^2 + x_2{}^2 < 1$, in hyperbolic coordinates?

6. Show that the lines $[u_1, u_2, u_3]$ and $[u_1, u_2, u_3']$, $u_3 \neq u_3'$, do not intersect.

## 12.5. MEASURE AND CONGRUENCE OF ANGLES

In Section 11.3 we first used the logarithm of a cross ratio and succeeded in assigning angle measure in similarity and Euclidean geometry. A similar approach succeeded in the preceding section in assigning measure to segments. Although duality is not appropriate in hyperbolic geometry, the results of the previous section do suggest a possible approach. Given two lines $a$ and $b$, why not consider the two lines through their intersection which are on the absolute regarded as a line conic? These two lines would just be the tangents to the conic from the point of intersection of the lines.

If we consider the special case of lines $u_1x_1 + u_2x_2 = 0$ and $u_1'x_1 + u_2'x_2 = 0$ meeting at the origin, the tangents to the conic will be tangent at $(1, i, 0)$ and $1, -i, 0)$, and all the preliminary analysis of Section 11.3 and even the result of Theorem 11.33 will be valid in this special case. This does not mean that hyperbolic angle measure will be the same as Euclidean angle measure, and we shall see shortly that the result of Theorem 11.33 is not generally valid in the hyperbolic plane. It does suggest, however, that the approach is promising and leads to a sequence of definitions and theorems similar to that of Section 11.3. We begin with

**Definition 12.51.** *The* measure of the intersecting lines *a and b is the number*

$$m(a, b) = \left|\frac{i}{2}\log_e R(a, b, j, k)\right|, \qquad 0 < m(a, b) \leq \frac{\pi}{2},$$

*where j and k are the tangents to the absolute through the point of intersection of a and b.*

As before we introduce

**Definition 12.52.** *Two intersecting lines are* perpendicular *if their measure is* $\pi/2$.

We can then prove

**Theorem 12.51.** *Through a point on a given line there passes a unique line perpendicular to the given line.*

PROOF. We apply Theorem 12.15, mapping the given point to (0, 0, 1) and the given line to $x_2 = 0$. It will then follow from the formula of Theorem 11.33 that the unique perpendicular is the unique line of the pencil through the given point mapped onto the line $x_1 = 0$.

In the next section we shall consider perpendicularity in more detail and arrive at an alternate and more sophisticated proof of this theorem.

We can now use the plane separation property of Theorem 2.22 to classify angles.

**Definition 12.53.** *If h and k are opposite rays on a line, (h, k) is a* straight angle. *If h and k lie on perpendicular lines , (h, k) is a* right angle. *If k lies in the same half plane as does h with respect to the line perpendicular to h through the vertex of the angle, (h, k) is an* acute angle. *If it lies in the opposite half plane, (h, k) is an* obtuse angle.

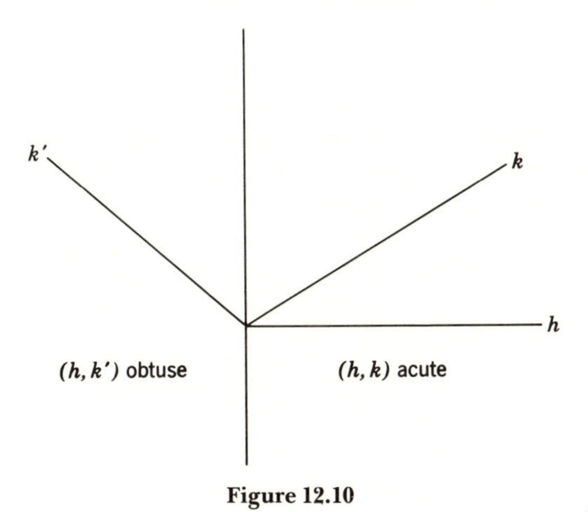


**Figure 12.10**

As in Section 11.3 we must prove

**Theorem 12.52.** *An acute angle is well defined.*

PROOF. We can "save" the proof of Theorem 11.35. We apply Theorem 12.15 to map the vertex of the angle to the origin so that we may use the formula of Theorem 11.33. In the proof of Theorem 11.35 we replace "parallel" by "parallel or nonintersecting." The points $I, J, M$, and $N$ are now to be the ultraideal points of the corresponding projective lines lying on $x_3 = 0$. The argument of the proof of Theorem 11.35 can then be carried through in the present case.

We now finally can define angle measure.

**Definition 12.54.** *If the rays h and k lie on the intersecting lines a and b and form an angle (h, k), then the* measure $m(h, k)$ of the angle *is*

(1) $m(a, b)$ *if* $(h, k)$ *is acute or right,*
(2) $\pi - m(a, b)$ *if* $(h, k)$ *is obtuse,*
(3) $\pi$ *if* $(h, k)$ *is straight.*

We have at once

**Theorem 12.53.** *If T is an isometry, then* $m(h, k) = m[T(h), T(k)]$.

As before we define congruence in terms of measure.

**Definition 12.55.** *The angles* $(h, k)$ *and* $(h', k')$ *are* congruent $[(h, k) \cong (h', k')]$ *if* $m(h, k) = m(h', k')$.

We find

**Theorem 12.54.** *Hyperbolic geometry satisfies Hilbert's axioms of congruence for angles.*

The proof is the same as for Theorem 11.37.

Finally, let us seek an analogy of equation (12.43) for the measure of the lines $[u_1, u_2, u_3]$ and $[v_1, v_2, v_3]$. The desired tangent lines of the pencil determined by these lines will have coordinates $u_i + kv_i$ for a suitable $k$, and as in Section 12.4 the cross ratio desired will be the ratio of these $k$ values. If the lines are to be tangent to the absolute, we must have, in view of the discussion of Section 12.1,

$$(u_1 + kv_1)^2 + (u_2 + kv_2)^2 = (u_3 + kv_3)^2.$$

Again we have a quadratic condition

$$ak^2 + bk + c = 0$$

where now

$$a = v_1{}^2 + v_2{}^2 - v_3{}^2,$$
$$b = 2(u_1v_1 + u_2v_2 - u_3v_3),$$

and

$$c = u_1^2 + u_2^2 - u_3^2.$$

Now the discriminant is negative, and we have

$$m(a, b) = \left|\frac{i}{2}\log_e \frac{b+\sqrt{4ac-b^2}\,i}{b-\sqrt{4ac-b^2}\,i}\right|.$$

The cosine of the desired angle will then be the absolute value of the real part of the numerator divided by the modulus $b^2 + (4ac - b^2)$ of this quantity. Thus

$$\cos m(a, b) = \frac{|b|}{\sqrt{4ac}}$$

or

$$\cos m(a, b) = \frac{|u_1v_1 + u_2v_2 - u_3v_3|}{\sqrt{(u_1^2 + u_2^2 - u_3^2)(v_1^2 + v_2^2 - v_3^2)}}. \tag{12.52}$$

This is in agreement with the result of Theorem 11.33 only in the special case where $u_3 = v_3 = 0$.

## Exercises 12.5

1. Compute the measure between each of the following pairs of lines and compare with the Euclidean measure:
   *a.* [2, 0, 1] and [0, 1, 0],
   *b.* [2, 0, 1] and [3, 0, 1],
   *c.* [1, 0, $a$] and [0, 1, $b$], $a < 1$ and $b < 1$,
   *d.* [3, 3, 1] and [3, 4, 1].
2. Prove that two lines perpendicular in Euclidean geometry are perpendicular in hyperbolic geometry if and only if one line passes through (0, 0, 1).

## 12.6 PERPENDICULARS AND PARALLELS

From Equation (12.52) we note that the lines $[u_1, u_2, u_3]$ and $[v_1, v_2, v_3]$ are perpendicular if and only if

$$u_1v_1 + u_2v_2 - u_3v_3 = 0.$$

If $A$ is the matrix of the absolute, it is apparent that $A = A^{-1}$. From Definition 9.53 it follows that the lines are perpendicular if and only if they are conjugate with respect to the absolute, or they are perpendicular if and only if each contains the pole of the other. We have proved

**Theorem 12.61.** *Two lines are perpendicular if and only if each contains the pole of the other.*

From this result we immediately deduce

**Theorem 12.62.** *There exists a unique line perpendicular to a given line and passing through a given point.*

The proof is left to the reader. We note that there is no restriction on the incidence relation between the given point and the given line, and indeed this relation is irrelevant in the proof. The theorem thus generalizes Theorem 12.51. This theorem is of course also a theorem in Euclidean geometry. The next is not, being peculiar to the hyperbolic plane.

**Theorem 12.63.** *There exists a unique line perpendicular to each of two non-intersecting lines.*

**Theorem 12.64.** *If two lines have a common perpendicular, they are non-intersecting.*

The proofs are left to the reader.

Let us now consider the configuration consisting of a line $l$, a point $P$ not on the line, and the parallels to $l$ through $P$. Let the perpendicular to $l$ from $P$ meet $l$ at $O$. We are interested in the angles $\alpha$ and $\alpha'$ determined by the parallels and the ray from $P$ to $O$. We take the rays on the parallels containing, in the sense of Definition 12.32, the ideal points on $l$. We claim that the angles so determined are acute, since they lie in the right angles determined by the ray from $P$ to $O$ and the rays on the line perpendicular to $OP$ through $P$. One can verify this by choosing coordinates so that $l$ is $x_2 = 0$ and $P$ has coordinates $(0, c, 1)$. Since the collineation with matrix

$$\begin{pmatrix} -1 & 0 & 0 \\ 0 & 1 & 0 \\ 0 & 0 & 1 \end{pmatrix} \tag{12.61}$$

is an isometry and carries the rays determining $\alpha$ onto the rays determining $\alpha'$, we conclude that $\alpha$ and $\alpha'$ are congruent. Moreover, if $A$-$B$ is any segment congruent to $O$-$P$ with line $l'$ through $A$ perpendicular

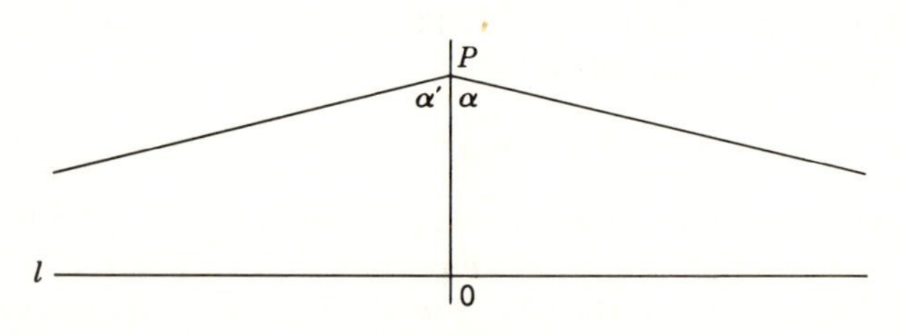


**Figure 12.11**

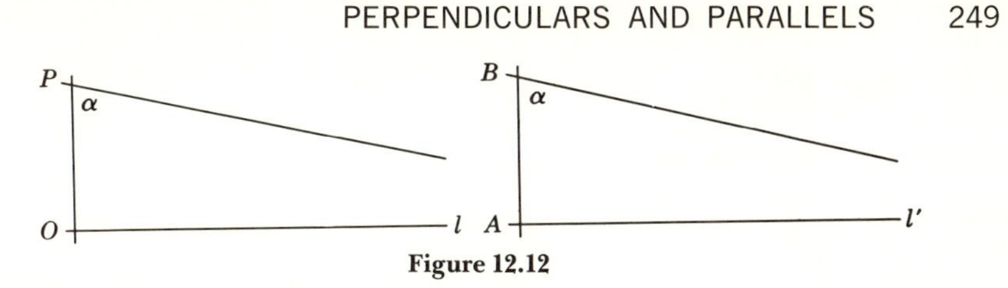


Figure 12.12

to $AB$, the angles at $B$ determined by the parallel rays to $l'$ are congruent to $\alpha$, since the configuration can be mapped onto the original configuration by an isometry. We summarize our analysis in the following theorem and definition.

**Theorem 12.65.** *The angle between a ray parallel to a given line and the coterminal ray perpendicular to the given line and containing a point of the given line is completely determined by the distance from the vertex of the angle to the given line.*

This means that congruent segments will determine congruent angles in accordance with the conditions of the theorem.

**Definition 12.61.** *The* angle of parallelism *of a segment measure is the angle determined by that measure and the conditions of Theorem 12.65.*

We turn now to the problem of finding an analytic expression for the angle of parallelism. If we use the choice of coordinates of the discussion leading to Theorem 12.65, the lines at issue will have coordinates $[1, 0, 0]$ and $[c, 1, -c]$. By equation (12.52), since the angle is acute, we find $\cos\alpha = c$. In terms of hyperbolic coordinates this becomes

$$\cos\alpha = \tanh a \tag{12.62}$$

where $\tanh a = c$ is just the change from hyperbolic to projective coordinates. We note that as $a \to 0$, $\alpha \to \pi/2$, so that the relation becomes Euclidean in the limit. If we let $P$ approach the ideal point on $l$, we find that the first projective coordinate of $P$ will decrease toward zero so that the angle of parallelism will approach $\pi/2$.

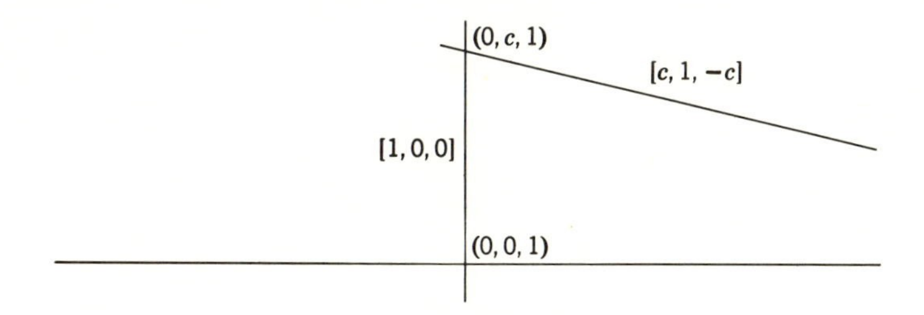


Figure 12.13

## Exercises 12.6

1. Prove Theorem 12.62.
2. Prove Theorem 12.63.
3. Prove Theorem 12.64.
4. Complete the proof of Theorem 12.65 by completely justifying all assertions made there.
5. Prove that the following are equivalent to equation (12.62):
   *a.* $\sin\alpha = \operatorname{sech} a$,
   *b.* $\tan\alpha = \operatorname{csch} a$,
   *c.* $\cot\alpha = \sinh a$,
   *d.* $\sec\alpha = \coth a$,
   *e.* $\csc\alpha = \cosh a$.

# 12.7 CONGRUENCE OF TRIANGLES

The one remaining congruence axiom of Hilbert's system is the axiom combining segment and angle congruence. We prove this by bypassing his axiom and proving directly the "side-angle-side" congruence theorem.

**Theorem 12.71.** *If two triangles have two sides of one and the included angle congruent respectively to two sides and the included angle of the other, the triangles are congruent.*

PROOF. We suppose that $A\text{-}B \cong A'\text{-}B'$, $A\text{-}C \cong A'\text{-}C'$, and $\angle A \cong \angle A'$. Let $T$ be the isometry mapping the ray from $A$ to $B$ onto the ray from $A'$ to $B'$. We may further require that $T$ map $C$ onto the same side of $A'B'$ as is $C'$. In view of Axiom 16, $T$ must map the ray from $A$ to $C$ onto the ray from $A'$ to $C'$. Then by Axiom 11 we conclude that $T(B, C) =$

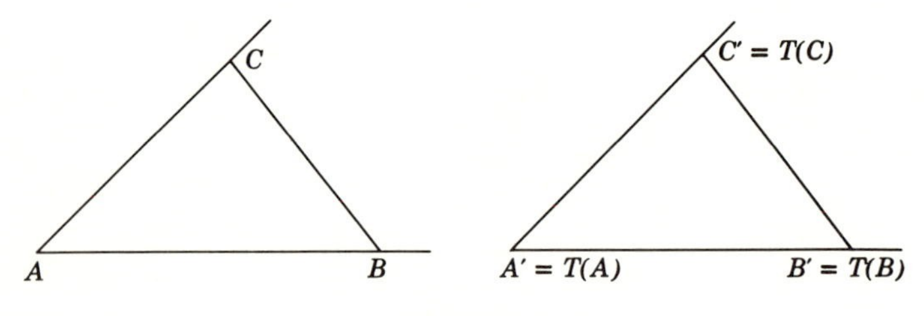


Figure 12.14

$B'$, $C'$. We conclude that $T$ thus maps the one triangle onto the other, hence that all corresponding parts are congruent.

We have now proved all of Hilbert's axioms except the one on parallels, which we clearly cannot prove, and the last two on continuity and completeness. We assert that the last two are consequences of the corresponding axioms in the projective plane in Chapter 10, but we shall not here go through the details of the proof.

With Theorem 12.71 proved, all remaining theorems in Section 2.3 can be proved in the hyperbolic plane; in particular, the "angle-side-angle" and "side-side-side" triangle congruence theorems can be proved. If we turn to Section 2.4 and consider Theorems 2.42 and 2.43, we assert that neither of these can be proved in the hyperbolic plane. We shall return later to Theorem 2.43. We claim that triangle congruence theory does lead to the following weakened version of Theorem 2.42.

**Theorem 12.72.** *If a transversal on two lines determines congruent alternate interior angles, the lines are nonintersecting.*

The proof is left to the reader.

Finally, although Theorem 2.43 is not valid, its weaker version is valid in both the Euclidean and hyperbolic planes, and we state it here, since its proof depends in part on triangle congruence.

**Theorem 12.73.** *(Exterior Angle Theorem) An exterior angle of a triangle has greater measure than either nonadjacent interior angle.*

The proof is left to the reader. Figure 12.15 may be helpful in constructing a proof.

## Exercises 12.7

1. Prove Theorem 12.72.
2. Prove Theorem 12.73.

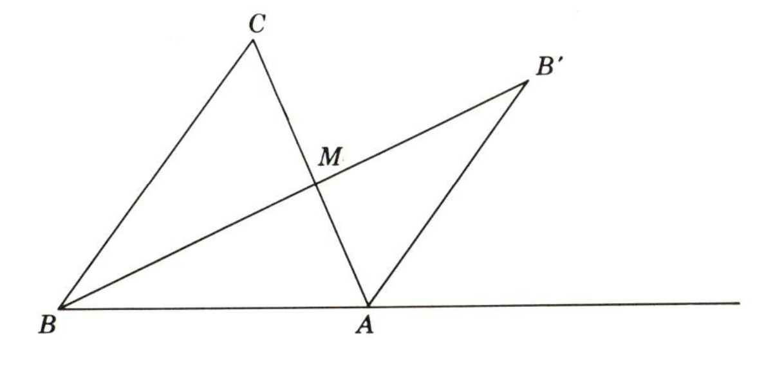


**Figure 12.15**

3. Let $A$ and $B$ be ordinary points. Consider the rays with these as terminal points and containing a common ideal point. The configuration consisting of $A$, $B$, $A$-$B$, and these two rays is called an asymptotic triangle. Prove that if any two corresponding pairs of parts of two asymptotic triangles are congruent, the third pair of corresponding parts must be congruent.

## 12.8 RIGHT TRIANGLE TRIGONOMETRY

As in the Euclidean case let us denote the nonright angles of a right triangle by $\alpha$ and $\beta$, the opposite legs by $a$ and $b$, respectively, and the hypotenuse by $c$. For a suitable choice of coordinates the triangle may be placed as in Figure 12.16. The coordinates in the figure are projective; hence we have $a' = \tanh a$ and $b' = \tanh b$. We also note that

$$\sqrt{1-a'^2} = \sqrt{1-\tanh^2 a} = \operatorname{sech} a$$

and

$$\sqrt{1-b'^2} = \operatorname{sech} b.$$

By equation (12.43) we have

$$\cosh c = \frac{1}{\sqrt{1-a'^2}\,\sqrt{1-b'^2}}$$

$$= \frac{1}{\operatorname{sech} a \operatorname{sech} b}$$

or

$$\cosh c = \cosh a \cosh b. \tag{12.81}$$

This formula gives us the length of the hypotenuse in terms of the lengths of the sides of the triangle and is the analogue in hyperbolic geometry to the theorem of Pythagoras in Euclidean geometry. We note that $\cosh x > 1$ for $x > 0$; hence in a right triangle $c > a$ and $c > b$ as in the Euclidean case.

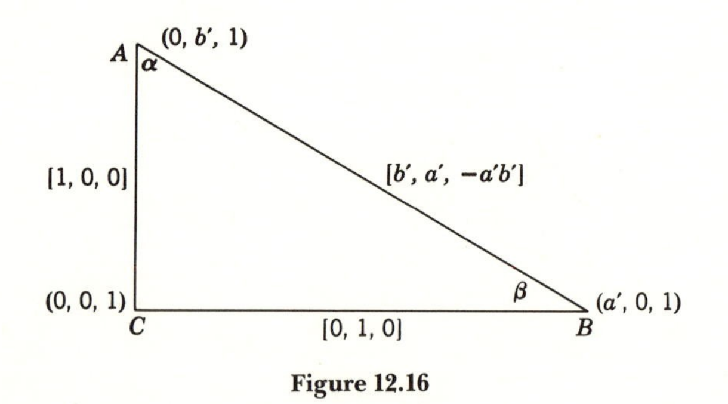


**Figure 12.16**

One readily verifies in Figure 12.16 that the nonright angles of the triangle are acute. The argument is similar to the one used in the preceding section in the discussion of angle of parallelism. By equation (12.52) we find

$$\begin{aligned}\cos\alpha &= \frac{b'}{\sqrt{b'^2+a'^2-a'^2b'^2}}\\ &= \frac{b'}{\sqrt{1-(1-a'^2)(1-b'^2)}}\\ &= \frac{\tanh b}{\sqrt{1-\operatorname{sech}^2 a\operatorname{sech}^2 b}}\\ &= \frac{\tanh b}{\sqrt{1-\operatorname{sech}^2 c}}\end{aligned}$$

or

$$\cos\alpha = \frac{\tanh b}{\tanh c}. \tag{12.82}$$

Since for an acute $\theta$, $\sin\theta = \sqrt{1-\cos^2\theta}$, we find

$$\begin{aligned}\sin\alpha &= \frac{a'\sqrt{1-b'^2}}{\sqrt{b'^2+a'^2-a'^2b'^2}}\\ &= \frac{\tanh a\operatorname{sech} b}{\tanh c}\\ &= \frac{\tanh a\operatorname{sech} b\cosh c}{\sinh c}\\ &= \frac{\tanh a\operatorname{sech} b\cosh a\cosh b}{\sinh c}\end{aligned}$$

or

$$\sin\alpha = \frac{\sinh a}{\sinh c}. \tag{12.83}$$

Finally,

$$\begin{aligned}\tan\alpha &= \frac{\sin\alpha}{\cos\alpha}\\ &= \frac{\sinh a\tanh c}{\sinh c\tanh b}\\ &= \frac{\sinh a}{\cosh c\tanh b}\\ &= \frac{\sinh a}{\cosh a\cosh b\tanh b}\end{aligned}$$

or

$$\tan\alpha = \frac{\tanh a}{\sinh b}. \tag{12.84}$$

By symmetry we also have

$$\cos\beta = \frac{\tanh a}{\tanh c}, \tag{12.85}$$

$$\sin\beta = \frac{\sinh b}{\sinh c}, \tag{12.86}$$

and

$$\tan\beta = \frac{\tanh b}{\sinh a}. \tag{12.87}$$

Equations (12.81) to (12.87) are sufficient for any right triangle computation although they by no means exhaust the relations among the parts of the triangle. The reader is referred to the exercises and the references at the end of the chapter for further examples.

Let us now consider

$$\tan(\alpha+\beta) = \frac{\tan\alpha + \tan\beta}{1 - \tan\alpha\tan\beta}.$$

The numerator of this expression is clearly positive. The denominator reduces to $1 - \operatorname{sech} c$, which is also positive. Hence $\tan(\alpha+\beta) > 0$ and $\alpha+\beta < \pi/2$. We have proved

**Theorem 12.81.** *The sum of the measures of the angles of a right triangle is less than $\pi$.*

From this we at once infer

**Theorem 12.82.** *The sum of the measures of the angles of a triangle is less than $\pi$.*

The proof is left to the reader. All one needs to do is drop a perpendicular from a vertex of the triangle to the opposite side. One must consider the possibilities that this perpendicular meets the line of the opposite side either internally or externally with respect to the segment which is the side.

**Theorem 12.83.** *The sum of the measures of the angles of a simple quadrilateral is less than $2\pi$.*

Again the proof is left to the reader. A simple quadrilateral is one whose sides (segments) do not intersect.

We may now prove an additional triangle congruence theorem not valid in the Euclidean plane.

**Theorem 12.84.** *If two triangles are such that the angles of one are congruent to the angles of the other, the triangles are congruent.*

PROOF. We suppose that $\angle A \cong \angle A'$, $\angle B \cong \angle B'$, and $\angle C \cong \angle C'$. As in the proof of Theorem 12.71, we consider the isometry mapping the

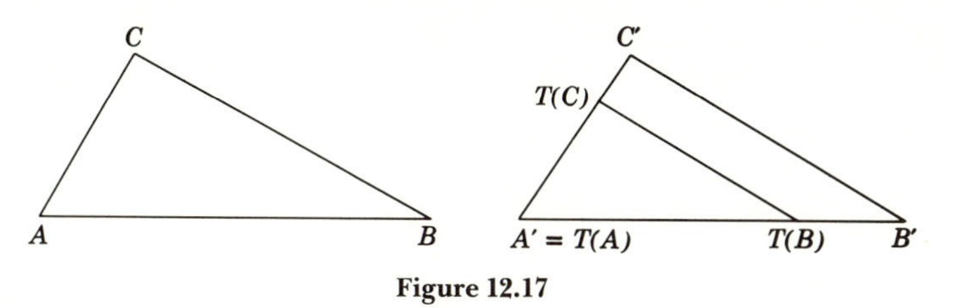


**Figure 12.17**

ray from $A$ to $B$ onto the ray from $A'$ to $B'$ and the ray from $A$ to $C$ onto the ray from $A'$ to $C'$. It will be sufficient to show that $T(B) = B'$. We suppose that this is not the case. We assume $T(B)$ to be between $A'$ and $B'$; a similar argument holds if $B'$ is between $A'$ and $T(B)$. We note that by Theorem 12.72 the segments $B'$-$C'$ and $T(B)$-$T(C)$ have no point in common. Since

$$\angle A'T(B)T(C) \cong \angle A'B'C'$$

and

$$\angle A'T(C)T(B) \cong \angle A'C'B',$$

the simple quadrilateral with vertices $B'$, $C'$, $T(C)$, and $T(B)$ would have the sum of the measures of its angles $2\pi$. This is impossible in view of Theorem 12.83.

Theorem 12.83 is in contrast to the result in Euclidean geometry; in the next chapter we shall see that the sum of the measures of the angles of a triangle in elliptic geometry must exceed $\pi$. One tends to differentiate between Euclidean, hyperbolic, and elliptic geometry on the basis of parallelism properties, but the distinction is just as sharp in terms of this property of the sum of the measures of the angles of a triangle.

Theorem 12.84 indicates that similarity and congruence are the same relation in the hyperbolic plane, unlike the Euclidean case. It also gives a strong indication as to why there is no intermediate "hyperbolic affine" geometry between projective and hyperbolic geometry. One can show in the hyperbolic plane that the amount by which the sum of the measures of the angles of a triangle falls short of $\pi$ can be taken as the measure of the area of a triangle; this result is clearly consistent with Theorem 12.84.

## Exercises 12.8

1. Prove Theorem 12.82.
2. Prove Theorem 12.83.
3. Find the remaining parts for each of the following right triangles:
   *a.* $a = 0.23$, $c = 1.31$;

*b*. $a = 0.23, b = 0.94$;

*c*. $a = 0.23, \alpha = \pi/3$;

*d*. $\alpha = \pi/4, \beta = \pi/6$.

4. Prove that $\tan\alpha \tan\beta = \operatorname{sech} c$.
5. Prove that $\cosh a = \cos\alpha/\sin\beta$.

## 12.9 GENERAL TRIANGLE TRIGONOMETRY

If we now consider a general triangle with side lengths *a, b,* and *c* and opposite angles $\alpha$, $\beta$, and $\gamma$, respectively, and let $h$ be the length of the altitude from the vertex $C$, we have by equation (12.83)

$$\sin\alpha = \frac{\sinh h}{\sinh b}$$

and

$$\sin\beta = \frac{\sinh h}{\sinh a}$$

from which we deduce

$$\frac{\sinh a}{\sin\alpha} = \frac{\sinh b}{\sin\beta}.$$

The generalization is obvious, and we have the *law of sines* for the hyperbolic plane

$$\frac{\sinh a}{\sin\alpha} = \frac{\sinh b}{\sin\beta} = \frac{\sinh c}{\sin\gamma}. \tag{12.91}$$

We suppose that the altitude from $C$ divides the segment $A$-$B$ into parts with lengths $d$ and $c-d$ as indicated in Figure 12.18. By equation (12.81) we have

$$\begin{aligned}\cosh a &= \cosh h \cosh(c-d)\\ &= \cosh h \cosh c \cosh d - \cosh h \sinh c \sinh d.\end{aligned}$$

Since

$$\cosh b = \cosh h \cosh d,$$

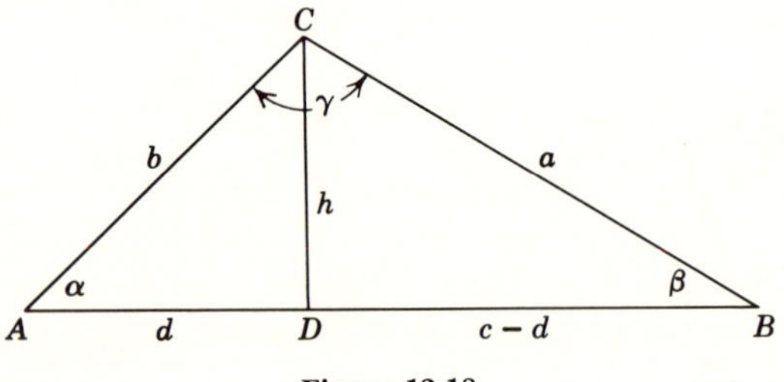


**Figure 12.18**

we have

$$\begin{aligned}\cosh a &= \cosh b \cosh c - \cosh h \sinh c \sinh d \\ &= \cosh b \cosh c - \frac{\cosh b}{\cosh d} \sinh c \sinh d \\ &= \cosh b \cosh c - \cosh b \tanh d \sinh c.\end{aligned}$$

Now by equation (12.82) we have

$$\cos \alpha = \frac{\tanh d}{\tanh b} = \frac{\tanh d \cosh b}{\sinh b}$$

or

$$\sinh b \cos \alpha = \cosh b \tanh d.$$

We thus find

$$\cosh a = \cosh b \cosh c - \sinh b \sinh c \cos \alpha, \tag{12.92}$$

the *law of cosines* in the hyperbolic plane. In a like manner we find

$$\cosh b = \cosh a \cosh c - \sinh a \sinh c \cos \beta$$

and

$$\cosh c = \cosh a \cosh b - \sinh a \sinh b \cos \gamma.$$

Since $\cos \gamma \geqslant -1$, we have

$$\begin{aligned}\cosh c &= \cosh a \cosh b - \sinh a \sinh b \cos \gamma \\ &\leqslant \cosh a \cosh b + \sinh a \sinh b \\ &= \cosh (a+b).\end{aligned}$$

We thus have the triangle inequality

**Theorem 12.91.** $m(A\text{-}B) \leqslant m(A\text{-}C) + m(C\text{-}B)$.

## Exercises 12.9

1. Justify equation (12.92) if $D$ is not between $A$ and $B$.
2. Find the remaining parts of each of the following triangles:
   *a*. $a = 0.45, \quad b = 0.23, \quad \gamma = \pi/6;$
   *b*. $a = 0.45, \quad b = 0.23, \quad \alpha = \pi/6;$
   *c*. $a = 0.45, \quad b = 0.23, \quad c = 0.62.$
3. The *defect* of a triangle is defined to be $\pi - (\alpha + \beta + \gamma)$. Show that if the interior of a triangle is subdivided into the interior of two triangles as in Figure 12.18, the sum of the defects of the resulting triangles is the defect of the original triangle. (This is an indication that the defect could be taken to be the measure of area of a triangle.)

## 12.10 SACCHERI AND LAMBERT QUADRILATERALS

It is evident from Theorem 12.83 that there can be no such thing as a rectangle in hyperbolic geometry. In their attempts to prove the parallel axiom of Euclidean geometry, Saccheri and Lambert each made use of certain quadrilaterals containing right angles. Although they were not, of course, able to use these quadrilaterals for the purpose intended, the resulting configurations are of considerable historical interest and of use in the further development of hyperbolic geometry. We shall note briefly their basic properties.

**Definition 12.101.** *A* Saccheri quadrilateral *is a quadrilateral with two adjacent right angles and two opposite congruent sides adjacent to these right angles. The side adjacent to both right angles is the* base *of the quadrilateral while the side opposite is its* summit.

The configuration is shown in Figure 12.19. The segment $A\text{-}A'$ is the base, while the segment $B\text{-}B'$ is the summit. If $T$ denotes the isometry of (12.61), it is apparent that

$$T(A, A', B, B', M, M') = A', A, B', B, M', M.$$

We conclude that the summit angles of a Saccheri quadrilateral are congruent and that the lines joining the midpoints of the base and summit is perpendicular to each. By Theorem 12.83 it follows that the summit angles at $B$ and $B'$ are acute. Finally, from equation (12.43) we have

$$\cosh \tfrac{1}{2}m(A\text{-}A') = \frac{1}{\sqrt{1-a^2}}$$

and

$$\cosh \tfrac{1}{2}m(B\text{-}B') = \frac{1-b^2}{\sqrt{1-b^2}\sqrt{1-b^2-a^2}}$$

$$= \frac{1}{\sqrt{1-[a^2/(1-b^2)]}}.$$

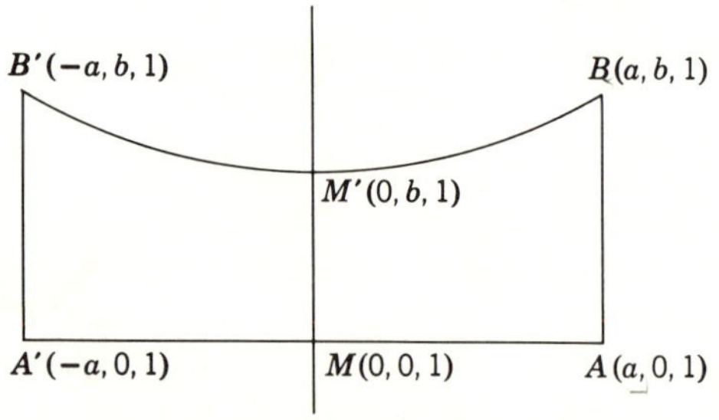

Figure 12.19

Since $0 < 1 - b^2 < 1$, we have

$$\frac{1}{\sqrt{1-a^2}} < \frac{1}{\sqrt{1-[a^2/(1-b^2)]}},$$

hence $m(A\text{-}A') < m(B\text{-}B')$. We summarize our results in

**Theorem 12.101.** *In a Saccheri quadrilateral*

*a. The summit angles are congruent and acute;*

*b. The line joining the midpoints of the base and summit is perpendicular to each;*

*c. The length of the base is less than the length of the summit.*

We now consider the Lambert quadrilateral.

**Definition 12.102.** *A* Lambert quadrilateral *is a quadrilateral with three right angles.*

We see that a Lambert quadrilateral is "one half" of a Saccheri quadrilateral in the sense that two congruent Lambert quadrilaterals can be so placed as to form a Saccheri quadrilateral; for example, in Figure 12.19 $ABM'M$ and $A'B'M'M$ are both Lambert quadrilaterals. From either Theorem 12.101 or 12.83 we have

**Theorem 12.102.** *A Lambert quadrilateral contains an acute angle.*

## Exercises 12.10

1. Give a synthetic proof of Theorem 12.101*a* and 12.101*b*.
2. Give a trigonometric proof of Theorem 12.101*c*.

## REFERENCES

Adler, Claire Fisher, *Modern Geometry, An Integrated First Course,* 2nd ed., McGraw-Hill Book Company, New York, 1967, Chapters 3, 11.

Artzy, Rafael, *Linear Geometry,* Addison-Wesley Publishing Company, Reading, Mass., 1965, Chapter 3.

Blumenthal, Leonard M., *A Modern View of Geometry,* W. H. Freeman and Company, San Francisco, 1961, Chapter 8.

Coxeter, H. S. M., *Introduction to Geometry,* John Wiley and Sons, New York, 1961, Chapter 16.

Coxeter, H. S. M., *Non-Euclidean Geometry,* 5th ed., University of Toronto Press, Toronto, 1965, Chapters 9, 10.

Eves, Howard, *A Survey of Geometry,* vol. I, Allyn and Bacon, Boston, 1963, Chapters 7, 8.

Gruenberg, K. W., and A. J. Weir, *Linear Geometry,* D. Van Nostrand Company, Princeton, 1967, Chapter 2.

Levy, Harry, *Projective and Related Geometries,* The Macmillan Company, New York, 1961, Chapter 5.

Meschkowski, Herbert, translated by A. Shenitzer, *Noneuclidean Geometry,* Academic Press, New York, 1964, Chapters 4 to 8.

Meserve, Bruce E., *Fundamental Concepts of Geometry,* Addison-Wesley Publishing Company, Reading, Mass., 1955, Chapter 8.

Moise, Edwin E., *Elementary Geometry from an Advanced Standpoint,* Addison-Wesley Publishing Company, Reading, Mass., 1963, Chapters 10, 24, 25.

Rainich, G. Y., and S. M. Dowdy, *Geometry for Teachers,* John Wiley and Sons, New York, 1968, Chapters 6, 8.

Springer, C. E., *Geometry and Analysis of Projective Spaces,* W. H. Freeman and Company, San Francisco, 1964, Chapter 9.

Tuller, Annita, *A Modern Introduction to Geometries,* D. Van Nostrand Company, Princeton, 1967, Chapters 1, 7.

Veblen, Oswald, and John Wesley Young, *Projective Geometry,* vol. II, Ginn and Company, Boston, 1910, Chapter 8.

Wolfe, Harold E., *Non-Euclidean Geometry,* Holt, Rinehart and Winston, New York, 1945, Chapters 1, 4, 5.

Wylie, C. R., Jr., *Foundations of Geometry,* McGraw-Hill Book Company, New York, 1964, Chapter 4.

## CHAPTER THIRTEEN

# ELLIPTIC GEOMETRY

We conclude by considering the geometry that results if we take as an absolute conic a nondegenerate imaginary one. We shall find that this leads to the other classical non-Euclidean geometry in which no parallels exist and that the analogies between this geometry and Euclidean geometry are weaker than those between hyperbolic and Euclidean geometries.

## 13.1 THE ELLIPTIC PLANE

Let us now choose as the absolute conic a nondegenerate imaginary one. For a suitable choice of coordinates in the projective plane the conic will have equation

$$x_1^2 + x_2^2 + x_3^2 = 0.$$

There are, of course, no real points on the absolute. Thus if we delete the points on it, we shall have deleted no points at all, and our plane will still be the projective plane. Moreover, it is evident that no analogy to Theorem 12.11 can hold in this case. It would appear, then, that there will be no classification of points in the real projective plane into ordinary and ideal. All points in the real projective plane will remain as points in the new geometry.

It is true, however, that it makes sense to ask that a collineation of the real projective plane leave this absolute fixed. Such a collineation has an associated matrix $A$ which is nonsingular and real. If $P$ is a point in the complex projective plane with associated column matrix $X$ of coordinates, the elements of $X$ being complex numbers, then the collineation can be extended to the complex projective plane by stipulating that it is to map the point $P$ onto the point $P'$ whose column matrix of coordinates is $AX$. If the collineation is to leave the absolute fixed, we are simply requiring that if $P$ is on the absolute, then $P'$ must be also. We thus can proceed in a manner similar to that at the beginning of Chapters 11 and 12.

**Theorem 13.11.** *The set of collineations leaving the absolute fixed is a subgroup of the collineation group.*

**Definition 13.11.** *The collineations leaving the absolute fixed are* elliptic isometries. *The real projective plane is the* elliptic plane. *The study of the elliptic plane and elliptic isometries is* elliptic geometry.

As in the preceding chapter, when no confusion results, we shall refer to an elliptic isometry as an isometry.

In projective geometry we found that the principle of duality provided us with a powerful tool in our analysis. In elliptic geometry we are concerned with the same set of points and lines found in the projective plane. We assert that this principle of duality can also be used in elliptic geometry if we are careful to dualize all definitions as we proceed, and we shall make use of duality in the remainder of this chapter.

As a first step we must inquire into the nature of the dual of an isometry. Even though we may think of an isometry primarily as a mapping of points onto points, we note that the isometry also maps lines onto lines, since it must carry collinear points onto collinear points; in other words, any collineation is truly a mapping of the points *and* lines of the plane onto the points and lines, respectively, of the plane. We now ask what the dual of an isometry is. Since it must preserve the tangent relation on the absolute, any isometry will map the tangents to the absolute onto the tangents to the absolute. If we regard these tangents as an absolute line conic, an isometry preserves this line conic and is a mapping of the lines of the plane onto the lines of the plane. In short, if we realize that a collineation is a mapping of points and lines, we see that an isometry is self-dual.

In the case of elliptic isometries we can get a simple characterization of the matrices $A$ which represent such isometries. We consider

$$\begin{aligned} x_1 &= a_{11}x_1' + a_{12}x_2' + a_{13}x_3' \\ x_2 &= a_{21}x_1' + a_{22}x_2' + a_{23}x_3' \\ x_3 &= a_{31}x_1' + a_{32}x_2' + a_{33}x_3' . \end{aligned}$$

If this is to represent an isometry, it must map the point $(1, i, 0)$ onto a point on the absolute.

This leads to the conditions

$$(a_{11}^2 + a_{21}^2 + a_{31}^2) - (a_{12}^2 + a_{22}^2 + a_{32}^2) = 0$$

and

$$a_{11}a_{12} + a_{21}a_{22} + a_{31}a_{32} = 0.$$

If we apply the same argument to the points $(1, 0, i)$ and $(0, 1, i)$, we are finally led to the conclusions

$$a_{1j}a_{1k} + a_{2j}a_{2k} + a_{3j}a_{3k} = 0, \qquad j \neq k,$$

and

$$\begin{aligned} a_{11}^2 + a_{21}^2 + a_{31}^2 &= a_{12}^2 + a_{22}^2 + a_{32}^2 \\ &= a_{13}^2 + a_{23}^2 + a_{33}^2 \\ &\neq 0. \end{aligned}$$

It is apparent that each of these sums can be made 1 for a suitable normalization of $A$. We can then summarize all the above in the condition

$$a_{1j}a_{1k} + a_{2j}a_{2k} + a_{3j}a_{3k} = \delta_{jk}$$

where $\delta_{jk} = 1$ if $j = k$ and $\delta_{jk} = 0$ if $j \neq k$. We now see that $AA^* = I$ whence $A^* = A^{-1}$. Now $A^{-1}$ must also represent an isometry, from which we conclude that

$$a_{j1}a_{k1} + a_{j2}a_{k2} + a_{j3}a_{k3} = \delta_{jk}.$$

We summarize our results in

**Theorem 13.12.** *An elliptic isometry is associated with a matrix* $A = (a_{ij})$ *for which*

$$\begin{aligned} \delta_{jk} &= a_{1j}a_{1k} + a_{2j}a_{2k} + a_{3j}a_{3k} \\ &= a_{j1}a_{k1} + a_{j2}a_{k2} + a_{j3}a_{k3}. \end{aligned}$$

**Theorem 13.13.** *A matrix* $A = (a_{ij})$ *for which*

$$\begin{aligned} \delta_{jk} &= a_{1j}a_{1k} + a_{2j}a_{2k} + a_{3j}a_{3k} \\ &= a_{j1}a_{k1} + a_{j2}a_{k2} + a_{j3}a_{k3} \end{aligned}$$

*is the matrix of an isometry.*

PROOF. Any collineation maps a conic onto a conic. The absolute contains the points $(1, i, 0)$, $(1, -i, 0)$, $(1, 0, i)$, $(1, 0, -i)$, and $(0, 1, i)$, no three of which are collinear. Such a matrix will map all these points onto points on the absolute, which is thus preserved under the transformation.

In the development of hyperbolic geometry we made use on occasion of the polarity induced by the absolute conic. In the present case we still have a polarity even if the conic is imaginary, and in fact the polar line of a real point is clearly a real line. In subsequent sections we shall make frequent use of this polarity, for it plays a key role in this geometry. We note that "polar line of a point" is dual to "pole of a line." The idea is used in the proof of the following theorem (which is self-dual), the elliptic version of Theorem 12.15.

**Theorem 13.14.** *If $l$ and $l'$ are lines containing the points $P$ and $P'$, respectively, there exists an isometry $T$ such that $T(P, l) = P', l'$.*

PROOF. Let $L$ be the pole (with respect to the absolute) of $l$, and let $M$ be the pole of $LP$. It is clear that $M$ is on $l$ and that $P$ is the pole of $LM$. We determine $L'$ and $M'$ in a like manner. Now any collineation $T$ for which $T(P, L, M) = P', L', M'$ has the desired property. It will suffice to show that there is a collineation $T$ which is an isometry.

Let such a collineation $T$ have matrix $A$. Let $P$, $L$, and $M$ have column matrices of coordinates $X, Y$, and $Z$, respectively. We so choose coordinates of $P$, $L$, and $M$ that

$$X^*X = Y^*Y = Z^*Z = (1).$$

Now $P', L'$, and $M'$ must have coordinates of the form $k_1AX$, $k_2AY$, and $k_3AZ$, respectively. We choose the $k_i$ so that

$$k_1^2X^*A^*AX = k_2^2Y^*A^*AY = k_3^2Z^*A^*AZ = (1).$$

We note that this is possible, since the matrices $X^*A^*AX$, $Y^*A^*AY$, and $Z^*A^*AZ$ represent sums of squares of coordinates of $P'$, $L'$, and $M'$, hence are not zero. We have by this condition put a restriction on $T$, for we are requiring that for the given coordinates of $P', L'$, and $M'$, $T$ must map $P+L+M$ onto $k_1P'+k_2K'+k_3M'$; that is, we have made our unit point choice. The matrix $A$ of a collineation $T$ satisfying this condition will be such that $P'$, $L'$, and $M'$ have coordinate matrices $AX$, $AY$, and $AZ$, respectively, where

$$X^*A^*AX = Y^*A^*AY = Z^*A^*AZ = (1).$$

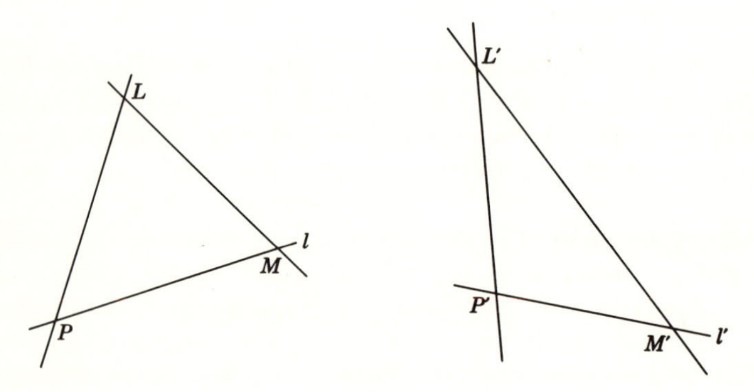


**Figure 13.1**

Now since the absolute conic has matrix $I$ and any two of the points $P$, $L$, and $M$ are conjugate with respect to it, we have

$$X^*IY = X^*IZ = Y^*IX = Y^*IZ = Z^*IX = Z^*IY = (0).$$

In a like manner we know that

$$X^*A^*AY = X^*A^*AZ = Y^*A^*AX = Y^*A^*AZ = Z^*A^*AX = Z^*A^*AY = (0).$$

Let

$$B = \begin{pmatrix} x_1 & y_1 & z_1 \\ x_2 & y_2 & z_2 \\ x_3 & y_3 & z_3 \end{pmatrix}.$$

Then in view of the above we find that $B^*B = I$ or $B^* = B^{-1}$. Similarly, we find that $B^*A^*AB = I$. We thus have

$$B(B^*A^*AB)B^* = BIB^* = I$$

and

$$B(B^*A^*AB)B^* = (BB^*)(A^*A)(BB^*) = A^*A.$$

Thus $A^*A = I$, and it is readily verified that such a matrix $A$ satisfies the hypotheses of Theorem 13.13.

## Exercises 13.1

1. Prove Theorem 13.11.
2. Complete the details in the proof of Theorem 13.13.
3. Complete the details in the proof of Theorem 13.14.
4. Find an isometry mapping (1, 1, 1) onto (0, 0, 1) and [1, −1, 0] onto [0, 1, 0].
5. Prove that each isometry has associated with it two matrices of the form of Theorem 13.13.
6. Prove that the number of isometries fulfilling the conditions of Theorem 13.14 is four.
7. Prove that the matrix $A = (a_{ij})$ of a hyperbolic isometry must satisfy the conditions

$$\begin{aligned} a_{11}^2 + a_{21}^2 - a_{31}^2 &= a_{12}^2 + a_{22}^2 - a_{32}^2 \\ &= -(a_{13}^2 + a_{23}^2 - a_{33}^2) \end{aligned}$$

and

$$a_{1j}a_{1k} + a_{2j}a_{2k} - a_{3j}a_{3k} = 0, \qquad j \neq k.$$

## 13.2 INCIDENCE AND PARALLELISM

In elliptic geometry we have deleted no points from the projective plane, but have only restricted the set of transformations we consider.

Since Hilbert's axioms of incidence are satisfied in the projective plane as well as in the Euclidean plane, we have at once

**Theorem 13.21.** *Hilbert's axioms of incidence are satisfied in the elliptic plane.*

It is evident that Theorems 2.11 and 2.12 are valid in the elliptic plane, but that Theorem 2.13 may be replaced by the stronger

**Theorem 13.22.** *Two distinct lines meet in one and only one point.*

For the sake of completeness we could restate Definition 2.41 as

**Definition 13.21.** *Two lines are* parallel *if they have no common point.*

In view of Theorem 13.22, though, we see that we must now replace Hilbert's axiom of parallelism by

**Theorem 13.23.** *Given a line and a point not on the line, there is no line containing the given point and parallel to the given line.*

Elliptic geometry is thus a geometry that disclaims parallelism and embodies the third alternative suggested by natural variations of the Euclidean axiom of parallelism. It is clear that Theorems 2.42 and 2.43 cannot be used in this geometry. We note that technically Theorem 2.41 is still valid, but its hypothesis is never satisfied so that it is of no use in the geometry.

## 13.3 METRIC CONCEPTS

If we were to follow the path taken in the development of affine and hyperbolic geometry, it would be appropriate next to consider the order relation. In each case we defined the betweenness concept in terms of separation and ideal points (points on the absolute). In the present case points on the absolute would have imaginary coordinates, and attempts to bring in separation in terms of cross ratio would fail because the resulting cross ratio would not be a real number. We here introduce the appropriate measure concepts first and use them to introduce the order and congruence relations for the elliptic plane.

In the hyperbolic plane we determined measure of a segment $A$-$B$ using $\log_e R(A, B, J, K)$ where $J$ and $K$ were points on the absolute and on the line $AB$. There is nothing to prevent consideration of this quantity in the present case. To be sure, $J$ and $K$ will have imaginary coordinates, but we have seen in our considerations of angle measure that such a phenomenon is not insurmountable.

Let us make use of the invariance of cross ratio under an isometry and Theorem 13.14 and carry out a preliminary investigation of $\log_e R(A, B,$

$J, K$). Let $A$ have coordinates $(0, 0, 1)$, $B$ have coordinates $(b_1, 0, b_2)$, $J$ have coordinates $(1, 0, i)$, and $K$ have coordinates $(1, 0, -i)$. It is clear that $J$ and $K$ are the points of intersection of the line $AB$ with the absolute. We find

$$R(A, B, J, K) = \frac{b_2 + b_1 i}{b_2 - b_1 i}$$

and we may take

$$\log_e R(A, B, J, K) = 2\theta i$$

where $\theta = \tan^{-1} b_1/b_2$ if $b_2 \neq 0$ and $\theta = \pi/2$ if $b_2 = 0$.

The quantity $\theta$ is the one we need to put a number scale on the line $AB$. Unlike the Euclidean and hyperbolic cases, that number scale does not use the whole set of real numbers, since we require $-\pi/2 < \theta \leq \pi/2$. We have already noted that the projective line in some sense seems "closed" or "circular," and this determination of the nature of $\theta$ confirms that notion. An appropriate picture for such a scaled elliptic (or projective) line is shown in Figure 13.2. It is already apparent that we are going to have difficulties in defining the order relation, hence in defining segment and ray. Let us define this measure formally and defer its use in defining order and congruence relations to later sections.

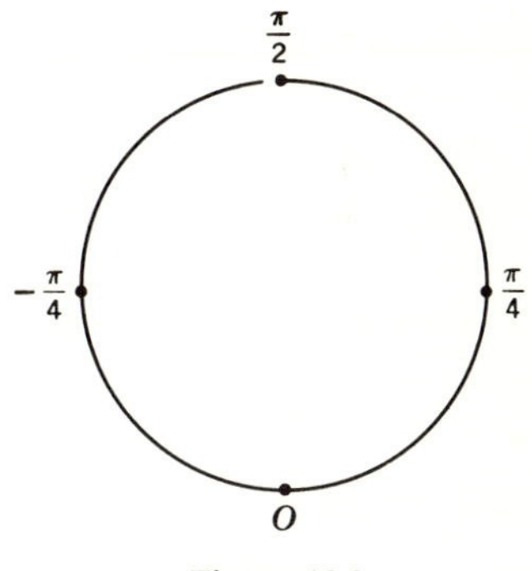


**Figure 13.2**

**Definition 13.31.** *The* measure of the points *A and B is the number*

$$m(A, B) = \left|\frac{i}{2} \log_e R(A, B, J, K)\right|, \qquad 0 \leq m(A, B) \leq \frac{\pi}{2}$$

*where J and K are the intersections of the line AB with the absolute.*

We have at once

**Theorem 13.31.** $m(A, B) \geq 0$ *and* $m(A, B) = 0$ *if and only if* $A = B$.

**Theorem 13.32.** *If T is an isometry,* $m(A, B) = m[T(A), T(B)]$.

Let the points $A$ and $B$ have coordinates $(x_1, x_2, x_3)$ and $(y_1, y_2, y_3)$. An argument similar to the one used to derive equation (12.52) leads to the formula

$$\cos m(A, B) = \frac{|x_1 y_1 + x_2 y_2 + x_3 y_3|}{\sqrt{(x_1^2 + x_2^2 + x_3^2)(y_1^2 + y_2^2 + y_3^2)}}. \tag{13.31}$$

From this we infer

**Theorem 13.33.** $m(A, B) = \pi/2$ *if and only if A and B are conjugate with respect to the absolute polarity.*

If we turn now to a consideration of the measure of two lines, we are led by our need to preserve duality to

**Definition 13.32.** *The* measure of the lines *a and b is the number*

$$m(a, b) = \left|\frac{i}{2} \log_e R(a, b, j, k)\right|, \qquad 0 \leq m(a, b) \leq \frac{\pi}{2},$$

*where j and k are the tangents to the absolute through the intersection of a and b.*

We have

**Theorem 13.34.** $m(a, b) \geq 0$ *and* $m(a, b) = 0$ *if and only if* $a = b$.

**Theorem 13.35.** *If T is an isometry,* $m(a, b) = m[T(a), T(b)]$.

If the lines have coordinates $[u_1, u_2, u_3]$ and $[v_1, v_2, v_3]$, we have

$$\cos m(a, b) = \frac{|u_1v_1 + u_2v_2 + u_3v_3|}{\sqrt{(u_1^2 + u_2^2 + u_3^2)(v_1^2 + v_2^2 + v_3^2)}} \cdot \qquad (13.32)$$

We can now use this equation to develop readily the theory of perpendicularity in the elliptic plane. We introduce

**Definition 13.33.** *The lines a and b are* perpendicular *if* $m(a, b) = \pi/2$.

We see that we have

**Theorem 13.36.** *Two lines are perpendicular if and only if each contains the pole of the other.*

**Theorem 13.37.** *If a given point is not the pole of a given line, there is a unique line perpendicular to the given line and passing through the given point.*

**Theorem 13.38.** *If a point is the pole of a given line, every line through the point is perpendicular to the given line.*

**Theorem 13.39.** *There exists a unique line perpendicular to each of two distinct lines.*

The polarity induced by the absolute enables us to prove the following theorem linking point and line measures.

**Theorem 13.310.** *If a and b are the lines joining the points A and B, respectively, to the pole of AB, then* $m(A, B) = m(a, b)$.

PROOF. We suppose that $AB$ meets the absolute at points $J$ and $K$. Let $O$ be the pole of $AB$, let $OJ$ be $j$, and let $OK$ be $k$. By Theorem 9.58[1] the lines $j$ and $k$ must be tangent to the absolute; that is, they are the $j$ and $k$ needed to apply Definition 13.32. Since $R(A, B, J, K) = R(a, b, j, k)$, we must have $m(A, B) = m(a, b)$.

[1]The argument leading up to the proof of this theorem is purely analytic and is as applicable in the complex projective plane as in the real projective plane.

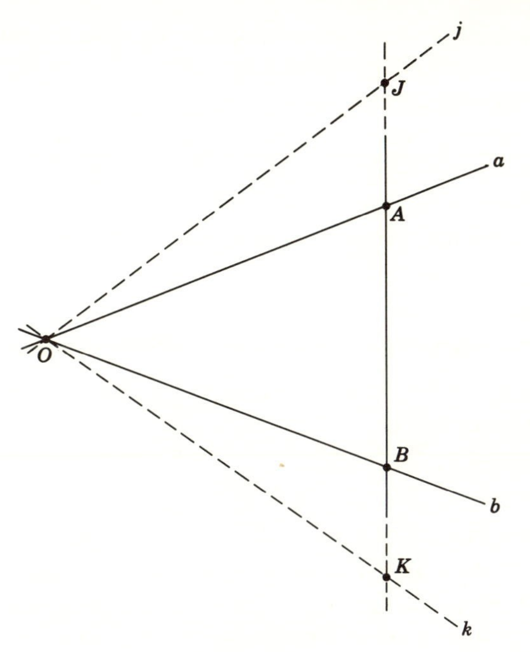


**Figure 13.3**

The "closed" or "circular" nature of a line in the elliptic plane has perhaps led the reader to think of spherical geometry, where lines are the great circles on a sphere. Certainly the last four theorems are valid in this spherical geometry, as is Theorem 13.23, which is the key theorem so far in our development. We assert that elliptic geometry cannot be spherical geometry, however, for in the former two distinct lines meet in a single point, whereas in the latter they meet in two distinct points.

One can obtain a model for elliptic geometry by taking the surface of a sphere and "identifying" opposite points; that is, we interpret the opposite points to be a single point. In some sense we thus confine ourselves to the surface of a hemisphere. Note that the opposite points on the boundary of the hemisphere must be identified; that is, we must take the hemisphere and sew it to itself in such a way that opposite points on the boundary are in contact (try it with half of a thin rubber ball!). The resulting very peculiar surface is a model for the elliptic or projective plane.

There are many similarities between elliptic and spherical geometry, and they are sometimes described as singly elliptic and doubly elliptic, respectively.

## Exercises 13.3

1. Prove Theorem 13.31.
2. Prove Theorem 13.32.
3. Derive equation (13.31).
4. Prove Theorem 13.33.
5. Derive equation (13.32).
6. Prove Theorem 13.37.
7. Prove Theorem 13.38.
8. Prove Theorem 13.39.
9. Find the measure of the following pairs of points:
   *a.* $(\frac{1}{4}, \frac{1}{4}, 1)$ and $(\frac{1}{2}, \frac{1}{2}, 1)$,
   *b.* $(\frac{1}{2}, \frac{1}{2}, 1)$ and $(\frac{3}{4}, \frac{3}{4}, 1)$.
10. Work Exercise 12.51 in the elliptic plane.
11. What is the dual of Definition 13.33?
12. One can build a model of the elliptic plane by considering all lines in Euclidean space through the origin. The line with direction numbers $a:b:c$ represents the point with coordinates $(a, b, c)$. Show that the measure of two points is the Euclidean measure of the angle between the corresponding lines. What can one associate with an elliptic line in this model? What corresponds in the model to the measure of two elliptic lines?

## 13.4 ORDER

We have seen that the definition of measure of two points has put a scale on an elliptic line which reinforces our feeling that it is "closed" and represents a circle. In such a case we certainly have no easy way of determining the meaning of betweenness. It is true, however, that with measures defined we can anticipate a need for the concepts of segment and angle, so that we can proceed to a consideration of the congruence relation. Let us consider possible ways of defining segment and see what this implies about the order relation.

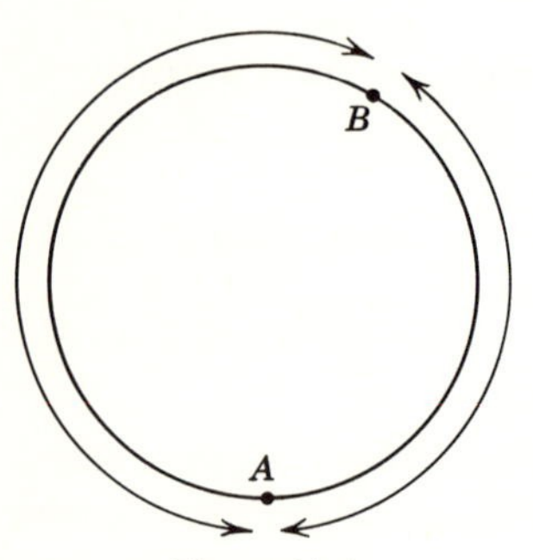


**Figure 13.4**

We consider distinct points $A$ and $B$ and place a number scale on their line so that $A$ has coordinate $O$. Suppose that $B$ has coordinate $b$. Intuitively we feel that the two points $A$ and $B$ have divided the line into two portions. We have two options as to how to proceed.

**I.** *Let $A$ and $B$ determine a unique segment on the line, if possible.*

In this case we wish to determine the shorter segment, if we can. If $b < 0$, we could take the segment to be the set of points whose coordinate $x$ satisfies $b < x < 0$. If $0 < b < \pi/2$, we could take the segment to be the set of points whose coordinate $x$ satisfies $0 < x < b$. If $b = \pi/2$, so that $B$ is conjugate to $A$, we could not make a decision, hence might refuse to let such a pair of points determine a segment.

If we accept the normal relation between segment and order, this would lead us to define $C$ to be between $A$ and $B$ if it were on the segment $A$-$B$. One could then prove that Hilbert's Axioms 5, 6, and 7 are satisfied. Axiom 8 would not be satisfied; if we take points with coordinates $-\pi/3$, 0, and $\pi/3$, no one of the points will be between the other two. Axiom 9 is also not satisfied. The simplest way to see this at the moment is to show that the plane separation property of Theorem 2.22 is not satisfied.

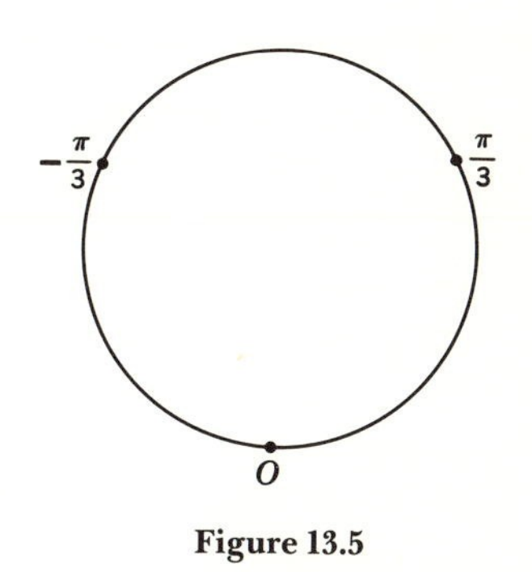


**Figure 13.5**

Let $l$ be a line containing the point $A$. We consider a line other than $l$ through $A$ with coordinates on the line chosen so that $A$ has coordinate 0. Let $B$, $C$, and $D$ on this line have coordinates $\pi/6, \pi/2$, and $-\pi/6$, respectively. Then $B$-$D$ meets $l$ while $B$-$C$ and $C$-$D$ do not. This would imply that $B$ and $D$ are in opposite classes with respect to the line while $C$ is in the same class with each of them, a clear contradiction of plane separation.

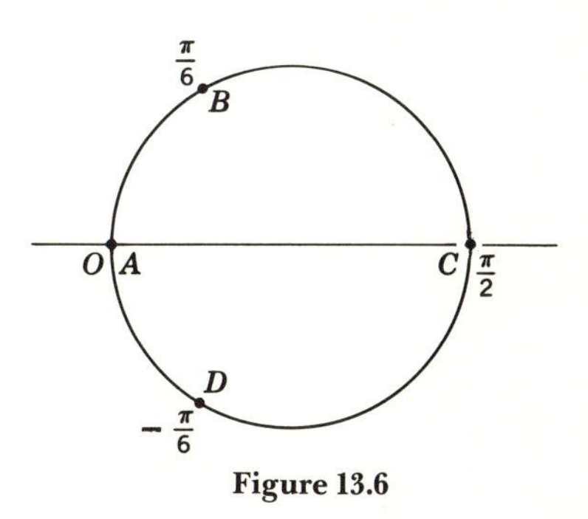


**Figure 13.6**

A second alternative is

**II.** *Let $A$ and $B$ determine two segments.*

It is this alternative that we shall pursue. Our first step must be a formal determination of the two portions into which $A$ and $B$ divide the line. This we do in

**Theorem 13.41.** *Two distinct points $A$ and $B$ divide the points other than $A$ and $B$ on the line $AB$ into two classes, two points $P$ and $Q$ being in the same class if and only if $AB \not\parallel PQ$.*

PROOF. We choose a point $C$ distinct from $A$ and $B$ on $AB$. Let one class consist of all points $P$ for which $AB \| CP$, while the other class consists of $C$ and all points $P$ for which $AB \nparallel CP$. We know that separation is equivalent to a negative cross ratio and from Theorem 8.55 that

$$R(A, B, C, P) R(A, B, P, Q) = R(A, B, C, Q).$$

From this it can be shown that $P$ and $Q$ are in the same class if and only if $AB \nparallel PQ$ and that the classes are independent of the choice of $C$. The details are left to the reader.

**Definition 13.41.** *Either of the two classes of Theorem* 13.41 *is a* segment *determined by $A$ and $B$. If $P$ is a point in one of these segments, the segment is denoted by $A$-$P$-$B$.*

We could now define betweenness as in alternative I by stating that $C$ is between $A$ and $B$ if it is on a segment determined by them. In this case Axiom 8 becomes ridiculous, for given any three distinct collinear points, each of the points will be between the other two. Axioms 5, 6, and 7 would be satisfied in a trivial sense and would be of no use. It would appear reasonable to conclude that the order relations in elliptic geometry are those of separation, as in the projective plane, and not the betweenness used in the Euclidean and hyperbolic cases.

We note that as far as Axiom 10 is concerned, it is automatically fulfilled if interpreted literally, for the given line will certainly meet one of the two segments determined by each pair of points. In the Euclidean and hyperbolic cases we tend to interpret this axiom in terms of a triangle, for in these geometries it can be stated in the form: if a line passes through no vertex of a triangle and meets a side, it must meet a second side. This interpretation in the elliptic plane leads to difficulties and suggests a more careful consideration of triangle.

If we were to agree that a triangle is a configuration consisting of three noncollinear points and three segments determined by pairs of these points, three distinct noncollinear points would determine eight triangles. Thus in Figure 13.7 the triangle whose sides are $A$-$N$-$B$, $B$-$L$-$C$, and $C$-$M$-$A$ would be distinct from the triangle whose sides are $A$-$N$-$B$, $B$-$L'$-$C$, and $C$-$M'$-$A$, yet each would have the same vertices. Suppose that a line meets $A$-$N$-$B$ and $B$-$L$-$C$. If it meets $C$-$M$-$A$, the "triangular" version of Axiom 10 is violated for the second triangle above. If it meets $C$-$M'$-$A$, the same remark holds for the triangle with sides $A$-$N$-$B$, $B$-$L'$-$C$, and $C$-$M$-$A$.

We can avoid some of these difficulties by a suitable redefinition of a triangle. We first prove

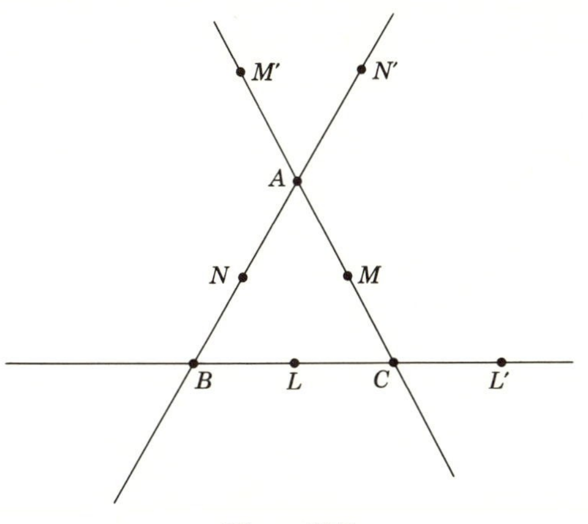


**Figure 13.7**

**Theorem 13.42.** *If $A$, $B$, and $C$ are noncollinear points and if two lines meet the same segments determined by $A$ and $B$ and by $B$ and $C$ on $AB$ and $BC$, they meet the same segment determined by $A$ and $C$ on $AC$.*

PROOF. Let the lines $l$ and $l'$ meet $AB$, $BC$, and $AC$ at $N$, $L$, $M$ and $N'$, $L'$, $M'$, respectively. If $M = M'$, there is nothing to prove. By hypothesis $N$ and $N'$ are on the same segment of $AB$ determined by $A$ and $B$ and $L$ and $L'$ are on the same segment determined by $B$ and $C$ on $BC$. Let $ML'$ meet $AB$ at $P$. If we consider the central perspectivity $T$ from $BC$ to $AB$ with center $M$, we have

$$T(B, L', L, C) = B, P, N, A.$$

Since separation, hence segments, are preserved by $T$, we conclude that $P$, $N'$, and $N$ are all on the same segment determined by $A$ and $B$ on $AB$. Now if $S$ is the perspectivity from $AB$ to $AC$ with center $L'$, we have

$$S(A, N', P, B) = A, M', M, C.$$

The preservation of separation and segments then indicates that $M$ and $M'$ are on the same segment on $AC$ determined by $A$ and $C$.

This theorem shows that of the eight potential triangles in Figure 13.7 four have the desired property embodied in Axiom 10; namely, if a line meets one side and no vertex, it will meet a second but not a third side. We shall restrict our use of the term triangle to make use of this result, and we introduce

**Definition 13.42.** *A* triangle *is a configuration consisting of three non-*

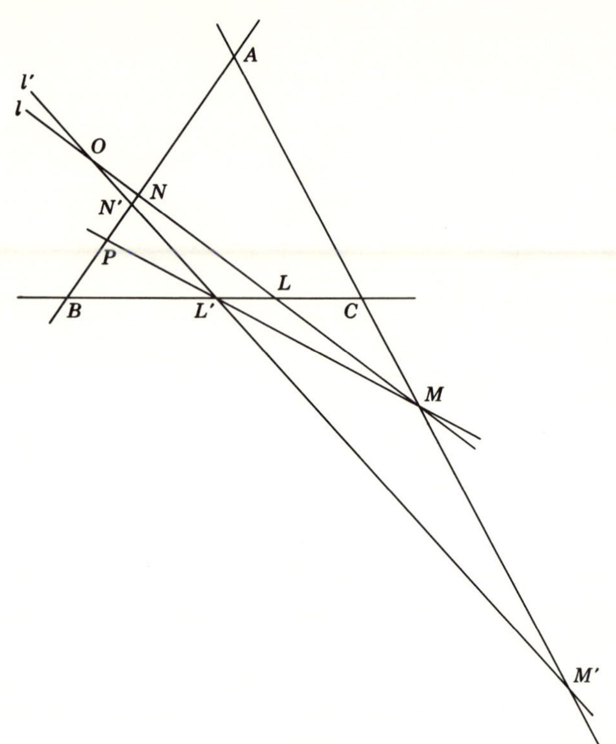


**Figure 13.8**

*collinear points* (vertices) *and segments* (sides) *determined by pairs of these points, the segments being so chosen that a line through no vertex meets exactly two or none of the sides.*

We have at once

**Theorem 13.43.** *A triangle is determined by its vertices and two of its sides.*

We have previously used the concept of ray to define angle. We now use duality instead and obtain the following by dualizing the previous results of this section.

**Theorem 13.44.** *Two distinct lines $a$ and $b$ divide the lines through their intersection, other than $a$ and $b$, into two classes, two lines $p$ and $q$ being in the same class if and only if $ab \not\parallel pq$.*

**Definition 13.43.** *Either of the two classes of Theorem* 13.44 *is an* angle *determined by $a$ and $b$. If $p$ is a line in one of these angles, the angle is denoted by $a$-$p$-$b$.*

**Theorem 13.45.** *If $a$, $b$, and $c$ are nonconcurrent lines and if two points lie on* *the same angle determined by a and b and by b and c, they lie on the same angle determined by a and c.*

**Definition 13.44.** *A trilateral is the configuration consisting of three non-concurrent lines (sides) and angles determined by pairs of these lines, the angles being so chosen that a point on no side lies on exactly two or none of the angles.*

**Theorem 13.46.** *A trilateral is determined by its sides and two of its angles.*

It is still possible to introduce the ray concept and associate an angle with a pair of rays. We define ray in

**Definition 13.45.** *If O is a point on line l, a ray on line l with terminal point O is the point O and one of the segments determined by O and its conjugate O′ on l.*

Let us consider the angle $a$-$p$-$b$ and suppose that $h$ and $h'$ are the rays on $a$ with terminal point $O$, the intersection of $a$ and $b$. Let $O'$ and $O''$ be the conjugates of $O$ on $a$ and $b$, respectively, and let $P$ be the intersection of $O'O''$ and $p$. The triangle with vertices $O$, $O'$, and $O''$ and sides $h$ and $O'$-$P$-$O''$ is uniquely determined, and its third side must be a ray on $b$ with terminal point $O$, say $k$. We can then associate the angle $a$-$p$-$b$ with the pair of rays $(h, k)$. It is obvious that if $k'$ is the other ray on $b$ with terminal point $O$, the pair $(h', k')$ is also associated with angle $a$-$p$-$b$.

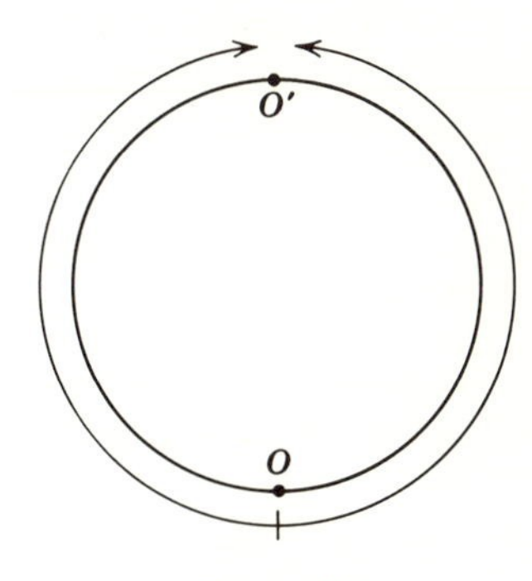


Figure 13.9

We should like to be able to determine the angles of a triangle, and we are now in a position to do this simply. Let the triangle have vertices $A$, $B$, and $C$ and sides $B$-$L$-$C$ and $A$-$M$-$C$. Let $AL$ and $BM$ meet at $P$. Let $PC$ and $LM$ meet $AB$ at $N$ and $Q$ respectively. It is immediate from Figure

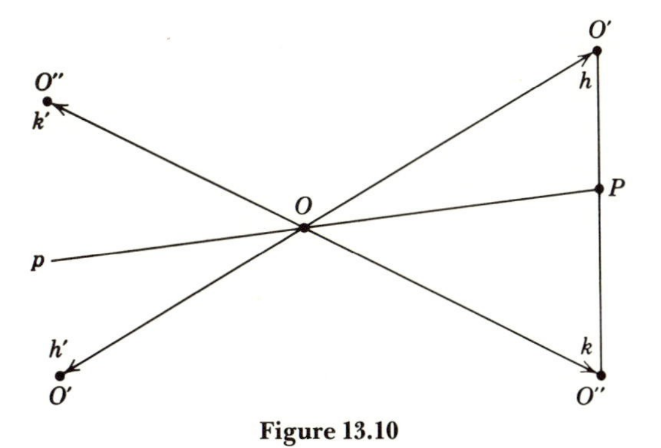


Figure 13.10

13.11 that $H(A, B; N, Q)$. In view of Definition 13.42 the point $Q$ is not on the third side of the triangle, hence the third side is $A$-$N$-$B$. The point $P$ lies on lines of one of the angles determined by $AB$ and $AC$, by $AB$ and $BC$, and by $BC$ and $AC$, and we agree to call these angles the angles of the triangle, denoting them by $A$, $B$, and $C$. We note that these are well determined in the sense that they are independent of the choice of $L$ and $M$ on the sides of the triangle. We also note that we can define the interior of the triangle to be the set of all points $P$ determined as above.

One can use duality to define the segments associated with a trilateral, and this is left to the reader. We note that ultimately either a triangle or trilateral is a configuration consisting of three vertices, three "line-sides," three "segment-sides," and three angles. It should be apparent from the foregoing discussion and Definition 13.44 that of the eight tentative triangles we anticipated when we began our discussion, four are triangles and four are trilaterals.

## Exercises 13.4

1. Prove in detail the stated relations between Hilbert's axioms of order and alternative I for segments.
2. Complete the proof of Theorem 13.41.
3. Prove that $H(A, B; C, D)$ implies that $A$-$C$-$B$ and $A$-$D$-$B$ are distinct.

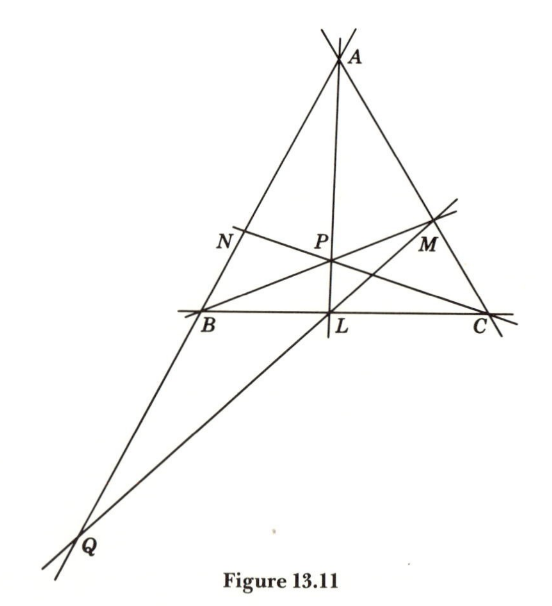


**Figure 13.11**

4. List the vertices, "segment-sides," and "line-sides" and describe the angles for each of the four triangles and four trilaterals of Figure 13.7.

5. Carry out the dual argument to determine the "segment-sides" of a trilateral.

6. Prove that the points of the plane not on the "line-sides" of a triangle are separated into four classes such that no two points of the same class are separated by any two points in which their line meets the "line-sides" of the triangle and any two points of different classes are separated by some pair of points in which their line meets the "line-sides" of the triangle. What are these four classes?

## 13.5 CONGRUENCE OF SEGMENTS AND ANGLES

We have completed the preliminaries on segment and angle to the point where we are now ready to define their measures and thus set up criteria for congruence. In these considerations we shall find it convenient to use the following lemma.

**Lemma 13.51.** *If $A$ and $B$ are distinct nonconjugate points and if their conjugates on $AB$ are $A'$ and $B'$, respectively, then $AA' \| BB'$.*

PROOF. We may assume without loss of generality that the line $AB$ has equation $x_3 = 0$. If $A$ has coordinates $(a_1, a_2, 0)$ and $B$ has coordinates $(b_1, b_2, 0)$, then $A'$ and $B'$ have coordinates $(a_2, -a_1, 0)$ and $(b_2, -b_1, 0)$, respectively. We find

$$R(A, A', B, B') = -\frac{(a_1b_2 - a_2b_1)^2}{(a_1b_1 + a_2b_2)^2} < 0,$$

whence $AA' \| BB'$. Note that the distinctness of $A$ and $B$ shows the numerator in the cross ratio does not vanish, while their nonconjugate property implies that the denominator does not vanish.

**Definition 13.51.** *The segment $A$-$M$-$B$ is* acute *if it does not contain the conjugate of $A$, right if $B$ is the conjugate of $A$, and* obtuse *if it contains the conjugate of $A$.*

We must show that acuteness and obtuseness of segments are well defined. If $A$-$M$-$B$ is acute, hence does not contain the conjugate $A'$ of $A$ on $AB$, we must show that it cannot contain the conjugate $B'$ of $B$ on $AB$. We know that $AB \| MA'$ and by the lemma that $AA' \| BB'$. It follows from Theorem 8.67 or Axiom 9d of Chapter 10 that $AB \| MB'$. We have proved

**Theorem 13.51.** *Acuteness and obtuseness of segments are well defined.*

We can now readily assign measures to segments by

**Definition 13.52.** *The* measure $m(A\text{-}M\text{-}B)$ of the segment $A\text{-}M\text{-}B$ *is*

(1) $m(A, B)$ *if* $A\text{-}M\text{-}B$ *is acute or right,*
(2) $\pi - m(A, B)$ *if* $A\text{-}M\text{-}B$ *is obtuse.*

Since an isometry must preserve separation, it clearly must carry acute segments to acute segments and obtuse segments to obtuse segments. Hence in view of Theorem 13.32 we have

**Theorem 13.52.** *If $T$ is an isometry,* $m(A\text{-}M\text{-}B) = m[T(A)\text{-}T(M)\text{-}T(B)]$.

We now define congruence by

**Definition 13.53.** *The segments* $A\text{-}M\text{-}B$ *and* $A'\text{-}M'\text{-}B'$ *are* congruent $(A\text{-}M\text{-}B \cong A'\text{-}M'\text{-}B')$ *if* $m(A\text{-}M\text{-}B) = m(A'\text{-}M'\text{-}B')$.

We at once have

**Theorem 13.53.** *Elliptic geometry satisfies Hilbert's Axioms* 12, 13, *and* 14 *of congruence for segments.*

Axiom 11 requires a slight modification and should now be taken in the form

**Theorem 13.54.** *Given a segment $A\text{-}M\text{-}B$ and a point $A'$ on a line $a$, there are two segments on $a$ with end points $A'$ which are congruent to $A\text{-}M\text{-}B$. If $B$ is not conjugate to $A$, their other end points are distinct. If $B$ is conjugate to $A$, their other end point in each case is conjugate to $A'$.*

The proof is left to the reader. Axiom 15 should now be stated in the form of

**Theorem 13.55.** *If $A\text{-}M\text{-}B$ and $B\text{-}N\text{-}C$ are segments on a line $a$ with no common point, if $A'\text{-}M'\text{-}B'$ and $B'\text{-}N'\text{-}C'$ are segments on a line $a'$ with no common point, if $A\text{-}M\text{-}B \cong A'\text{-}M'\text{-}B'$, and if $B\text{-}N\text{-}C \cong B'\text{-}N'\text{-}C'$, then $A\text{-}B\text{-}C \cong A'\text{-}B'\text{-}C'$.*

The proof is left to the reader.

The corresponding results below for angles now follow by duality.

**Lemma 13.52.** *If $a$ and $b$ are distinct nonconjugate lines and if their conjugates through the intersection of $a$ and $b$ are $a'$ and $b'$, respectively, then $aa' \| bb'$.*

**Definition 13.54.** *The angle $a\text{-}m\text{-}b$ is* acute *if it does not contain the conjugate of $a$ through the intersection of $a$ and $b$,* right *if $b$ is a conjugate of $a$, and* obtuse *if it contains the conjugate of $a$ through the intersection of $a$ and $b$.*

**Theorem 13.56.** *Acuteness and obtuseness of angles are well defined.*

**Definition 13.55.** *The* measure $m(a\text{-}w\text{-}b)$ of the angle $a\text{-}w\text{-}b$ *is*

(1) $m(a, b)$ *if* $a\text{-}w\text{-}b$ *is acute or right,*
(2) $\pi - m(a, b)$ *if* $a\text{-}w\text{-}b$ *is obtuse.*

**Theorem 13.57.** *If $T$ is an isometry, $m(a\text{-}w\text{-}b) = m[T(a)\text{-}T(w)\text{-}T(b)]$.*

**Definition 13.56.** *The angles $a\text{-}w\text{-}b$ and $a'\text{-}w'\text{-}b'$ are* congruent *($a\text{-}w\text{-}b \cong a'\text{-}w'\text{-}b'$) if $m(a\text{-}w\text{-}b) = m(a'\text{-}w'\text{-}b')$.*

**Theorem 13.58.** *Elliptic geometry satisfies Hilbert's Axioms* 17, 18, *and* 19 *for congruence of angles.*

**Theorem 13.59.** *Given an angle $a\text{-}m\text{-}b$ and a ray $h$, there are two rays $k$ such that $(h, k)$ is congruent to $a\text{-}m\text{-}b$. If $b$ is not conjugate to $a$, these rays are on distinct lines. If $b$ is conjugate to $a$, these rays are each on a line conjugate to the line of $h$ and through the terminal point of $h$.*

**Theorem 13.510.** *If $a\text{-}m\text{-}b$ and $b\text{-}n\text{-}c$ are angles with vertex $A$ having no common line, if $a'\text{-}m'\text{-}b'$ and $b'\text{-}n'\text{-}c'$ are angles with vertex $A'$ having no common line, if $a\text{-}m\text{-}b \cong a'\text{-}m'\text{-}b'$, and if $b\text{-}n\text{-}c \cong b'\text{-}n'\text{-}c'$, then $a\text{-}b\text{-}c \cong a'\text{-}b'\text{-}c'$.*

## Exercises 13.5

1. Prove Theorem 13.54.
2. Prove Theorem 13.55.

## 13.6 CONGRUENCE OF TRIANGLES

The three standard triangle congruence theorems hold in elliptic as well as in Euclidean and hyperbolic geometry. We shall have to take some care in their proof, however, for the plane separation properties used in Sections 2.3 and 12.7 are no longer available to us.

Let us reinvestigate Theorem 13.14. We have noted in Exercise 13.16 that there are at least four isometries fulfilling the theorem. The key conditions here are our choices of the $k_i$ such that

$$k_1{}^2X^*A^*AX = k_2{}^2Y^*A^*AY = k_3{}^2Z^*A^*AZ = (1).$$

Clearly, there are two possible choices for each $k_i$, or eight choices in all. A change in sign of all the $k_i$ will not change the transformation determined, however. There are thus four different choices of transformations associated with different sign choices of the $k_i$. Referring to Figure 13.1, we note that if we choose points $P' + rM'$ and $P' - rM'$, we have

$$R(P', M', P' + rM', P' - rM') = -1,$$

showing that the points are on opposite rays with terminal point $P'$. It is clear that an isometry will preserve separation, hence map the rays with terminal point $P$ onto rays with terminal point $P'$. The effect of a change of sign in $k_1k_2$ will be to interchange $P'+rM'$ and $P'-rM'$ as far as their antecedents under $T$ are concerned. We thus conclude that we may not only produce an isometry $T$ such that $T(P,l)=P',l'$, but that we may also specify the mapping with respect to the two rays with terminal point $P$, stating which ray is to be mapped onto which ray on $l'$ with terminal point $P'$.

Suppose now that $(h,k)\cong(h',k')$. We know we may produce an isometry $T$ such that $T(h)=h'$. We know from Theorem 13.57 that $(h,k)\cong(h',T[k])$. Now from Theorem 13.59 there are two rays $k''$ such that $(h,k)\cong(h',k'')$. An argument similar to that of the preceding paragraph shows that the two different choices of sign of $k_1k_3$ in the proof of Theorem 13.14 lead to two different isometries for which $T(k)$ is one of the rays $k''$. For one of these $T(k)=k'$. We have proved

**Theorem 13.61.** *If $(h, k)\cong(h', k')$, there exists an isometry $T$ such that $T(h)=h'$ and $T(k)=k'$.*

We are now ready to prove

**Theorem 13.62.** *If two triangles have two sides of one and the included angle congruent respectively to two sides and the included angle of the other, the triangles are congruent.*

PROOF. We suppose that $A\text{-}N\text{-}B\cong A'\text{-}N'\text{-}B'$, $A\text{-}M\text{-}C\cong A'\text{-}M'\text{-}C'$, and $\angle A\cong\angle A'$. Let $B$ lie on a ray $b$ with terminal point $A$, $C$ lie on a ray $c$ with terminal point $A$, $B'$ lie on a ray $b'$ with terminal point $A'$, and $C'$ lie on a ray $c'$ with terminal point $A'$. Then $(b,c)\cong(b',c')$, for these will either be the angles $A$ and $A'$ or their supplements. We now apply Theorem 13.61, letting $T$ map $b$ onto $b'$ and $c$ onto $c'$. Thus $T$ maps $\angle A$ onto $\angle A'$, $B$ onto $B'$, and $C$ onto $C'$. By Theorem 13.43 $T$ must map the third side of one triangle onto the third side of the other triangle; hence these sides are congruent. $T$ will also map the lines joining $C$ to the points $A\text{-}N\text{-}B$ onto the lines joining $C'$ to the points of $A'\text{-}N'\text{-}B'$; hence it maps

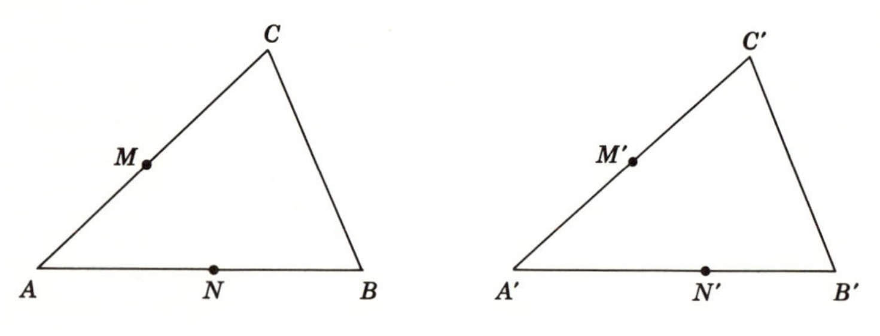


**Figure 13.12**

$\angle C$ onto $\angle C'$, and these angles are congruent. A similar argument establishes the congruence of $\angle B$ and $\angle B'$.

By duality the "angle-side-angle" congruence of trilaterals now follows. Since a triangle is associated with a trilateral with one side being the supplement of the triangle side, the other two sides coinciding, the same "angle-side-angle" result holds for the triangle inasmuch as we can establish one side congruence, then apply the previous theorem. We thus have indirectly by duality

**Theorem 13.63.** *If two triangles have two angles and the included side of one congruent respectively to two angles and the included side of the other, the triangles are congruent.*

The lack of plane separation makes our previous proof of the "side-side-side" theorem suspect. We state the theorem here but defer its proof until we have developed elliptic trigonometry, at which point the proof is extremely simple.

**Theorem 13.64.** *If two triangles have three sides of one congruent respectively to three sides of the other, the triangles are congruent.*

By an indirect duality argument this leads to

**Theorem 13.65.** *If two triangles have three angles of one congruent respectively to three angles of the other, the triangles are congruent.*

Theorems 12.72 and 12.73 cannot be proved in elliptic geometry. In the case of Theorem 12.72 this is obvious because of the lack of parallelism. Theorem 12.73 is not so obvious. We leave it to the reader to show the theorem does not hold for any of the triangles whose vertices are $(1, 0, 0)$, $(0, 1, 0)$, and $(0, 0, 1)$. This theorem is of historical interest, for it is the first theorem in the order in which Euclid developed his geometry which is invalid in the elliptic plane.

We have now completed our development of Hilbert's axioms in the plane. The order axioms have in effect been discarded, but with slight modifications we have been able to save the congruence ones. It is apparent by now that the analogies between elliptic and Euclidean geometry are indeed weaker than those between hyperbolic and Euclidean geometry.

## Exercises 13.6

1. Disprove Theorem 12.73 in the elliptic plane.
2. Criticize the proof of Theorem 2.36 as applied to the elliptic plane.

## 13.7 RIGHT TRIANGLE TRIGONOMETRY

We develop the theory for a particular right triangle of Figure 13.13. There are four right triangles associated with this figure, and we leave it to the reader to verify the validity of the formulas for the other three triangles.

We assume in the figure that $a' > 0$ and $b' > 0$. As segment on the line $BC$ we take those points that have coordinates $(x, 0, 1)$, $0 < x < a'$; as segment on the line $AC$ we take those points with coordinates $(0, y, 1)$, $0 < y < b'$. The line determined by $(a'/2, 0, 1)$ and $(0, b'/2, 1)$ will meet $AB$ at the point with coordinates $(a', -b', 0)$. From this result it follows readily that the desired segment on $AB$ consists of those points $(x, y, 1)$ on the line for which $xy > 0$. The details of these segment verifications are left to the reader as an exercise. We assert that these segments are all acute. The conjugate of $C$ on $BC$ is the point $(1, 0, 0)$, which is not on the segment chosen on $BC$, verifying its acuteness. A similar argument shows that the chosen segment on $AC$ is acute. The conjugate of $B$ on $AB$ is the point with coordinates $(-a', b'[a'^2+1], a'^2)$, which is not on the segment determined on $AB$.

We now consider the angles in the chosen triangle. It follows at once from equation (13.32) that $\angle C$ is a right angle. The desired angle $\alpha$ at $A$ is the set of lines joining $A$ to the points on the segment on $BC$. We have just seen that $B$ and $C$ separate such points and the conjugate of $C$ on $BC$. Hence the lines $AC$ and $AB$ separate the lines of the angle and the line through $A$ conjugate to $AC$. Therefore $\alpha$ is acute. A similar argument shows that $\beta$ is acute.

With the acuteness verified, we may now use Definition 13.52 and equation (13.31) to find

$$\cos a = \frac{1}{\sqrt{1+a'^2}},$$

$$\cos b = \frac{1}{\sqrt{1+b'^2}},$$

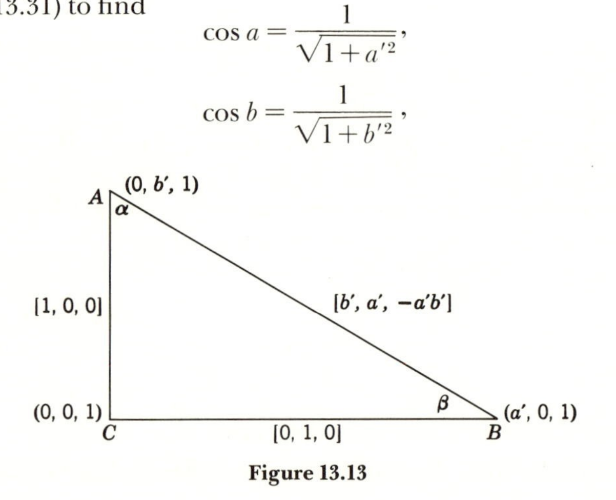


**Figure 13.13**

and

$$\cos c = \frac{1}{\sqrt{(1+a'^2)(1+b'^2)}},$$

where $a$, $b$, and $c$ are the measures of the sides of the triangles. We thus have

$$\cos c = \cos a \cos b. \tag{13.71}$$

The analogy with the beginning development in Section 12.8 is striking and holds throughout the derivation of the other formulas. We state the results, leaving derivations as exercises. Note that the acuteness of the angles is needed here to enable us to use Definition 13.55. We find

$$\cos \alpha = \frac{\tan b}{\tan c}, \tag{13.72}$$

$$\sin \alpha = \frac{\sin a}{\sin c}, \tag{13.73}$$

$$\tan \alpha = \frac{\tan a}{\sin b}, \tag{13.74}$$

$$\cos \beta = \frac{\tan a}{\tan c}, \tag{13.75}$$

$$\sin \beta = \frac{\sin b}{\sin c}, \tag{13.76}$$

and

$$\tan \beta = \frac{\tan b}{\sin a}. \tag{13.77}$$

If we now consider

$$\tan(\alpha+\beta) = \frac{\tan\alpha + \tan\beta}{1-\tan\alpha\tan\beta},$$

where $\alpha$ and $\beta$ are acute, we know that the numerator is positive. The denominator reduces to $1-\sec c < 0$. Hence $\tan(\alpha+\beta) < 0$ and $\alpha+\beta > \pi/2$. This inequality is strengthened for any of the other right triangles of Figure 13.13, for they will contain some combination of $\alpha$ and $\beta$ and their supplements, and the angle sum will be even greater. Thus we have

**Theorem 13.71.** *The sum of the measures of the angles of a right triangle exceeds $\pi$.*

From this we find

**Theorem 13.72.** *The sum of the measures of the angles of a triangle exceeds $\pi$.*

## Exercises 13.7

1. Show in detail that the analytic conditions imposed in the argument above did indeed define segments on the lines $AB$, $AC$, and $BC$ with appropriate end points.

2. Derive equations (13.72), (13.73), and (13.74).

3. Work Exercise 12.83 in the elliptic plane.

4. Prove Theorem 13.72.

5. Prove $\cot \alpha = \sin b / \tan a$.

6. Prove $\cos \alpha = \cos a \sin \beta$.

7. Prove $\cos c = \cot \alpha \cot \beta$.

8. Verify that equations (13.71) to (13.77) are valid for the other three right triangles of Figure 13.13.

9. Define a right trilateral and find the basic formulas relating its angles and "segment-sides."

10. State the duals of Theorems 13.71 and 13.72.

# 13.8 GENERAL TRIANGLE TRIGONOMETRY

We now consider a general triangle with sides having measures $a$, $b$, and $c$ and opposite angles $\alpha$, $\beta$, and $\gamma$, respectively. The derivations for the laws of sines and cosines are similar to those of Section 12.9 and are left as exercises. We have

$$\frac{\sin a}{\sin \alpha} = \frac{\sin b}{\sin \beta} = \frac{\sin c}{\sin \gamma}, \tag{13.81}$$

$$\cos a = \cos b \cos c + \sin a \sin b \cos \alpha,$$
$$\cos b = \cos a \cos c + \sin a \sin c \cos \beta,$$

and

$$\cos c = \cos a \cos b + \sin a \sin b \cos \gamma. \tag{13.82}$$

We note that the three versions of the law of cosines assert that if side measures are known, the angles are known, thus giving a simple proof of Theorem 13.64. The argument of Section 12.9 can also be modified to give the triangle inequality

**Theorem 13.81.** *If $a$, $b$, and $c$ are measures of the sides of a triangle, then $c < a + b$.*

## Exercises 13.8

1. Derive equation (13.81).

2. Derive equation (13.82).

3. Prove Theorem 13.81. Why can this theorem not be stated in the form of Theorem 12.91?

4. Work Exercise 12.92 in the elliptic plane.

5. State an analogue in the elliptic plane of the result of Exercise 12.93, and prove it.

6. State the laws of sines and cosines for trilaterals.

7. State the dual of Theorem 13.81.

8. Prove Theorem 13.64.

## REFERENCES

Adler, Claire Fisher, *Modern Geometry, An Integrated First Course,* 2nd ed., McGraw-Hill Book Company, New York, 1967, Chapters 3, 11.

Blumenthal, Leonard M., *A Modern View of Geometry*, W. H. Freeman and Company, San Francisco, 1961, Chapter 8.

Coxeter, H. S. M., *Non-Euclidean Geometry*, 5th ed., University of Toronto Press, Toronto, 1965, Chapters 5, 6, 7, 14.

Gans, David, "An Introduction to Elliptic Geometry," *American Mathematical Monthly*, vol. 62, no. 7, part II, Aug.–Sept. 1955.

Levy, Harry, *Projective and Related Geometries*, The Macmillan Company, New York, 1961, Chapter 5.

Meschkowski, Herbert, translated by A. Shenitzer, *Noneuclidean Geometry*, Academic Press, New York, 1964, Chapter 9.

Meserve, Bruce E., *Fundamental Concepts of Geometry*, Addison-Wesley Publishing Company, Reading, Mass., 1955, Chapter 8.

Springer, C. E., *Geometry and Analysis of Projective Spaces*, W. H. Freeman and Company, San Francisco, 1964, Chapter 9.

Tuller, Annita, *A Modern Introduction to Geometries*, D. Van Nostrand Company, Princeton, 1967, Chapters 1, 7.

Veblen, Oswald, and John Wesley Young, *Projective Geometry*, vol. II, Ginn and Company, Boston, 1910, Chapter 8.

Wolfe, Harold E., *Non-Euclidean Geometry*, Holt, Rinehart and Winston, New York, 1945, Chapter 7.

# ANSWERS AND HINTS FOR EXERCISES

**Exercises 1.2** Page 4

1. Solve $x^2+y^2=a^2$ and $(x-a)^2+y^2=a^2$ simultaneously.

3. Solve $x^2+y^2=a^2$ and $(x-c)^2+y^2=b^2$ simultaneously and examine the discriminant of the resulting quadratic equation.

**Exercises 2.1** Page 10

1. Apply Axioms 1 and 2.

3. If $l$ and $l'$ meet at $P$, they contain points $Q$ and $R$, respectively, each distinct from $P$ by Axiom 2. The line of $Q$ and $R$ does not contain $P$ by Axiom 4.

5. Three.

**Exercises 2.2** Page 15

3. The points are $A$, $B$, and $C$ of Axiom 7. We cannot be sure that $B$ and $D$ are distinct.

5. Suppose the line meets $AB$ at $D$, $AC$ at $E$, and $BC$ at $F$. Apply Pasch's axiom to $\triangle BDF$.

6. Draw a diagonal.

7. *a*. $CAB$ and $ABD$, or $CBA$ and $BAD$.

10. Consider the half planes determined by the lines on which the sides of the angle lie.

12. The interior is convex because any two points in the interior lie in each of the half planes used to define the interior.

**Exercises 2.3** Page 20

2. Follows from Axiom 16 and Theorem 2.34. If $(h, l)$ is a straight angle with $k$ in its interior and if $(h, k) \cong (h', k')$, the unique $l'$ on the opposite side of $k'$ from $h'$ for which $(k, l) \cong (k', l')$ must be such that $(h, l) \cong (h', l')$, and $(h', l')$ is a straight angle.

5. By Theorem 2.31 it is sufficient to prove $A\text{-}C \cong A'\text{-}C'$. Suppose they are not, choose $C''$ on the same side of $A$ on the line of $A$ and $C$ as is $C$, and consider triangles $ABC''$ and $A'B'C'$.

**Exercises 2.4** Page 21

2. Let $l$ meet $m$ and $n$ at $A$ and $B$, respectively. If $m \nparallel n$, let them meet at $C$. Choose $D$ on $n$ on the opposite side of $l$ from $C$ such that $B\text{-}D \cong A\text{-}C$ (we assume the congruent alternate interior angles are $\angle CAB$ and $\angle ABD$). The congruence of triangles $CAB$ and $ABD$ implies $D$ is on $m$ and $n$, which is a contradiction. The converse follows by the foregoing and Playfair's axiom.

4. This is a statement about *measure* of angles, not their congruence.

**Exercises 2.5** Page 23

1. If $a > 0$, then $(a, b) > (0, nc) = n(0, c)$ for all $n$.

**Exercises 2.6** Page 24

1. If $A$, $B$, and $C$ are coplanar and $D$ is not in this plane, consider the planes determined by $A$, $B$, and $C$, by $A$, $B$, and $D$, and by $A$, $C$, and $D$.

2. Consider also the plane of $B$, $C$, and $D$.

5. Choose a point $P$ not on the plane. One class consists of those points $Q$ for which $P\text{-}Q$ does not meet the plane; the other consists of $P$ and those $Q$ for which $P\text{-}Q$ meets the plane. Apply Axiom 10 or Theorem 2.22 in any plane containing $P\text{-}Q$ not parallel to the given plane.

**Exercises 4.2** Page 34

1. If the lines are ordinary and not parallel, they have distinct ideal points and meet only in their common ordinary point. If they are ordinary and parallel, they meet in a unique ideal point. This is also the case if one line is ordinary and one is ideal.

**Exercises 4.4** Page 39

1. *a*. The set of all lines through a point; *d*. the point of intersection of a given line and the line determined by two given points; *g*. distinct collinear points lie on only one common line.

**Exercises 4.5** Page 42

2. Do not consider cases. If $\pi$ meets $\pi'$ at $m$ and $\pi''$ at $n$ (Theorem 4.51), then $m$ and $n$ are coplanar. Apply Theorem 4.54.

4. Choose a point in the plane and consider the plane determined by this point and the line (two points on the line). This plane meets the given plane in a line. Apply Theorem 4.54 to the resulting lines.

6. Apply Theorem 4.57.

**Exercises 4.6** Page 43

2. *a.* The set of planes through a line; *d.* the set of points on a plane; *f.* the set of lines through a common point; *h.* the set of all lines in space.

**Exercises 5.1** Page 50

4. In Figure 5.5 choose an appropriate $\triangle A''B''C''$ not in $\pi$, show that $P$ and $P'$ are then determined, and that the necessary incidence relations then hold to give the desired line concurrence at $O$.

5. The center of perspectivity is the intersection of the common side with the line joining the vertices not lying on the common side.

6. *a.* The axis of perspectivity is the ideal line; *d.* see Exercise 5.15.

9. See Figure 5.3.

11. Suppose the triangles are $ABC$, $A'B'C'$, and $A''B''C''$. Let $AB$ meet $A'B'$ at $X$, $AB$ meet $A''B''$ at $Y$, $A'B'$ meet $A''B''$ at $Z$, $BC$ meet $B'C'$ at $X'$, $BC$ meet $B''C''$ at $Y'$, and $B'C'$ meet $B''C''$ at $Z'$. Apply the theorem of Desargues to triangles $XYZ$ and $X'Y'Z'$.

13. Find Desarguean triangles with center of perspectivity the point of concurrence of $AA'$, $BB'$, and $CC'$.

14. Apply the theorem of Desargues to corresponding faces, and consider the incidence relations of the resulting axes of perspectivity.

**Exercises 5.2** Page 56

2. $BACD$, $ABDC$, $BADC$, $CDAB$, $CDBA$, $DCAB$, and $DCBA$.

3. See Theorem 5.21.

4. Carry out a harmonic sequence construction avoiding $R$.

7.–8. Note that if $P$ and $Q$ are vertices of a complete quadrangle and are collinear with diagonal point $D$ and $E$ (intersection of $D'D''$ and $PQ$), then $H(D, E; P, Q)$. The quadrangle is not unique in Exercise 5.27 but is in Exercise 5.38.

9. Use Theorems 5.23 and 5.22 to show that any two of the conjugate points are collinear with the given point on the opposite side of the triangle, then apply the theorem of Desargues to the given triangle and the triangle determined by the three conjugate points.

11. Carry out the construction for the fourth harmonic point choosing $P$ such that $PC \perp AB$.

13. Apply Exercise 5.211 and Theorem 5.23.

**Exercises 5.3** Page 61

4. *a.* No; *c.* yes; *e.* yes; *g.* no; *i.* no.

**Exercises 5.4** Page 66

2. If $T$ is the projectivity and $S$ is any central perspectivity for which $ST$ is defined, then $S^{-1}ST = T$.

4. If $T$ maps $l$ onto $l'$, then $S$ must map $l'$ onto $l$.

5. Consider $T(A, B, C, D)$ and apply Theorem 5.21.

8. If $T(A, B, C) = A, B', C'$, by the Fundamental Theorem $T$ must be the central perspectivity with center at the intersection of $BB'$ and $CC'$.

11. Suppose $W$ is a central perspectivity for which $W(A, B, C, D) = A', B', C', D$. Let $V$ be a central perspectivity with center $A$ such that $V(A', B', C', D) = A'', B'', C', C$. Let $U$ be a central perspectivity with center $B'$ such that $U(A'', B'', C', C) = B, A, D, C$. Then $S = UVW$.

**Exercises 6.1** Page 76

1. *a.* $(0, 0, 1)$; *c.* $(0, 1, 0)$; *e.* $(3, -7, 1)$; *g.* $(1, 1, 1)$.
2. *a.* $(0, 1)$; *c.* $(1, 1)$.
3. *a.* Euclidean line $2x - 3y + 4 = 0$; *c.* $y$-axis; *e.* ideal line.
5. *a.* $x_2 = 0$; *c.* $4x_1 - x_2 - 5x_3 = 0$; *e.* $3x_1 - 2x_2 + 5x_3 = 0$.
7. The equation is linear in the $x$'s and is satisfied by $(a_1, a_2, a_3)$ and $(b_1, b_2, b_3)$.

**Exercises 6.2** Page 78

1. *b.* $[0, 0, 1]$; *d.* $[3, 2, -6]$; *f.* $[0, 1, -3]$.
2. *b.* $u_1 = 0$; *d.* $3u_1 + 7u_2 + u_3 = 0$.
3. $a^2 + b^2 + c^2 = 0$ for real $a$, $b$, and $c$ if and only if $a = b = c = 0$.

**Exercises 6.3** Page 82

1. *a.* $[8, 23, 13]$; *c.* $[0, 0, 1]$; *e.* points are not distinct.
4. $[-29, 8, -34]$.
5. Expand the determinant using cofactors of the first row.

**Exercises 6.4** Page 85

1. *a.* $(0, 0, 0, 1)$; *d.* $(4, -6, -1, 1)$.
2. *b.* $[0, 0, 0, 1]$; *d.* $[2, 7, 0, -8]$.
3. *a.* $[11, 29, -30, -13]$; *c.* $[3, 2, -1, -9]$; *e.* points are collinear.
5. $[-3, -54, 9, 34]$.
6. See hint to Exercise 6.17.

**Exercises 7.1** Page 89

1. *b.* $(-2, 1, 12)$; *d.* $(7, 1, 1, -12)$.
2. *b.* $(19, -41, 21)$; *d.* $(-4, 8, -22, 18)$.

**Exercises 7.2** Page 91

1. *b.* Independent; *d.* $(4, 6, 10) = 2(2, 3, 5)$; *f.* independent
3. $c_0 + c_1x + c_2x^2 = 0$ identically only if $c_0 = c_1 = c_2 = 0$.
5. Show that if, say, $\mathbf{v}_m$ is a linear combination of $\mathbf{v}_1, \mathbf{v}_2, \cdots, \mathbf{v}_{m-1}$, then $c_m\mathbf{v}_m$ is a linear combination of $c_1\mathbf{v}_1, c_2\mathbf{v}_2, \cdots, c_{m-1}\mathbf{v}_{m-1}$. You must consider both $c_i = 0$ and $c_i \neq 0$.

**Exercises 7.3** Page 94

1. *a.* $\begin{pmatrix} 16 & 4 \\ 3 & 8 \end{pmatrix}$; *d.* impossible; *g.* $\begin{pmatrix} 28 & 11 \\ 42 & 18 \end{pmatrix}$; *j.* impossible;

*m.* $(6\ 2\ 8)$; *p.* $\begin{pmatrix} -3 & 0 & -7 \\ 7 & 3 & 12 \\ 1 & 8 & 1 \end{pmatrix}$; *s.* $(0)$

**Exercises 7.4** Page 97

2. Rank of $A, B, C$, and $D$ is 2; of $E, F$, and $K$, 1; of $G$ and $H$, 3.
4. If the given matrix is $(a_{ij})$, consider the rank of $(b_{ij})$ where $b_{ij} = a_{ji}$.

**Exercises 7.5** Page 99

2. $(1, 1, 1) = (-\frac{1}{7})(3, 2, 1) + (\frac{3}{7})(2, -1, 0) + (\frac{4}{7})(1, 3, 2)$.
4. See Exercise 7.52.

**Exercises 7.6** Page 102

1. *a.* Nonsingular; *b.* singular; *c.* nonsingular; *d.* singular.
3. Note that the row vectors of $B$ are linear combinations of the row vectors of $C$.

**Exercises 8.1** Page 109

1. *a.* Independent; *b.* independent; *c.* dependent; *d.* dependent.
3. *a.* $(0, 19, 15) = (6, 21, 27) + (-6, -2, -12)$,
   $(0, \frac{19}{3}, 5) = (2, 7, 9) + (-\frac{2}{3})(3, 1, 6)$.
7. Noncollinearity and Theorem 8.15 imply

$$c_1P_1 + c_2P_2 + c_3P_3 + c_4P_4 = \theta$$

with no $c_i$ zero.

10. *b.* The collinearity of the points will imply that in each $p_{ij}'$ each row is the *same* linear combination of the rows of $p_{ij}$.

**Exercises 8.2** Page 115

2. Using the notation of Theorem 8.22, show that the points in question are $P_1 - P_2$, $P_2 - P_3$, and $P_3 - P_1$.

**Exercises 8.3** Page 123

1. $\begin{pmatrix} 2 & 3 & 0 \\ 2 & 3 & 4 \\ 0 & 6 & 16 \end{pmatrix}$, $(20, -16, 5)$.

3. $\begin{pmatrix} 3 & -4 \\ 6 & -6 \end{pmatrix}$, $(-13, -12)$.

5. $U' = UA$ where $U$ and $U'$ are row matrices.

7. $\begin{pmatrix} 1 & 2 & 1 & 0 \\ 1 & 0 & 0 & 0 \\ 0 & 2 & 0 & 2 \\ 0 & 0 & 1 & 2 \end{pmatrix}$

**Exercises 8.4** Page 127

2. $\begin{pmatrix} -32 & 58 \\ 45 & -96 \end{pmatrix}$.

4. Impossible, not one-to-one.

**Exercises 8.5** Page 130

1. *a.* $-\frac{17}{144}$; *c.* 0; *e.* $-\frac{1}{2}$.
3. Choose coordinates for $A$ and $B$ wisely!
6. Clearly $R(A,B,D,C) = 1/t$. Use Exercise 8.53 to show that $R(D,B,A,C) = 1-t$. The others follow readily from these two.

**Exercises 8.7** Page 137

1. *a.* Parabolic; *c.* elliptic.
2. *a.* $\begin{pmatrix} 26 & -25 \\ 22 & -26 \end{pmatrix}$; *c.* $\begin{pmatrix} 2 & 1 \\ -1 & -2 \end{pmatrix}$; *e.* $\begin{pmatrix} 63 & 36 \\ 72 & -63 \end{pmatrix}$.
4. Consider $R(T[M], T[N], T[A], T^2[A])$.
6. Show the center has nonhomogeneous coordinate $a_{11}/a_{21}$. The product of the distances is $(a_{12}a_{21} + a_{11}^2)/a_{21}^2$.
8. Consider the powers of the matrix of a parabolic transformation mapping the origin ($x = 0$) onto itself.
10. If $T(A,B,C) = A', B', C'$, consider the involutions $U$ and $V$ for which $U(A,B,C) = B', A', C''$ and $V(A', C'') = B', C'$.

**Exercises 9.1** Page 149

2. Apply Theorem 9.12.
3. *a.* Let $T_1$ and $T_2$ be the projectivities which map the lines of the pencils defining the conic onto the points at which they meet the given line, and let $T_3$ be the projectivity between the pencils. Then $T = T_2T_3T_1^{-1}$.

**Exercises 9.2** Page 154

2. Plane of points, or one or two lines, one of which is the line of the five points. Plane of points, or two lines, one containing the four collinear points and the other through the remaining point. Plane of points, or two lines.
4. Apply Theorems 8.76, 8.77, and 9.22.

**Exercises 9.3** Page 159

1. Yes.
5. Theorem 8.82, Definitions 9.11 and 9.12, and Theorem 9.39 are sufficient.

**Exercises 9.4** Page 167

1. *a.* (1) $x_1^2 - 2x_1x_3 + 3x_2^2 - 2x_3x_1 - x_3^2 = 0$;
   (3) $x_1^2 - 2x_1x_3 + 3x_2^2 - 2x_3x_1 + 4x_3^2 = 0$;
   *b.* (1) $x_1^2 + x_2^2 - x_3^2 = 0$; (3) $x_1^2 + x_2^2 = 0$.
2. *a.* $4x_1x_2 - x_1x_3 - 3x_2x_3 = 0$;
   *c.* $-7x_1^2 + 12x_1x_2 + 6x_1x_3 - 5x_2^2 - 4x_2x_3 + x_3^2 = 0$ (determine pencils defining the conic).
3. $3x_1^2 + 5x_1x_2 - 16x_1x_3 - 4x_2x_3 + 8x_3^2 = 0$.
5. Compute the determinant of $(a_{ij})$.
8. Use the canonical form for a nondegenerate conic and an arbitrary line. Solve simultaneously.

**Exercises 9.5** Page 172

2. *a.* $x_3 = 0$; *b.* $[1, -5, -8]$; *c.* $(9, -5, 2)$; *d.* $[1, 0, 0]$ and $[1, 2, 0]$.
6. Use the canonical form for the conic.
10. Use Theorems 9.37 and 9.58.
11. Use Exercise 9.510.

**Exercises 10.1** Page 178

2. Use Axioms 2, 5, and 4.
4. Given the point, show there is a line not on this point, then apply Axioms 2 and 4.
6. Use a point on the axis of perspectivity as a center of perspectivity for an appropriate pair of triangles.
7. Denote by $mn$ the point of intersection of the lines $m$ and $n$. Apply Axiom 7 to $a'c'$, $ac'$, $bc'$ and $bc$, $b'c$, $a'c$, where $a$, $b$, $c$ and $a'$, $b'$, $c'$ are the two sets of concurrent lines.

**Exercises 10.2** Page 186

1. $P, Q, R, Q \rightarrow A + O = A$.
5. $K, L, M, O \rightarrow A \cdot O = O$ and $K, L, M, L \rightarrow A \cdot U = A$.
10. *a.* No; *c.* yes; *e.* yes; *g.* no.
11. Apply Theorem 8.77 to quadrangle *PQRS*.
14. Apply Theorem 8.77 to quadrangle *KLMN*.

**Exercises 10.4** Page 197

2. $a + a = a(1+1) = 0$, $a^2 = -(a+1) = a+1$, $(a+1)^2 = a$, $a(a+1) = 1$, etc.
4. Three.

**Exercises 10.5** Page 199

2. *a.* $[1-i, -1-i, -7-9i]$; *c.* $[15+10i, -25, -9+4i]$; *e.* $[1, -i, 0]$.
4. (1) $x_1^2 + x_2^2 + x_3^2 = 0$; (3) $x_1^2 + x_2^2 = 0$.

**Exercises 10.6** Page 205

1. Let the line and circle be coplanar. Join all points on the line to a fixed point within the circle by lines. The pencil of lines will determine a one-to-one correspondence.

**Exercise 11.1** Page 210

2. Use Theorem 11.13.

**Exercises 11.2** Page 217

2. $T$ must map the intersection of $m$ and $n$ onto an ideal point.
4. Consider the complete quadrangle whose vertices are the vertices of the parallelogram.
7. Use Theorem 11.13.
9. The axes are translated, and the scales are changed on the axes.
11. *a.* Conic degenerate; *c.* $x^2 + y^2 = 1$; *e.* $x^2 + y^2 = -1$.

**Exercises 11.3** Page 223

5. The triangles have corresponding angles with the same measure in view of Definition 11.31 and Theorem 11.36.

6. Apply Theorem 11.36.

7. Apply Exercise 11.36.

**Exercises 12.1** Page 234

5. Note that $H$ is the matrix of the absolute. If a point with column matrix $X$ of coordinates satisfies the matrix equation of the absolute, so must $AX$.

7. Consider the sign of $a^2+b^2-1$ as discussed in the proof of Theorem 12.13.

**Exercises 12.2** Page 235

4. Consider lines joining ideal points on the two given lines.

**Exercises 12.3** Page 239

2, 3. Generalize the cross ratio computation in the proof of Theorem 12.31.

**Exercises 12.4** Page 244

3. 0.88.

5. $u_1 \tanh z_1 + u_2 \tanh z_2 + u_3 = 0$.

**Exercises 12.5** Page 247

1. *b.* Fail to intersect in each case; *d.* $\cos^{-1} 20/\sqrt{408}$, hyperbolic; $\cos^{-1} 21/\sqrt{450}$, Euclidean.

2. Compare equation (12.52) and the equation of Theorem 11.33.

**Exercises 12.6** Page 249

1. Consider the line joining the given point to the pole of the given line.

3. See Exercises 12.17 and 12.18.

**Exercises 12.7** Page 251

2. In Figure 12.15 let $M$ be the midpoint of $A$-$C$ and choose $B'$ so that $M$ is midpoint of $B$-$B'$. Be careful in your consideration of the *interior* of an angle.

3. Show that there must be an isometry mapping the one asymptotic triangle onto the other.

**Exercises 12.8** Page 255

1. Drop a perpendicular from one vertex to the opposite side. Consider cases.

3. *a.* $b=1.28, \alpha=0.14$ (radians), $\beta=1.31$; *c.* $b=0.13, \beta=1.06, c=0.27$.

**Exercises 12.9** Page 257

2. *b.* $\beta=1.32, c=$ impossible; *c.* $\gamma=0.91, \beta=0.28, \alpha=0.59$.

**Exercises 13.1** Page 265

4. $\begin{pmatrix} 1/\sqrt{6} & 1/\sqrt{6} & -2/\sqrt{6} \\ 1/\sqrt{2} & -1/\sqrt{2} & 0 \\ 1/\sqrt{3} & 1/\sqrt{3} & 1/\sqrt{3} \end{pmatrix}$. Others possible.

6. See proof of Theorem 13.61.

7. The points $(1, i, 0)$, $(1, 0, 1)$, $(1, 0, -1)$, $(0, 1, 1)$, and $(0, 1, -1)$ must be mapped onto points on the absolute.

**Exercises 13.3** Page 270

3. See the derivation of equation (12.52).
6. See hint to Exercise 12.61.
8. Consider the polar line of the point of intersection of the two lines.
9. *a.* $\cos^{-1} 5/\sqrt{27}$.
12. A line corresponds to a set of coplanar lines. Line measure corresponds to the measure of the dihedral angle between two planes.

**Exercises 13.4** Page 276

3. Use Theorem 8.54.
6. Draw a figure for Theorem 13.45.

**Exercises 13.5** Page 279

1. Choose coordinates on $a$ so that $A'$ has coordinate 0.

**Exercises 13.6** Page 281

2. Consider all of the occasions when order principles are used in the proof.

**Exercises 13.7** Page 284

3. *a.* $b = 1.30, \alpha = 0.24, \beta = 1.51$; *c.* $c = 0.27, b = 0.14, \beta = 0.54$.
8. In one of the other triangles the parts (with respect to the given triangle) are $b, \beta, \pi - a, \pi - \alpha$, and $\pi - c$. If these changes are made in the basic formulas, they will remain unchanged or undergo sign changes on both sides of the equation.
9. Interchange sides and angles in the basic formulas.

**Exercises 13.8** Page 285

3. Note that for acute $x$ and $y$, $x < y$ will imply $\cos x > \cos y$. The difficulty in the statement of this result traces back to the fact that two points determine two different segments.
4. *b.* $\beta = 0.20, c = 0.64, \gamma = 2.39$; *c.* $\gamma = 2.26, \beta = 0.30, \alpha = 0.60$.
6. Interchange sides and angles.

# INDEX